THE APPLE REVOLUTION

ABOUT THE AUTHOR

Luke Dormehl is a journalist and award-winning documentary filmmaker. With a particular focus on pop culture, his writing has appeared extensively in dozens of online and print publications, while his films have been screened at the Cannes festival and on Channel 4. This is his second book.

THE APPLE REVOLUTION

STEVE JOBS, THE COUNTER CULTURE AND HOW THE CRAZY ONES TOOK OVER THE WORLD

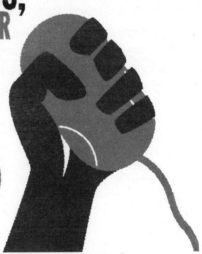

LUKE DORMEHL

2 4 6 8 10 9 7 5 3

First published in the UK in 2012 by Virgin Books, an imprint of Ebury Publishing

A Random House Group Company

Copyright © Luke Dormehl 2012

Luke Dormehl has asserted his right under the Copyright, Designs and Patents
Act 1988 to be identified as the author of this work.

www.randomhouse.co.uk

Addresses for companies within The Random House Group Limited can be found at
www.randomhouse.co.uk/offices.htm

The Random House Group Limited Reg. No. 954009

A CIP catalogue record for this book is available from the British Library.

The Random House Group Limited supports The Forest Stewardship Council (FSC®),
the leading international forest certification organisation. Our books carrying the FSC
label are printed on FSC® certified paper. FSC is the only forest certification scheme
endorsed by the leading environmental organisations, including Greenpeace. Our paper
procurement policy can be found at www.randomhouse.co.uk/environment

MIX
Paper from
responsible sources
FSC® C016897

Text design and typesetting by carrdesignstudio.com
Printed and bound in Great Britain by Clays Ltd, St Ives PLC

ISBN: 9780753540626

To buy books by your favourite authors and register for offers,
visit www.randomhouse.co.uk

To Andy Palmer

"In the beginning (of the Information Age) was the void. And the void was digital. But lo, there came upon the land, the shadow of Steven Jobs (and Stephen Wozniak). And Steven (Stephen) said, 'Let there be Apple.' And there was Apple. And Steven (Stephen) beheld Apple. And it was good. And Apple begat Macintosh. And it was good. And soon upon the land there began to appear, The Cult of Macintosh. For they had tasted of Apple. And it was good."

Russell W. Belk and Gülnur Tumbat,
'The Cult of Macintosh',
Consumption, Markets and Culture,
Vol. 8, No. 3, September 2005

❏

"Ready or not, computers are coming to the people.
That's good news, maybe the best since psychedelics."

Stewart Brand,
'*Spacewar: Fanatic Life and Symbolic Death among the Computer Bums*',
Rolling Stone magazine, 7 December 1972

❏

"Never trust a hippie."

Malcolm McLaren, former manager of the Sex Pistols

CONTENTS

AUTHOR'S NOTE & ACKNOWLEDGEMENTS

As a lifelong fan of high-tech and a similarly longtime fascinated follower of industry trends, I could easily make the argument that the groundwork for this book dates back the best part of three decades. More officially I began work on this project proper in the early part of 2011, several months before Walter Isaacson announced his exhaustive authorised biography of Steve Jobs. For more than a year I conducted over a hundred interviews with former Apple employees, as well as those who knew Steve Jobs, Steve Wozniak and the other major players in this story, in an effort to piece together an important chapter in Silicon Valley history. I thank all of them for their time and willingness to be cross-examined, often on multiple occasions and about the minutiae of events that, in some instances, occurred almost forty years ago.

This is not a business book and nor is it a biography of either of Apple's co-founders. Instead this is a history of Apple, Inc. as social artefact: a look at how the countercultural revolution of the sixties and seventies (and beyond) imbued the company with much of its original identity, and continues to do so almost half a century later. There are, naturally, a few tangents. To set the scene I have attempted to paint a picture of the changing face of Silicon Valley cyberlibertarianism. Because the Apple revolution is just one prominent example of other, interlinked revolutions, this is an effort to fill in some of the space surrounding Apple.

As such I have also opted, in individual chapters, to tell the story of NeXT, the computer company founded by Jobs during his decade away from Apple, where he had his first experience as CEO, and Pixar, a company which shares much of Apple's DNA and which, ultimately, turned Jobs into a billionaire. Since both entities impacted upon Apple in some way, I feel that both of these stories are key to building up that trickiest of things: the bigger picture.

From here simply to add a note of special thanks to all those that have helped me along the way. Ed Faulkner, my tireless editor at Virgin Books, had a seemingly bottomless well of enthusiasm for this book from the start and his good advice continued throughout the writing process. Jake Lingwood, of Ebury Publishing, who helped make this project happen in the first place. My agent, Margaret Hanbury, whose unwavering support was always greatly appreciated. My good friends Tim Matts and Alex Millington, who both read over various sections of the book and offered detailed and incredibly useful feedback and suggestions. Vicky Thomas, who proved a far speedier transcriber than I could ever hope to be. In addition much love to Clara, as well as assorted friends and family members too numerous to list.

A note of gratitude, too, to those intrepid Apple chroniclers who came before me. Several books proved particularly helpful in researching *The Apple Revolution*. John Markoff's fabulous *What the Dormouse Said* may very well be viewed as a spiritual prequel to this tome, dealing with the social history of the personal computer up to the point at which Mr Jobs and Mr Wozniak first arrived on the scene. Michael Moritz's *Return to the Little Kingdom* (originally published under the title, *The Little Kingdom*) laid the groundwork for all Apple books to have come after, and is required reading for anyone wanting more information about Apple's early years. Frank Rose's *West of Eden* is something of an underrated gem, dealing with the company's inner-workings in the mid-1980s. Both Randall Stross' *Steve Jobs & the NeXT Big Thing* and Alan Deutschman's *The Second Coming of Steve Jobs* perform different, but equally enthralling, functions regarding Jobs' lesser-known wilderness years outside of Apple.

Last but certainly not least, I reiterate my sincerest gratitude to all those who agreed to speak with me as part of my own research (both on and off the record), along with those working at Apple and other such high-tech companies today. Without their constant attempts to invent the future, there would be nothing for us tech writers to write about in the first place.

PART I

THE TWO STEVES

Prologue

TALKiNG ABOUT iGENERATiON

'I think hackers ... are the most interesting and effective body of intellectuals since the framers of the US Constitution. No other group that I know of has set out to liberate a technology and succeeded. They not only did so against the disinterest of corporate America, their success forced corporate America to adopt their style in the end ... High tech is now something that mass consumers do, rather than just have done to them ... The quietest of the 60s sub-subcultures has emerged as the most innovative and most powerful - and most suspicious of power.'

Stewart Brand,
Whole Earth Review (1985)

A New Yorker driving homewards one day in late March 1999, who just so happened to glance upwards while travelling down Manhattan's East 14th Street, might well have received something of a surprise. Between Avenue A and B, a giant 246-square-foot billboard bearing Apple's familiar 'Think Different' slogan was bolted to the side of a six-storey tenement building. The existence of the billboard itself was nothing out of the ordinary. Nor was the line of copy, which the Cupertino-based computer company had been using to great acclaim since 1997, the year in which co-founder Steve Jobs had returned to the fold after a decade in the high-tech wilderness. The tagline put into ungrammatically exact words what Apple's bosses, employees and fans had been saying since day one. Apple thought different(ly).

The words were paired with images of some of the most recognisable figures of the previous hundred years: Albert Einstein, Martin Luther King, Jr, John Lennon, Muhammad Ali, Pablo Picasso, and others. They had been selected by Jobs and Apple's ad agency, TBWA/Chiat/Day, for the profound, indelible contribution each had made upon a variety of aspects of life in the twentieth century. And yet all were pointedly controversial in some regard. As the 'Think Different' commercial noted, these were figures that had been both revered and reviled during their lifetimes. As a member of the Beatles, John Lennon had been part of a group that shook pop music to its very core. '[N]o one will ever be more revolutionary, more creative and more distinctive,' former *Rolling Stone* associate editor Robert Greenfield once wrote. However, Lennon had also spoken wishfully of his hope for a world free of the violence that comes from ideology, to the point where his famous 1971 song 'Imagine' is sometimes performed in America with the phrase 'no religion too' changed to 'and *one* religion too'. Meanwhile, Muhammad Ali might have commanded awe and respect as boxing's heavyweight champion of the world, but he was also stripped of that title and had his licence suspended in 1967, after refusing to be conscripted into the US military to fight in Vietnam, telling the Johnson administration that 'No Viet Cong ever called me nigger'.

Ironically, many of the reasons these people are loved and admired today is for the same reason they were once vilified. To a degree, the Apple billboards represented the notion that, given enough time, everything becomes respectable.

Which was why this particular billboard appeared so brazen. When they saw it, some of those driving on East 14th Street stopped their cars and got out to take a look. This being two years before digital cameras came into popular use, several people took photographs on their one-use disposable cameras for later developing. No matter what, it was certainly an attention-grabbing image. Older people might have thought it was vulgar; attention-craving in the worst possible taste. Younger people may have found it edgy; the appropriation of yet another piece of retro pop culture in the same year that *Boogie Nights*, a film about the adult entertainment industry in 1970s California, starring A-list actor Mark Wahlberg, was raking in money at the box office. No matter how one responded, however, the choice of image was undoubtedly controversial. As people looked up at the grimacing face leering down at them from the billboard, they asked themselves and one another the same question: was cultist murderer Charles Manson *really* the best choice of spokesperson for the new, improved Apple Computer?

❏ ❏ ❏

As it transpired, Manson – as differently as he may have thought – was not really the latest representative for the world's most alternative computer company. The image was the creation of Ron English, a guerrilla street artist and cultural jammer who took it upon himself to 'liberate' billboards to transmit his own subversive messages. He referred to it as 'subvertising'. One of English's billboards which appeared in New York in 1990 bore the likeness of the All-American icon Uncle Sam, with the legend 'Censorship is good because ███ ██████ ██████'. Another mirrored the look of a Camel cigarettes advertisement, and depicted one cartoon camel asking another, 'Hook any new kids today?'

English did not pick his targets at random. All concerned either issues that he felt strongly about, or companies which had grievously offended him in some way. The aspect of Apple's 'Think Different' campaign that English had taken issue with was its appropriation of some of the world's greatest unconventional achievers for corporate-boosting means. Had he still been alive, would Gandhi, for example – the anti-materialist, who saw modern civilisation's greed for wealth and worldly pleasures as its greatest downfall – have been first in line to act a spokesperson for a new computer? 'The religion of our culture is commercialism,' said English, 'but the first time I saw the "Think Different" ads it freaked me out. I couldn't believe that those people's images had been sold. It was like today's revolutionary is tomorrow's corporate shill. I thought, you want different: *here you go.*'

The billboard transformation had occurred that morning, right under (or, more accurately, above) people's noses. English had set off early from his Tribeca loft, where he lived with his wife, Tarssa Yazdani, and his two children, Zephyr and Mars. He was accompanied on his mission by Don Goede, the editor for a Lower East Side publishing company; Janet O'Faolain, a member of a modern dance troupe; and the owner of an independent record label named Jake Szufnarowski. The foursome caught a train from City Hall subway station to Union Square, and walked the rest of the way to East 14th Street. While O'Faolain separated off to set up a video camera she used to record the action, English, Goede and Szufnarowski ascended six flights of stairs to reach the roof of a nearby building. There they mixed a bucket of cornstarch paste and then, using stiff-bristled brushes with long wooden handles, applied the gloopy cement to the existing billboard's surface. Once this had been done they carefully unrolled English's specially prepared 22x11-foot 'Think Different' image and manoeuvred it into place. The whole process lasted no longer than twenty minutes. Years of billboard liberation had taught English that the best way to get away with this brand of activity was to do it in clear view of everyone. No New York City dweller batted an eyelid at the sight of a new billboard being erected during daylight hours. People tended to become more suspicious when it was done under cover of darkness. Even in a city that never sleeps, finding workmen that dedicated was a little unusual.

The clues that this was not an official Apple advertisement presented themselves upon closer inspection. Scrutiny revealed that the rainbow-coloured stripes which then made up the Apple logo – bands of green, yellow, orange, red, purple and blue – were instead a mass of interlocking salamanders; the emblem English had used as his signature since his days as a high-school student in Decatur, Illinois. Given the rabble-rousing nature of his work, English was used to extreme reactions. But while few would jump to the aid of cigarette companies, or McDonald's, or governmental censorship, Apple was different. It was the province of artists, designers, musicians and freedom-fighting journos. Naomi Klein, author of the bestselling anti-globalisation manifesto *No Logo: Taking Aim at the Brand Bullies*, used a Mac. So, for that matter, did English. 'I had so many people say to me, "Hey, Apple is a great product. Why do you have to go after them?"' he recalls.

To his credit he didn't back down. If anything, his next bit of 'agit-pop' was even more offensive to the Apple faithful.

It was a 'Think Different' advertisement featuring Microsoft CEO, Bill Gates.

ロ ロ ロ

When Steve Jobs died on 5 October 2011, at the age of fifty-six, numerous publications and media outlets made mention of his roots as a different-thinking child of the sixties. Growing up in Cupertino, on America's West Coast, he was a short drive from

San Francisco, where a seismic cultural shift was taking place that, at least on the surface, seemed to have nothing to do with the manufacture of luxury personal computers or portable music players. What connected it, however, was the scope of ambition. This was a generation hell-bent on bringing about revolution: *sexual, political, ecological, technological* ... 'It's remarkable what's come out of Haight-Ashbury,' U2's front man Bono told *Rolling Stone* magazine the month that the Apple CEO passed away. 'The children of the sixties are seriously changing the world. Steve Jobs is right up there. He is, in many ways, the Bob Dylan of machines; he's the Elvis of the kind of hardware-software dialectic. He's a creature of quite progressive thinking, and his reverence for shape and sound and contour and creativity did not come from the boardroom.'

Despite the ubiquity of such technologies today, the rise of the personal computer from its humble, garage-born roots to the vast industry that it currently represents is, in a very real sense, the forgotten part of the counterculture story. While the music of Bob Dylan, Jimi Hendrix, John Lennon, Joan Baez, the Rolling Stones and the Grateful Dead; the fashion for flared trousers and tie-dyed bandannas; and a healthy appreciation for psychedelics will forever conjure up an image of youthful insurrection, it is all too easy to pass over the contributions of the wireheads, phreakers, cyberpunks and other techno-geeks who seized upon the countercultural revolution as their opportunity to claim high-tech for the masses. If, as has been often stated, the three groups which brought about such rebellion in America were the dope dealers, the rock musicians and the underground artists and writers, then the computer engineers and technological soothsayers were right behind them. In the words of business historian Peter Drucker, the so-called 'bards and hot-gospellers of technology keepers' have become the countercultural torch bearers. 'The people who built Silicon Valley were engineers,'

Jobs told *Wired* in 1996, shortly before his return to Apple. 'They learned business, they learned a lot of different things, but they had a real belief that humans – if they worked hard with other creative, smart people – could solve most of humankind's problems. I believe that very much.'

Jobs' own sensibilities were forged at a critical juncture in US history; a period when, for the briefest of moments, the gates of bureaucracy came crashing down and the infidels flooded the castle. He was the CEO with a difference; the one who claimed that taking LSD was one of the two or three most important things he had done in his life, and that there were aspects of himself that modern corporate America could never truly understand for this very reason. 'Psychedelics and computers overlapped, and Steve was certainly a clear case of someone who dove very early and often into the computer world, and likewise dove early and often into the LSD world,' says Stewart Brand, founder of the seminal countercultural document *The Whole Earth Catalog*. 'He referred to his "LSD perspective" on things throughout his life.'

Along with fellow Apple co-founder Steve Wozniak, Jobs served as a notable reminder of the time when the hippies took over the high-tech industry. 'It's no accident that Steve Jobs and Steve Wozniak were both long-haired, barefoot freaks,' Timothy Leary, an ex-Harvard professor, LSD proponent and self-appointed spokesperson for the underground, once said. 'They were the original cyberpunks. Their names will go down with great libertarians in history.'* Long after the hippie communes came down, LSD gave way to new designer drugs, and popular music graduated from protest songs to arena

---

* When the term 'libertarian' is used in this context, please read as neoliberal, rather than in any sense anarcho-communist. The same refers to anytime I use the word myself throughout the book.

rock, Apple continued to wage its wars against conformity, the Establishment and standardised file types.

And in the process, something unexpected happened to the underdog, renegade computer company: it became the biggest tech company in the world.

The question of Apple's changing role is at the centre of this book. The past decade and a half has seen Apple undergo a reversal in its fortunes which will go down as one of the most remarkable turnarounds in corporate history. In 1996, the company was on the ropes; haemorrhaging $740 million in the first quarter of the year alone. When asked what he would do if he were Apple CEO, Michael Dell commented: 'I'd shut it down and give the money back to the shareholders.' Today that notion seems patently absurd. Apple's shares have shot up almost 6,000 per cent since 2003.* Meanwhile, the company has changed everything from the way we buy and listen to music (iTunes has around a 70 per cent share in legal music downloads); the means by which we make telephone calls; and the method by which we browse the Internet and thus, by extension, how we gather much of our information on a daily basis. In April 2011, Apple posted profits over and above those of one-time enemy Microsoft. Several weeks later it was reported that Apple now had a bigger market cap than both Microsoft and Intel combined; the so-called Wintel alliance which had almost put it out of business a decade earlier. At times it has had more money than the US Treasury, with $76 billion to the Treasury's meagre $74 billion. Apple, at least at the time of writing, was the world's second most valuable company overall behind Exxon, which it has briefly pipped to the post on several occasions, to

---

* As a somewhat disquieting thought, if someone had bought Apple shares in 1984 with the money they might have otherwise used to purchase a Macintosh they would be sitting on around a quarter of a million dollars' worth of stock today.

momentarily sit astride the globe as the all-conquering colossus Steve Jobs always saw it could be.

Despite Ron English's best efforts, Apple is proof positive of the fact that the culture cannot be jammed. A Fordist corporation can become the world's biggest company *and* keep its countercultural cool. It's everything that the hippie generation dreamed of – and more. Ultimately this is a story about how a garage business started by two long-haired college dropouts rose to become the tech giant that Apple, Inc. is today. It's also the story of the rise of the all-conquering Apple ideology; the revolution it came from, the one it helped start and the one that it left behind.

THE SECRET HiSTORY OF COMPUTERS

'The plot ... thickened when countercultural code-cowgirls and code-cowboys, combining the insights and liberated attitudes of beats, hippies, acidheads, rock 'n' rollers, hackers, cyberpunks, and electronic visionaries, rode into Silicon Valley and foiled the great brain robbery by developing the great equalizer: the Personal Computer.'

Dr Timothy Leary,
Chaos and Cyberculture (1994)

In the early 1970s, before the Macintosh, the iPhone and the iPad were the faintest of gleams in Steve Jobs' eye, a wonderful new invention shook the technological world to its very core. Its appearance would change the course of high-tech history and bring to the masses what had previously been available only to a select few. It was the handheld calculator, and for a short while it was the greatest thing that the geeks of Silicon Valley had ever seen. To this rabid fan base – of whom, at the time, there was a greater population on a *per capita* basis in Silicon Valley than anywhere else in the world – the calculator's appeal extended far beyond simply being a portable adding device, or even the obvious benefit of being a piece of relatively high-tech equipment that one could buy and take home. It was programmable. In other words, one could purchase such a calculator (a low-end model might cost a couple of hundred dollars, something superior closer to a thousand), take it home and within several hours – and provided one knew what one was doing – be using the right combination of the INT, STO, GTO and TAN keys to enter in simple programs. These could range from simple games of noughts and crosses to, in the case of one ambitious NASA scientist, a complete model of the solar system, capable of producing predictions of planetary positioning. As programming tools, calculators had their faults, of course. The number of 'steps' that could be entered was severely restricted by the machine's tiny data store. Worse, if the adding device was turned off or its battery died, the program that had been keyed in was lost for ever – or until one painstakingly reprogrammed it. In the words of one calculator hobbyist, this repetitious and rather cruel and unusual activity amounted to 'medieval punishment'. Overall, however, the positives far outweighed the negatives – and until something demonstratively better came along the geeks of Silicon Valley knew this.

The mad rush for calculators created an entirely new market, and manufacturers dutifully rushed to respond to demand. In addition to the established likes of Hewlett-Packard and Texas Instruments, smaller operations began getting in on the act. Two men in Albuquerque, New Mexico, named Ed Roberts and Forrest Mims III, first entered the hobbyist market in 1970, co-founding a company named MITS (Micro Instrumentation and Telemetry Systems), through which they hoped to sell model rockets to space enthusiasts. But sales were disappointing and MITS turned its attentions to calculators as another potential money-spinner. This time they proved correct, and by 1973 the company was selling every calculator it could run off its production line, with 110 employees working back-to-back shifts to fulfil orders. Eventually competitive pricing drove Roberts and Mims out of that business, too, when they discovered that they could buy a fully assembled calculator in an electronics store for less money than the individual parts were worth. MITS was out of luck for a second time. Having lost millions of dollars, nobody would have blamed either man for throwing in the towel, but they refused to do so. In 1975, they re-emerged with a fantastic piece of kit called the Altair 8800: the machine considered by many to be the world's first real personal computer. When it came to pricing the Altair, Roberts and Mims struck upon another piece of good fortune. While the cost of a fully assembled machine came to $621, a great number of would-be customers were happy to instead pay the cheaper price of $439 for a kit of parts. Where previously the assemblage of a piece of electronics was merely an accepted part and parcel of the whole pursuit, both men found to their amazement that there was an entire subset of geeks who relished building things; and the more complex, unwieldy and infuriating, the better. In less than a month, MITS's bank balance went from minus figures to $250,000 in the black.

The Altair was, as both technical and consumer product, fairly basic. It had no built-in display for one thing, other than a series of blinking lights. Nor any existing software. Nor did it come packaged with a keyboard. Nor did it have – until two bespectacled university dropouts by the name of Bill Gates and Paul Allen were able to cobble one together – a language comprehendible by any more than a select few. What it did represent was a blank slate, upon which could be projected whatever the user could envisage to program. It was also a real, genuine computer, and for a generation of nerds previously left out in the technological cold, that was enough. This first collective of fully realised computer geeks flocked around the

Altair with an interest that bordered on mania. Some started up fanzines with quirky names like *Dr. Dobb's Journal of Computer Calisthenics and Orthodontia*, through which they were able to share their wildly optimistic predictions about the direction in which computers were headed. Others took jobs at upstart dealerships such as ComputerLand and the Byte Shop. The jobs themselves often paid barely enough to cover the rent on a tiny studio flat, but that didn't matter. 'A lot of us took our pay in merchandise anyway,' admits Michael Holley, one early devotee.

What was surprising about the ferocity with which geeks flocked to computers in the seventies was the suggestion that it made about the tentative first steps in what would eventually become a full-on public re-evaluation of the medium. Just one decade earlier, computers had been eyed suspiciously as sinister tools of governmental oppression; giant mainframes sequestered away in labyrinthine chambers at the back of laboratories and overseen by highly qualified people in white lab coats. Key players speak of the technological priesthood which existed around such machines – being separated off from mainstream society and requiring a lengthy period of seminary training before an individual was considered able to interact with one. A story that neatly encapsulates these Howard Hughes levels of secrecy, security and cleanliness tells of the mainframe manager who posted up a sign ordering that nobody be admitted to the computer room without wearing specially designed cloth coverings over their shoes. The manager admitted that there was no essential need for these precautions, 'but it sure kept people out of the computer room'.

MITS's Altair, meanwhile, accomplished an altogether more impressive feat than simply giving a handful of ragtag, technology-starved hobbyists something to do on an otherwise empty Friday night: it kept the dwindling spirit of the counterculture ticking over, in the same way that an electric current might keep a heart beating long after it would normally have ceased to do so. Those who were in early enough to remember the heady thrills of the Altair 8800 became the intrepid frontiersmen of the personal-computer revolution, just as the hipsters and Beat poets had been the founding fathers of the sixties hippie spirit.

Although the idea has gained acceptance over the years, the temptation is still to see the cyber-ideologues as a curious sideline to the overall countercultural revolution; nervously standing at the edge of the party, undancing and looking distinctly uncomfortable, while the musician likes of the Grateful Dead and Jimi Hendrix take centre stage. This may be true in terms of the force of personality of many of its representatives, but it is decidedly not the case when it comes to the targeting of their attacks, or their lasting influence. At its root the counterculture was all about striking back at the conformist hegemony of the 1950s, in which long-standing traditions of individualism disappeared with the rise of giant corporations. In the words of political activist Abbie Hoffman it was about rejecting 'a system that channelled human beings like so many laboratory rats with electrodes rammed up their asses into a highly mechanized maze of class rankings, degrees, careers, neon supermarkets, military-industrial complexes, suburbs, repressed sexuality, hypocrisy, ulcers and psychoanalysts'. The man widely considered to have coined the term 'counterculture' is an American academic named Theodore Roszak. In his 1969 book *The Making of a Counter Culture*, Roszak argued that the youth rebellion at every level – technological or otherwise – was united in some manner against the so-called 'technocracy': a governmental order which, like a giant mainframe computer, operated by increasing efficiency and devising automated solutions to whatever myriad problems might be thrown at it. He suggested that it was the computer's ability to function as an emotionless, calculating logic machine – hardwired for rationality – that made it so impressive to America's power elite, who would openly brag about their country's 'overkill' ability to wipe out any of its enemies ten times over. 'Man's infatuation with the machine is frequently misunderstood as being a love affair with mere power,' Roszak writes. '... Is it not the machine's capacity to be severely routinized that we admire quite as much as its sheer strength?'

Several of the older Bay Area computer geeks (read: anyone over thirty) had been studying at University of California, Berkeley, in December 1964, when protesting students marched for Free Speech. Refusing to be thought of by the university bureaucracy as simply dehumanised bits of data, they seized upon the computational metaphor to register their anger. Many took blank punch cards (this in the days when computer programs were stored on pieces of stiff paper in which digital information was conveyed through the presence or absence of punctures in predefined positions), filled them with strange new configurations of holes, reading 'Strike!' and 'Free Speech', which were unlikely to render workable programs, and hung them around their necks. One particularly creative protestor – showing the blend of progressive politics and snarky humour that can only be found in students – augmented his outfit with a sign spoofing the punch card's user instructions. 'I am a UC student,' it read. 'Please do not fold, bend, spindle or mutilate me.'*

The suggestion that technology might not always be in man's best interest is not mutually exclusive with high levels of Cold War paranoia, of course. Techno-scepticism dates back at least

---

* Over the coming years this critique of modern man as robot, or the human brain as hardware, gained some measure of critical mass. In *The Greening of America*, author Charles A. Reich referred to it as *Consciousness II*: in which workers endured 'a robot life, in which man is deprived of his own being, and he becomes instead a mere role, occupation, or function'. Unlike their parents, however, the young countercultural generation (afforded the perspective of *Consciousness III*) felt anger at their government's betrayal, and would kick against it with a newfound 'capacity for outrage'.

as far as the Romantic poets' vision of Dark Satanic Mills. The Berkeley students were not wrong, though. Computers at the time were essentially tools of war: designed primarily to crack secret military codes and to aid with missile guidance. The IBM Corporation, for example, only built its first programmable digital computer after its doing so was requested by the United States Defense Department during the Korean War. Resituating the Dark Satanic Mainframes of the military-industrial complex as benevolent humanitarian aids would take time, and the kind of full-on public makeover that would unsettle even the most determined of PR agencies. It started gradually. At MIT in Massachusetts, a group of sympathetic academics formulated an intensely meritocratic set of principles, which would later be referred to as the 'hacker ethic' by author Steven Levy in his book *Hackers: Heroes of the Computer Revolution*. The hacker ethic included the following tenets:

- Access to computers – and anything which might teach you something about the way the world works – should be unlimited and total;
- All information should be free;
- Mistrust authority – promote decentralization;
- Hackers should be judged by their hacking, not bogus criteria such as degrees, age, race, or position;
- You can create art and beauty on a computer;
- Computers can change your life for the better.

When these ideals spread westwards they were met with approval by the countercultural techno-nerds. 'There was a feeling that the sky was the limit,' remembers Marc LeBrun, a programmer who was then working at the Stanford Artificial Intelligence Laboratory. 'We could redesign our consciousness, redesign society – why not redesign technology?' From here it began to snowball. Several geeks latched on to the ideas presented in Norbert Wiener's book *The Human Use of Human Beings*, in which the celebrated mathematician argued for the use of cybernetics in a way that would benefit humankind. 'These kids are explorers,' said an unnamed interviewee quoted in one of a series of articles that appeared in the *San Francisco Chronicle* during the summer of 1967 under the title 'I Was a Hippie'. 'They're trying to find creative ways to live in a leisure age – ten years away, when machines and computers will do most of the work.' Over at the California Institute of Technology, a 32-year-old poet in residence named Richard Brautigan wrote and hand-distributed a poem entitled 'All Watched Over by Machines of Loving Grace' in which he shared his increasingly popular dream for a world in which technology and small woodland creatures might cheerfully coexist.

Earlier critics of technology might have argued that simply pursuing efficiency through technological advancement did untold damage to both human beings and nature. For the philosopher Martin Heidegger this was most certainly the case. Taking on the level of abstraction he adopted in one 1949 lecture, mechanised agriculture was said to be no different from the gas chambers of the Holocaust, which was no different from electricity, which was no different from the atomic bomb. Many flower children readily accepted this view. The techno-hippies, however, had grown up as the first generation of rabid electronic consumers. For them, the better approach was to appropriate ill-gotten technology from its government overlords, to undermine its top-heavy power structure and democratise it for the masses. One of the loudest dissenting voices came from a philosopher by the name of Ted Nelson, who would become famous for coining the terms 'hypertext' and 'hypermedia'. In 1974 Nelson published what has become known as the world's first personal computer-specific book:

a Janus-like codex which joins two volumes back-to-back, entitled *Computer Lib / Dream Machines*. In it Nelson makes a convincing argument for rallying against Central Processing in all its myriad manifestations, and stresses that 'You can and must understand computers NOW':

> *Knowledge is power and so it tends to be hoarded.*
> *Experts in any field rarely want people to understand*
> *what they do, and generally enjoy putting people down*
> *... Doctors, lawyers and construction engineers are*
> *the same way. But computers are very special, and we*
> *have to deal with them everywhere, and this effectively*
> *gives the computer priesthood a stranglehold on the*
> *operation of all large organizations, of government*
> *bureaux, and everything else that they run ... It is*
> *imperative for many reasons that the appalling gap*
> *between public and computer insider be closed. As*
> *the saying goes, war is too important to be left to the*
> *generals.*

Thus, in the years that followed the UC Berkeley riots, computers were rethought of not as simply a cool thing to build – an Airfix model for those with high IQs and a fondness for soldering irons – but as the key to a utopian future; one in which cybernetics could bring people together rather than driving them apart. 'I had no problem with technology itself,' says Alvy Ray Smith, a wild-haired hippie who would later go on to co-found the computer-animation giant Pixar. 'I had problems with how the government *used* technology. I grew up in New Mexico with rockets going off. I heard the first atom bomb exploding when I was two, and there were rockets across the mountains from me. I lived near White Sands Missile Range, so after the Second World War, the

Americans brought Wernher von Braun and his V-2 rockets to White Sands to learn how to build rockets from the Germans. High-tech was just everywhere. I learned about computation there. I was still a short-haired, politically naive kid from the sticks, madly in love with technology and the future.'

Viewing it through the prism of the early 1970s the future looked to be a funny place. The sixties had ended with man walking on the moon, seeming to usher in a new age of high-tech utopianism. When President John F. Kennedy announced in 1961: 'I believe that [the United States] should commit itself to achieving the goal, before this decade is out, of landing a man on the Moon and returning him safely to Earth', the sentiment proved to be infectious. Big business jumped to take part. At that decade's New York World's Fair, visitors to the General Motors exhibition tent were invited to partake in a theme-park ride through a promised future of multi-lane motorways, glittering skyscrapers, moving pavements, underwater communities and holiday destinations in space. At the same time as this was going on, the spectre of nuclear holocaust remained hanging ominously overhead, as generations of schoolchildren were being taught to 'duck and cover' in the event of a nuclear attack by the Soviet Union.* Either way, science fact was starting rapidly to close the gap with science fiction, and the only question the public asked themselves was which mode of science fiction it would most resemble. Here the geeks – consuming, as they did, large quantities of science-fiction novels, and thus knowing the genre like the back of their sweaty hands – remained confident that it was to be a future they would approve of. 'We all read Robert

---

* Sample narration from a 1951 civil-defence film shown in schools: 'The bomb might explode when there are no grownups near us. Paul and Patty remember this, and are always ready to take care of themselves. Here they are on their way to school on a beautiful spring day. But no matter where they go or what they do, they always try to remember what to do if the atom bomb exploded right then.'

Heinlein's epic *Stranger in a Strange Land* as well as his libertarian screed-novel, *The Moon Is a Harsh Mistress*,' techno-hippie entrepreneur Stewart Brand recalled in a 1995 *Time* magazine article entitled 'We Owe It All to the Hippies'. 'Hippies and nerds alike revelled in Heinlein's contempt for centralised authority ... Ever since the 1950s, for reasons that are unclear to me, science fiction has been almost universally libertarian in outlook.'

Space rockets, of course, now seem strangely old-fashioned; appearing as one last Promethean burst at the tail end of the Age of Industry. The geeks would be proven right about the future, and the scale that it was to be enacted upon. The key to freedom was not to be found in the high-powered blast of a Saturn V rocket which would liberate man from the confines of his home planet, but in a smaller, more personal freedom, that had its roots in autonomy rather than astronomy. 'There was a huge desire for liberation on all fronts, and the personal computer was seen as being a tool of liberation,' says John Perry Barlow, a former lyricist for the Grateful Dead and a self-described cyber-libertarian. 'It was, in a way, a kind of unwitting Marxism where the means of production were being seized by the workforce and were no longer going to be something that was strictly the province of large institutions. That was directly connected with everything that had happened culturally in the sixties.' Something quasi-cohesive started to emerge – not entirely political, but not wholly technological either. It was the beginning of a breed of hippie ideology: sort of *Woodstock for wireheads, the New Left with better networking capabilities.* The widely accepted manifesto – as laid out in the locally produced periodical for a band of hackers called PCC – described the grand plans for the movement, filtered through the hackneyed, 'gee-whiz' dialogue of a pulp comic book: 'Computers are mostly used against people instead of for people; used to control people instead of to free them. Time to change all that – we need a ... People's Computer Company.'

If the language itself was revolutionary, it should come as little surprise. In 1976, the United States celebrated its bicentennial, marking the 200th anniversary of the establishment of the country as its own independent republic. The word 'revolution' was everywhere one looked, and changes that may previously have been viewed as simple progression began to be discussed in terms tantamount to a full-on insurrection.

Even discounting unnecessary hyperbole, though, there was every reason for the techno-geeks to feel that the future would be forged from digits instead of steel. From 1972, America's post-war Age of Affluence came to a sharp, jolting close as a series of oil-shortage crises hit the United States, one after the other. For almost the first time since the Great Depression it became evident that resources might not be quite as limitless as popular wisdom perceived. At the same time, the counterculture – which was more economically dependent than many wanted to believe – began to diffuse. Those who had been all too happy to follow Dr Timothy Leary's mantra of 'tune in, turn on, drop out' began, in the words of rockabilly singer Billy Craddock, to 'freak out, fuck up [and] crawl back'. In Silicon Valley, meanwhile, a little theorem known as Moore's Law suggested that this didn't have to be the case. Moore's Law, named after Intel co-founder Gordon Moore, theorised a long-term trend in computing in which the number of transistors that could be placed inexpensively on an integrated circuit would double every two years. The same year that those like Roszak were solemnly toasting the recently departed youth revolution, Intel released its 8008 chip, the world's first 8-bit microprocessor. The 8008's more popular successor, the Intel 8080, was what became the basis for MITS's Altair.

More than just the memory store of individual microchips – that led to the so-called 'memory race' among semiconductor manufacturers in Silicon Valley – the 1970s saw another knock-on effect which would transform the personal-computer scene and help launch dozens of companies. While Intel and the like rushed to produce bigger and better chips, the price of individual units inevitably shrank as, much like the calculator wars of the early part of the decade, semiconductor companies vied for a larger share of the market. A chip, such as the Intel 8080A, which sold for $110 in 1975, would cost just $20 two years later. By the end of the seventies, a standard 8-bit microprocessor might sell for anywhere between $5 and $8. To offset the inevitable losses semiconductor companies faced, larger and larger batches of microchips were ordered into production, since economies of scale meant that the longer a production run, the cheaper the price of manufacture. This also had the all too pleasant impact of flooding the streets with affordable microchips. Intel co-founder Andy Grove referred to them as 'high technology jelly beans'. *All the better to build personal computers with!*

In the same manner that Muhammad Ali would effortlessly dance out of the way of a larger, more lumbering opponent, only to dart in to pepper him with punches before once again retreating to safety, the geeks – armed to the teeth with cheap microchips, and not afraid to use them – knew that they had one major advantage over large computer-manufacturing corporations. They were smaller. They were quicker. And they were certainly more in tune with what was going on. Indeed, despite America's overall dominance in the global marketplace there was never a time when the corporation was perceived as being less fashionable. In an article entitled 'Love Is Dead' which appeared in the *New York Times Magazine*, Earl Shorris took great pleasure in highlighting the repressed company man as the barely living antithesis of the carefree hippie. He describes how 'a man grown weary of singing company songs at IBM picnics, feeling guilty about the profits he has made on defense stocks, who hasn't really loved his wife for 10 years, must admire, envy and wish for a life of love and contemplation, a simple life leading to a beatific peace. He soothes his despair with the possibility that the hippies have found the answers to problems he does not dare to face.'

For their part, giant corporations saw personal computers as a dead end. They told themselves that they were too niche; too small; too grubby; incapable of carrying out any task the general public might actually be interested in. 'I blew them off altogether,' admits Alvy Ray Smith, who was then working in the R&D wing of Xerox. 'I thought they were just silly little toys.' Their own business models called not for the manufacture and sale of myriad small, low-cost machines but in producing a relatively small number of high-end computers and then charging clients through the nose to maintain them. (Let us also not forget that this was at a time when any businessman worth his salt wouldn't be caught dead typing out his own correspondence.) Sizing up their potential new enemies, IBM and the rest saw little to be concerned about. 'They just thought we were a bunch of crackpots out playing in the garage,' says Randy Wiggington, who went on to become an early Apple employee. This perception was more or less shared by the general public, for whom there was an undoubted grubbiness attached to a world inhabited by electronics hobbyists. The names of the products themselves didn't help matters a great deal. One week a local Bay Area parts distributor, Marty Spergel, who was nicknamed 'The Junkman', had a shipment of computer joysticks – packed into boxes labelled 'for hobbyist usage' – confiscated by US customs on the basis that they appeared to be some sort of sex toy. Nor did the typical computer geek make the pursuit appear any less seedy.

'I was not very socially adept,' says Lee Felsenstein, a hobbyist who was heavily involved with the scene. 'I didn't know until I was in my early twenties that when somebody said, "How are you?" that they weren't asking an objective question. I thought they really wanted to know how I *was*, what my condition was. Someone had to sit me down and explain that it was just a kind of ritual.' Another person, who went on to make millions of dollars in the tech industry, told me the story of a young woman rumoured to have attended a hackers' conference at this time, hoping to meet eligible young men and fully aware that nearly all the attendees would be male. Her verdict: 'the odds were good, but the goods were odd'.

Shunned by the general public, corporations and more than a few of their peers in academia, it became apparent that the geeks and techno-hippies needed somewhere to talk. That place became known as the Homebrew Computer Club. (Some initially suggested the 'Eight Bit Byte Bangers' or the 'Midget Brains' as better choices of name, but Homebrew stuck; largely as tribute to one of the club's spiritual founders, who was a homebrew-beer aficionado.) 'This kind of allure of the computer, and denial of access to it, created a pent-up demand within Silicon Valley,' recalls one former member. 'The Homebrew Club was the place where you could begin to scratch that itch.' Homebrew was a social gathering for the socially maladroit, which functioned as both regular instructional seminar and support group. It was started by two men, Gordon French and Fred Moore, who had first met at the Community Computer Centre in nearby Menlo Park. The inaugural meeting took place in March 1975 and French volunteered his garage as a venue. A posted-up bulletin-board notice advertising the event invited interested parties to 'come to a gathering of people with like-minded interests. Exchange information, swap ideas, help work on a project, whatever.'

'It was a rainy afternoon,' remembers Bob Lash, an attendee who was among the first to answer the club's siren call. 'We just meandered around the garage and talked to each other. The big event was going into Gordon's house to see the computer he was building – an 8008-based minicomputer which he called The Chickenhawk.' Thirty-two people turned up at that first meeting. Before long, attendance had ballooned to around three hundred. Within a few years the club's mailing list (you needed to attend in person at least once to get on it) had around 3,000 names and addresses dutifully noted down. As this happened, the club's premises moved first from Gordon French's house to the second floor of a creaky private school, and then, when it outgrew this too, to the auditorium at Stanford's Linear Accelerator Centre. The techno-geeks had their tools, they had amassed an army and now they had a home base worth crowing about. *Let the revolution begin!*

Lee Felsenstein, in his early thirties at the time, took on the role of club chairman. Felsenstein is an important figure in the history of the personal-computer revolution. Born in 1945, he spent his childhood and teenage years in a Jewish neighbourhood in Philadelphia. 'I was brought up in a family which very much believed in the machine as a model for human society,' he recalls. Felsenstein's parents were prototypical beatniks-cum-communists, who were 'basically thrown out of the party for being freethinkers'. By his own admission he was a red-diaper baby, brought up in a house in which communist ideals thrived. At the age of eighteen he left home to attend UC Berkeley, a relatively inexpensive public college on the opposite side of the country, which was welcoming students from places other than just the West Coast in an effort to enrich the student population. When he arrived in California in 1963, he was subject to extreme culture shock, finding the whole scenario impossibly glamorous. 'My impression of it was that it was the land of storybooks,' he remembers. Everywhere he turned was some larger-than-life reminder of the difference between America's East and West Coasts. One of his first actions was to open an account at the local bank, which turned out to be Wells Fargo. Felsenstein couldn't believe it. To him, the name was synonymous with the Wells Fargo stagecoach of Wild West legend. It stopped him dead in his tracks. 'It was just a bank, but it had a great name,' he says. He felt as if he had left his old safe Jewish neighbourhood in the middle of the twentieth century and somehow arrived back on the American frontier. Overtly politicised from his communist upbringing,* he lost no time in getting involved with the various student protests which were going on at the time. The year he started at UC Berkeley he spent Hiroshima Day, along with a group of like-minded youths, 'trying to spoil everyone's Sunday by reminding them that this was the anniversary of dropping the Bomb'. In addition he got involved with the local civil-rights organisation, CoRE (The Congress of Racial Equality) and threw himself headlong into the Free Speech movement, becoming one of the 768 arrested in the climactic 'Sproul Hall Sit-In' of December 1964. Soon Felsenstein began working as the 'military editor' for the *Berkeley Barb*, one of the Bay Area's leading underground newspapers.

This protest impulse led directly into Felsenstein's growing interest in the world-changing possibilities offered by computers. In 1970 he was sent on a course to learn how to program. 'Its instructors were guys approximately my age, but they were wearing three-piece suits and strutting about most pompously,' he notes. When one of them commented that the reason the mainframe terminal Felsenstein was using had suddenly slowed down was because the network had switched over from one based in Los Angeles to one in Kansas City, he came to the

---

* Years later, Lee Felsenstein's therapist would help him stumble upon the revelation that his older brother, the mathematical-biologist Joe Felsenstein, was named after Joseph Stalin. 'My therapist mentioned "Uncle Joe" in talking about my childhood,' Lee recalls. 'I said, "no, no, Joe's my brother." And then I realised ... But you forget, in 1942 Stalin was on the front cover of *Time* magazine as their Man of the Year. He was a very big person in my parents' lives.'

snap realisation that all the machines were networked together. Since the computers were constantly speaking to one another, Felsenstein theorised that they could be used to help organise the kinds of mass protests he was involved with. 'The UC Berkeley campus had 20,000 students and several thousand more working at the university,' he says. 'One of the salient features of that was the alienation that the students used to feel there. They were alone in the midst of a huge crowd, and the administration counted on that. But I figured that if we could provide the tools by which people could identify themselves as members of a community, the barriers to lateral communication would come down ... When people began communicating laterally instead of hierarchically, magic things could happen.' From here he developed the Community Memory Project: a series of teletype computers placed in public areas, by which locals could communicate with one another. At first Felsenstein got a bit of flak for the idea.

'Lee, why do you want to put computers between people?' one person asked. 'If I want to talk to Pearl over there, why don't I just go over and say, "Pearl, I want to talk to you"?'

Felsenstein considered the point for a few seconds.

'Well,' he said, 'what if you don't know that it's Pearl that you want to talk to?'

There was a pause.

'I think I get what you mean,' came the response.

After a while, he moved on to the idea of personal computing. Several of the public Community Memory computers were damaged, and he realised that it would be altogether more sensible if people had their own machines through which to communicate. Felsenstein's approach to the technology would echo the same concerns as the computer utopians, although with something more of a nod to the apocalyptic. He dreamed of eventually creating a personal computer capable of surviving a third world war, when the world's peasants would inevitably storm into America to liberate it from the forces of capitalism. He was convinced that 'the industrial infrastructure might be snatched away at any time, and the people should be able to scrounge parts to keep their machines going in the rubble of the devastated society; ideally, the machine's design would be clear enough to allow users to figure out where to put those parts.'

At the Homebrew Club, Lee Felsenstein took his position very seriously and maintained order with an iron fist. When he spoke, he would gesticulate with the aid of a lecturer's stick, which he used to single out members of the audience. If people weren't paying attention he would clamp the stick to the podium with one hand and then pull it up and release it with the other to make a loud *thwacking* sound. In his own mind he was doing a pitch-perfect parody of the authoritarian professor. Most of the Homebrew attendees just thought of him as strict.

Meetings followed a regular format. After calling people to order, Felsenstein would theatrically announce the commencement of 'the Homebrew Computer Club, which does not exist'. After the applause had died down, he began the evening with a 'mapping period', during which club members could get up and ask a question, speculate on a bit of techno-gossip they had heard, or speak about an item of interest – often in the form of a presentation. After this there would be 'random access', in which attendees could wander around informally, trading information and helping one another. As word got out about the captive audience of computer-hungry geeks gathering fortnightly, local business owners began showing up, each hoping that they might be able to turn a profit on some obscure hardware component or unload a batch of new Intel microchips. The promise of new personal computers, each more impressive than the last, was a regular event. Inevitably most of these would never see the light of day. 'People would try and sell computers that were never designed,' recalls Randy Wiggington. 'These were pretty early days. A new computer would be announced, and then it would disappear within a few weeks and no one would ever hear of it again. There was a lot more snake oil than anything else.'

Although everyone was keen to get to the 'random access' part of the evening, the earlier 'mapping period' demonstrations often proved invaluable and, occasionally, ground-breaking. One particularly memorable showing occurred at Homebrew's 16 April 1975 meeting when a club member named Steve Dompier stood up in front of the group and used his Altair to play a tune on a nearby radio thanks to the magic of electrical interference. The music he chose was the Beatles' 'The Fool on the Hill' from their *Magical Mystery Tour* album. 'What do you do with a machine that so far has no input/output boards or peripherals?' Dompier asked when writing up his discovery. 'Well, there's always the front panel switches and machine language, so I was soon busy making up programs to test all of the 8080's functions, and getting a good set of calluses on my ten input devices. There's a lot of 8080 instructions! I had just finished setting in a "sort" program, and at the same time I was listening to a weather broadcast on a little transistor low-frequency radio, which was sitting next to the Altair. I hit the "run" switch on the computer and it took off running the same list of numbers over and over again. At the same time my radio also took off. The computer was sorting numbers and the radio was going *zzziiipp! zzziiipp! zzziiipp!!!*' After some investigation Dompier discovered that the radio was intercepting radio wave interference sent by the Altair. He tried several different programs to see how the radio would react, and found that he could enter code that, in turn, would produce differing tones. 'I had myself a program that could actually make music,' he recalled. 'Of a sort,' he added.*

This was, in the minds of those present, the democratisation of technology in action. When Dompier performed his musical symphony of radio interference, what made it so hair-raisingly exciting was not just the fact that he had devised a way to play music on an AM radio using a computer with 256 bytes of memory. It was that he was someone *completely untrained as a computer scientist* who had devised a way to play music on an AM radio using a computer with 256 bytes of memory. 'For him to be able to show it to the group was a *tour de force*,' says Felsenstein. 'It was motivational. If he could do it, then anyone could do it, something that astonishing.'

Randy Wiggington was just fourteen years old when he started attending the Homebrew Club. Because he was still several years away from being able to legally drive a car, it was difficult for him to get to and from the Stanford Linear Accelerator Center, where meetings were held. One evening when Felsenstein asked the group if anyone had any news they would like to share, Wiggington stuck his hand up, told everyone where he lived, and asked whether anyone in the room might be able to offer him a lift on future occasions. Later that night, Stephen Wozniak

---

* One person among those impressed by Dompier's music synthesiser was the nineteen-year-old Bill Gates. Less than a year later, Gates would again appear on the radar of those at Homebrew when he sent the club newsletter an indignant 'open letter to hobbyists', complaining that the frequent piracy of his Altair BASIC software had put his company Microsoft (then stylised as Micro-Soft) out of pocket by $40,000 in paid computer time. 'I laughed at it along with everyone else at the club,' says Lee Felsenstein. 'I don't think they were paying $40,000 for [the computer time]. They didn't have that. What everyone else would have done would be to show up after hours at a computer centre and to talk their way into being able to use the computer – and that was what (Microsoft co-founder) Paul Allen was doing. So when Bill said $40,000 of computer time everyone knew that it was disingenuous, and that this monetary number was funny-money. He lost all credibility at that point. Calling us names didn't help either.'

ambled over and introduced himself. It transpired that Woz, as he was known to friends, lived in an apartment about a mile from Wiggington's parents in nearby Sunnyvale. He said that he would be more than happy to pick Wiggington up on his way through, and to drop him back home afterwards. 'That's the way he was,' Wiggington says. 'Always very helpful, very kind. He wasn't pretentious or anything. He would give me a ride each time and we would talk all the way there and all the way back. I would just ask questions non-stop. That was sort of my education.' At meetings Wozniak, Wiggington and a growing group of teenaged hangers-on would occupy the centre back two rows of the hall. 'The SLAC auditorium was really set up geographically, and people would go to the same places,' recalled Chris Espinosa – today a senior employee at Apple but then a young hobbyist who knew Wiggington from summer school. 'It was like a high-school cafeteria: you sat with the "right people". It was very amusing. During the Random Access section of the meeting there was a lot more mixing, but in the Mapping section, you really knew where to sit, unless you were new, in which case you migrated around.' Espinosa's mother, who also went on to work for Apple for ten years, drove her son to and from meetings. 'She'd sit in the back, reading, while all of these slightly dirty and dangerous people discussed [computer] things,' Espinosa said.

On one of Wozniak's and Wiggington's fortnightly journeys to the Homebrew Club, Wozniak casually brought up the fact that he was building, from scratch, a computer of his own. Even at Homebrew, a group whose entire *raison d'être* was the building of personal computers, this was somewhat unusual. Most members – if they owned a computer at all – were likely to have bought one in kit form, like the Altair 8800. 'It was kind of like a Ferrari car club,' Wiggington recalls. 'You go along and *maybe* 10 per cent of the members actually own a Ferrari, and 90 per cent want one or want to hang out with those who do have one.'

The machine that Wozniak was building was a large part of the reason for his growing entourage. Its folksy moniker neatly summed up the homespun, organic, *friendly* approach to computing that the Homebrew Club epitomised. The fact that he added a roman numeral to the end of its name also hinted at the fact that this would be the first of several. For now, though, this one would have to do. Wozniak called it the Apple I.

2

ANGEL-HEADED HiPSTER HACKERS

'People on the East Coast were hearing a lot about California surfers, California bikers, hot rodders, car customizers, California hippies, and political protesters, and the picture they got was of young people in jeans and T-shirts who were casual, spontaneous, impulsive, emotional, sensual, undisciplined, and obnoxiously proud of it. So these semi-conductor outfits in the Silicon Valley with their CEOs dressed like camp counsellors struck them as the business versions of the same thing. They couldn't have been more wrong.'

Tom Wolfe,
'*The Tinkerings of Robert Noyce*' (1983)

The common refrain is that nobody is born in California. People *go* to California. This certainly proved to be the case in the 1960s, when tens of thousands of young people descended upon San Francisco – some wearing flowers in their hair, others not wearing very much of anything – to 'turn on, tune in and drop out'. These were not, however, the first wave of émigrés to arrive on the West Coast. Since the middle of the nineteenth century the area had been a desirable destination for those wishing to embrace the more relaxed California climes over the buttoned-down rigidity of America's East Coast. Like the country's Founding Fathers, virtually all of the early Silicon Valley arrivals were white men, none with roots in the area, and all expressing a disdain for the 'old-line' way that industry had operated where they had come from.* They brought with them the spirit of innovation. Although the Spanish had established missions in the region in the 1770s, the modern era for Silicon Valley began in 1850, when Irish immigrant Martin Murphy, Jr, and his family established the Bay View ranch. From the start, Murphy set the tone for entrepreneurship and different thinking. He granted permission to the San Jose and San Francisco Railroad Company to build a railway line across his lands en route to San Francisco. The track became the first such completed line in the state and fuelled economic growth in the vicinity. In the 1920s some of the Murphy estate was sold to Albert W. Besset, an inventor who had patented a nifty design for new, safer chicken coops. Besset used the money from his Jubilee Incubator Company to provide the start-up funds to his son's radio-shop business, which subsequently became the area's first electronic-manufacturing business. So it has continued since. 'There's a tilt in the continent and the kooks and innovators flow west until they can't flow any further and come up against the Pacific Ocean,' says Stewart Brand. 'Then

---

* The 'valley' in Silicon Valley refers to the Santa Clara Valley, located at the southern end of San Francisco Bay. Silicon, meanwhile, refers to the high concentration of companies involved in the semiconductor industry, in which silicon is used in the creation process. The term Silicon Valley itself was coined by Ralph Vaerst, a successful California entrepreneur. Its first published use was in 1971, in the weekly trade newspaper *Electronic News*, which ran a series of articles entitled 'Silicon Valley in the USA'. Nevertheless, I use this phrase to describe the area throughout its history to avoid confusion.

they start innovating at the edge.' This is particularly true in high-tech. Take a look at the names dominating Silicon Valley culture today and it becomes immediately apparent that these, too, are outsiders, driven to America's new Land of Opportunity by the promise of a new way of life. Intel's Andy Grove, for example, was born in communist Hungary, before arriving in California in 1963. Adobe Systems' John Warnock called Salt Lake City, Utah, his first home. Hewlett-Packard's David Packard came from Colorado, while his partner Bill Hewlett hailed from Michigan. So did Google's Larry Page, while co-founder Sergey Brin began life in Moscow, Russia. Even Silicon Valley's newest superstar, Facebook's Mark Zuckerberg, was born in an upscale part of New York and founded his multi-billion dollar social-networking service – originally called 'Thefacebook' – from a dorm room in Harvard before making the decision to move out to California.

By contrast, Steve Jobs and Steve Wozniak, the co-founders of Apple, Inc., were born in the area. In a part of the country dominated by émigrés seeking a new life for themselves, they were Californian through and through, and grew up with the region's disparate influences drummed into them from childhood.

They were outsiders in other ways.

□ □ □

Stephan Gary Wozniak was born on 11 August 1950. The most memorable item of local news that month was the crashing of a bomb-laden B-29 Superfortress into a residential area of California, a tragedy which killed seventeen people and injured a further sixty-eight. The son of a Lockheed engineer, Wozniak spent his early childhood building various electronic kits and becoming an avid ham-radio operator. 'I never did know for sure what my dad did for a living,' Wozniak commented in his memoir. '... I managed to put together little bits and pieces. I remember seeing NASA-type pictures of rockets, and stuff related to the

Polaris missile being shot from submarines or something, but he was just so closemouthed about it, the door slammed there.' As a crew-cut eleven-year-old, Wozniak was pictured on the front page of the *San Jose Mercury News*, posing next to Richard Nixon – then engaged in his 1962 California gubernatorial race – representing Nixon's unanimous support among members of Cupertino's Serra School ham-radio operators' group. In fact, there was no such group, and Wozniak was probably the only ham-radio operator for miles around; almost certainly the youngest. While other children had pictures of Wilt 'The Stilt' Chamberlain or the Beverly Hillbillies lining their bedroom walls, Wozniak had an image of a rack-mounted 4K SuperNOVA Computer posted above his bed.* One day, he told himself, he would own one. When his father commented that a mainframe computer would cost as much as a house, the young Wozniak replied, 'That's okay, I'll live in an apartment.'

Like many of the people drawn to electronics hobbyism, Wozniak was something of a loner in his youth; fretfully shy and introverted. 'He was very square at high school,' his mother once said. 'He wasn't too much with the girls.' The main attention that he received from his peers growing up was for his work as a prankster, which was exceptionally imaginative as such high jinks go. He may have pushed it too far on one occasion, however, when he fashioned what looked like several sticks of dynamite from pieces of scrap metal, attached an oscillator which made a loud ticking sound, and put the whole lot into a friend's locker, being extra careful to leave some incriminating wires hanging artistically out of the door. The next thing anyone knew, the school principal was rushing heroically to the football

---

* Amusingly, one newspaper journalist failed to hear Wozniak correctly during a 2008 interview and instead misreported this as an 'Iraqi super computer'. Such errors and their rapid online dissemination are what Internet dreams are made of.

field, holding Wozniak's 'bomb' at arm's length, before launching a furious investigation into who on earth had tried to blow up Homestead High School. When Wozniak was fingered as the culprit (initially he thought he was being called into the principal's office to be congratulated on winning a maths competition) he was sent for an overnight stay at the local juvenile hall. Walking into class the following day, he was given a rousing ovation by his classmates. Then he went back to being invisible again.

Rather than focus on making friends, Wozniak spent hours building electronics projects such as a machine he nicknamed the Cream Soda Computer, which served as a precursor to the Apple I. To complete the Cream Soda Computer (so-called because of the industrial quantities of Cragmont cream soda drunk while building it) Wozniak drafted in the services of his neighbour, Bill Fernandez, to build a functioning power supply. After it was finished, Wozniak's mother put her son in touch with a reporter she knew from the local paper, who agreed to come and write a story about the two boys. The reporter duly turned up and asked Wozniak and Fernandez to pose for a photograph by their creation. When this happened the Cream Soda Computer began billowing smoke and promptly expired with a loud bang. Fernandez discovered that the cause was an unmarked chip he had used for the power supply, given to him as thanks for some gardening he had done.

Following high school, Wozniak briefly attended UC Berkeley for what turned out to be a short college experience. Again his most memorable moments were prank-related. This time he sneaked into the university computer room and rigged a couple of programs to churn out endless reams of paper reading FUCK NIXON and GOOD SCRAP PAPER. After one year he dropped out altogether, moved home and started working at Hewlett-Packard, designing calculator circuits. It was during this time that he began attending the Homebrew Computer Club, being among those intrepid attendees present at the first meeting. 'We had a lot of interest and enthusiasm,' he would recall of the prevailing attitude at Homebrew. 'The rewards that drove us were all intrinsic ... It's not like you get a better salary, or a better title, or more respect at work, or a new car. We had the autonomy of creators. We could decide what was going to make a neat computer. We could implement it, and we could show it off. We also had excellent feedback from our peers.' He bought absolutely into the cultural freedom-fighting ideology that underpinned the club. 'We were revolutionaries,' he wrote in his later memoirs. 'Big companies like IBM and Digital Equipment didn't hear our social message. And they didn't have a clue how powerful a force this small computer could be. They looked at what we were doing – small computers, hobby computers – and said they would just remain toys ... They didn't imagine how they could evolve.'

□ □ □

Steven Paul Jobs was born five years later than Wozniak, on 24 February 1955. His parents were 23-year-old German-American Joanne Carole Schieble and Abdul Fattah Jandali, a young Syrian Muslim immigrant. Since Schieble's white, conservative Christian family would not accept her marrying Jandali, she took herself off from Wisconsin to the more liberal San Francisco in order to have her child and put it up for adoption without her parents finding out. Her child was subsequently adopted by Clara and Paul Jobs, a middle-class San Francisco couple. Childhood for Jobs, like Wozniak, occurred at a politically turbulent juncture in American history. 'I was walking across the grass at my schoolyard going home at about three in the afternoon when somebody yelled that the President had been shot and killed,' he recalled of the death of John F. Kennedy. 'I must have been about seven or eight years old, I guess, and I knew exactly what it meant. I also remember very much the Cuban Missile Crisis. I probably didn't sleep for three or four nights because I was afraid that if I went to sleep I wouldn't wake up. I guess I was seven years old at the time and I understood exactly what was going on. I think everybody did. It was really a terror that I will never forget, and it probably never really left. I think that everyone felt it at that time.' For someone who would show such an interest in the American counterculture throughout his life, it is interesting to note just how closely Jobs' life coincided with the movement. Several months after he was born, Allen Ginsberg would write his extraordinary work 'Howl', which went on to become the most popular poem of the Beat Generation, at a coffeehouse in Berkeley, an hour's drive from Cupertino.

Later in life Jobs further ruminated on the period which marked his childhood. 'It was a very interesting time in the United States,' he noted of the late 1950s and early sixties. 'America was sort of at its pinnacle of post World War II prosperity and everything had been fairly straight and narrow, from haircuts to culture in every way, and it was just starting to broaden into the sixties where things were going to start expanding out in new directions. Everything was still very successful, very young. America seemed young and naive in many ways to me, from my memories at that time.'

Of course, it is worth noting that America at the time might have seemed young and naive to Jobs because he was, in reality, young and naive himself. Despite his biographical overlap with the era loosely termed 'the counterculture' (yet more loosely termed as 'the sixties', despite clearly lasting far longer than ten years), Jobs was a decade too young to have been a meaningful part of those first few waves of countercultural activity, including the Beats, the hipsters and even the main thrust of the hippies. 'We wanted to be hippies but we were too young,' says Daniel Kottke, one of Jobs' later close friends from Reed College, who went on to become Apple's first paid employee. 'Too young to go to Woodstock.' In fact, Jobs was still in junior high at the time of the legendary music festival. Despite this, throughout his life he continued to place himself spiritually as a figure of the sixties wherever possible. The reasons for this can be endlessly debated. Certainly Jobs was never adverse to a bit of myth-making, particularly when the actual details of his early life appear to be both, respectively, traumatic and strikingly average.*

A student prone to the occasional bout of misbehaviour, he talked about how, had it not been for a kindly fourth-grade teacher, he 'would absolutely have ended up in jail'. Similarly, in the way of highly intelligent children who struggle to make friends of their own age, he gravitated to people older than himself. 'I was a little bit more mature for my age,' he said. Like the best Apple solutions his childhood was tweaked, simplified, redesigned until it resembled not the story of a middle-class kid growing up in a middle-class neighbourhood, but the Great American Novel. With Jobs, naturally, as its protagonist. 'Steve Jobs ... burst [into] the world's consciousness as a unique and

---

* Despite the psycoanalytical temptation to frame later idiosyncrasies, ranging from Jobs' desire for total control to his binary relationship patterns of total acceptance followed by harsh rejection, as characteristics arising from his knowledge of early parental abandonment, I will avoid doing so at length. More interesting is to contrast Jobs' childhood with that of Bill Gates', the figure most commonly seen as Jobs' opposite number throughout his career. Jobs, adopted at birth, grew up in a middle-class home with parents who were employed as a machinist (his father) and an accountant (his mother). Gates, meanwhile, was the son of a prominent lawyer, had a mother on the board of directors for First Interstate BancSystem and the United Way, and a grandfather who was a national bank president. Steve Jobs persuaded his parents to take him out of his first school because, in the words of a classmate, 'fights were a daily occurrence; as were shakedowns in the bathroom ... The year before [Jobs] arrived, a group of eighth-grade boys had gone to jail for gang rape. During [Jobs'] seventh-grade year, the school lost a wrestling match to Sunnyvale's Mango Junior High and proceeded to demolish that school's team bus.' Gates, for his part, attended the Lakeside School, an exclusive preparatory school in Seattle, Washington, where one quarter of alumni go on to Ivy League schools and 99 per cent to college. Jobs' background would favourably underpin Apple's underdog ideology, as much as Gates' upper-class upbringing would seem to cement Microsoft's image as the Establishment.

fully formed character – half hippie shaman, half tech savant,' wrote Silicon Valley journalist Michael S. Malone in his memoir. Malone, two years older than Jobs, had attended the same Mountain View primary, Monta Loma School, as the future tech billionaire. Years later, in 1984 – several weeks before the launch of the Macintosh computer – he interviewed the Apple co-founder for a magazine article. By sheer coincidence, Malone had recently found himself in possession of Jobs' fourth-grade geography report, which had somehow been mixed in with his own childhood work two decades earlier. As a joke, Malone decided to open the interview by presenting the report to Jobs, thinking that he would find it amusing. He didn't.

'Where did you get this?' Jobs snapped, as though Malone had stumbled across the plans for a hitherto unannounced next-generation Apple computer.

'He was already busy constructing that persona that night,' Malone wrote, '... and the last thing he needed was a reporter who seemed to have a dossier on his entire life.' Like Wozniak, Jobs also had an interest in electronics growing up, although with nothing like the passion of his eventual business partner. Upon the advice of an engineer neighbour, Larry Lang, Jobs joined the Hewlett-Packard Explorers Club and later – trying to get hold of spare parts to build a frequency counter – phoned HP founder, Bill Hewlett, and found himself being offered a summer job.

Jobs' teenaged years were marked by a continuance of the low-level rebellion he had exhibited as a child. 'Mr. Jobs may have experimented with illegal drugs, having come from *that generation* [emphasis added],' notes his 1991 FBI profile. Certainly he smoked marijuana, and began experimenting with LSD by his senior year at high school. This was often done in the fields around Cupertino, leading to a moment Jobs would later recount as one of the most profound that he ever experienced. 'All of a sudden the wheat field was playing Bach,' he recalled. 'It was the most wonderful experience of my life up to that point. I felt like the conductor of this symphony with Bach coming through the wheat field.'

□ □ □

The Two Steves met in the early 1970s, courtesy of mutual acquaintance Bill Fernandez, and quickly became close. 'We were introduced because we had two things in common: electronics and pranks,' Wozniak has said. Jobs – who later in life modestly described his aspirations as the need to 'put a ding in the universe' – cast their friendship in more galactic terms. 'We're sort of like two planets in their own orbits that every so often intersect,' he later told *Playboy* magazine. 'It wasn't just computers, either.' Jobs had grown up listening to the Beatles, who formed as a group when he was five and released their last original album, *Let It Be*, when he was fifteen. Wozniak turned him on to Bob Dylan, whose music proved to be a constant love throughout Jobs' life. The two spent long teenage hours discussing and comparing the finer points of both artists. They concluded that, as great as the Beatles were, their catalogue was made up of radio-friendly hits, while Dylan's music carried greater resonance. 'Dylan's songs struck a moral chord,' Wozniak recalled. 'They kind of made you think about what was right and wrong in the world, and how you're going to live and be.' Jobs' favourite Dylan album was 1975's *Blood on the Tracks*, widely seen as one of the artist's best – and most heartbreakingly plaintive. 'We used to listen to it over and over again,' Jobs' friend, Dan Kottke says. Years later, when *Rolling Stone* magazine interviewed Jobs about the launch of iTunes, journalist Jeff Goodell was able (after several abortive attempts) to engage him in a brief discussion about Dylan's work. 'He was a very clear thinker and he was a poet,' Jobs stated, picking up on two characteristics that he himself consciously strove to embody throughout his career. 'I think he wrote about what he saw and thought. The early stuff is very precise. But, as he matured, you know, you had to unravel it a little bit. But once you did, it was just as clear as a bell.'

'Steve was into everything hippie,' Wozniak recalled. 'He ran around shouting "free love, man" and eating seeds.' 'This was California,' Jobs noted. 'You could get LSD fresh made from Stanford. You could sleep on the beach at night with your girlfriend. California has a sense of experimentation and a sense of openness – openness to new possibilities.' Wozniak himself was a far closer fit for the archetypal image associated with the computer geek of the day, although he, too, would make the occasional nod to hippie haute couture. 'I would wear this little Indian headband, and I wore my hair really long and grew a beard,' he recalled in his autobiography. 'From the neck up, I looked like Jesus Christ. But from the neck down, I still wore the clothes of a regular kid, a kid engineer.' Wozniak also refused to take LSD, fearing that it would impact negatively on his engineering ability. According to Wozniak, he didn't even get drunk for the first time until the age of thirty.

There were few better places to be a disaffected teen in the sixties and seventies than Silicon Valley. Thanks to the rebellious spirit which coursed through the region, a prevailing attitude of intense meritocracy had developed. A person could drop out of college (or even high school), stumble into one of the numerous places in the area in which ground-breaking work was being carried out and, provided they were intellectually up to snuff, find themselves gainfully employed. One such research centre was the Stanford Artificial Intelligence Laboratory, better known as SAIL. Headed up by two MIT refugees, John McCarthy and Les Earnest, SAIL was the epitome of Silicon Valley geekoid utopia. Its rooms were named after locations from J. R. R. Tolkein's Middle Earth saga, while its printer was rigged in such a way that it could produce three (!) different Elven font types. Despite this, SAIL wore its renegade countercultural idealism on its sleeve like a badge of honour. As an intense intellectual, John McCarthy had little time for the dull rigidity of management. Hardly a micromanager, his standoffish nature cultivated an atmosphere of extreme freedom which, by extension, attracted an excitingly eclectic band of scientists. Les Earnest, meanwhile, came to represent the anarchic spirit of the lab, which itself owed much to the hacker ethic of MIT. The stories of his clashes with the Establishment were legion. In the early sixties, as an employee of the MITRE Corporation, Earnest had once been 'loaned' out to the CIA to help integrate several military computer systems. As part of his security clearance he was asked to fill out a personal profile. Under 'race' he put 'mongrel'. He was hauled in front of various higher-ups and given a dressing-down, before they eventually agreed to accept his chosen race only on condition that he sign a sworn affidavit to that effect.

The Two Steves were never employed at SAIL, but they would occasionally visit in order to hang out with a mutual friend, Allen Baum, who worked there for a time. Regardless, the palpable excitement which existed at SAIL added to Wozniak's longing for a computer of his own. Jobs, for his part, claimed that the 'vibrations' he felt setting foot into the laboratory remained with him his entire life.

□ □ □

If the Stanford AI lab had an impact on Jobs' developing sensibilities, then it was nothing compared to that of *The Whole Earth Catalog*, a do-it-yourself manual produced in nearby Menlo Park, which married the hippie interest in ecology with the cyber-libertarian ideals of the techies – for those whose idea of back-to-nature came with a hefty number of microchips. Along with *Rolling Stone*, *The Whole Earth Catalog* became one of the two most memorable countercultural documents to emerge from the Haight-Ashbury movement. It was founded by a 29-year-old

Bay Area hippie named Stewart Brand, who funded its creation with an inheritance from his father. Born in 1938, Brand had graduated from Stanford University in 1960 with a degree in Biology. After a stint as a parachutist in the US military, he hooked up with *One Flew Over the Cuckoo's Nest* author Ken Kesey and his group of Merry Pranksters, whose early travels around America in a converted school bus are chronicled in Tom Wolfe's seminal book *The Electric Kool-Aid Acid Test*. Brand went on to play a key role in the narcotic-heavy Haight-Ashbury scene; both partaking in and helping to organise Acid Tests for willing participants eager to expand their minds.

'How do you feel?' he was asked after taking LSD himself for the first time.

'Very *thing*!' he replied.

On a separate occasion after dropping acid, Brand conceived himself voyaging down into his dinner plate to walk among the vegetables. One particular piece of potato appeared to transform itself into a heroic visage of himself. Meanwhile, Brand's interest in the mind-expanding medium of computers blossomed. In 1968 he assisted the pioneering computer scientist Doug Engelbart in what is now remembered as The Mother of All Demos, during which Engelbart demonstrated the computer mouse, video conferencing, email, hypertext and word-processing to the world for the first time.* That same year Brand launched his *Whole Earth Catalog*. Its first mimeographed issue, limited to an initial print run of 1,000 copies, featured a NASA photograph of the earth from space on the cover, and introduced the familiar 'Access to Tools' subheading. It was a telling image to select; one which

---

* Doug Englebart was once given LSD under test conditions, with the rationale that *if he can come up with this stuff straight, imagine what he could do stoned*. He spent several hours staring fascinated at a wall, and dreamed up a miniature water wheel – which he called a 'tinkle toy' – that could sit in a toilet bowl and spin when urinated upon to aid with potty-training.

took the latest in space-age technology and reframed it within an anti-industrial, or even altogether counter-modern, paradigm. While the moon landings had the vast majority of people staring out in wonderment at the universe and considering the possibilities of intergalactic travel, *The Whole Earth Catalog* opted to turn its gaze back upon the earth.

Inside the publication was a smorgasbord of items, split across seven separate categories: *Understanding Whole Systems; Shelter and Land Use; Industry and Craft; Communications; Community; Nomadics;* and *Learning*. There were articles on such headscratchers as 'Should sportsmen take dope?', a proposal for 'A Libertarian Nomadic Association in Southern California', letters from Merry Pranksters Ken Kesey and Ron Bevirt, and notes on where to buy items ranging from camping equipment to programmable calculators. It was, in essence, everything that the self-respecting techno-hippie might need to truly do-it-themselves. 'DIY had been this unhip thing that pathetic daddies did in the backyard, the attic, the basement and the garage,' recalls Brand. 'We somehow made all of that sound hip and young and generational. It became a hippie libertarian bible. This was shortly after Jack Kennedy famously said, "Ask not what your country can do for you ..." and we all said, "Right, do it yourself."' As Andrew Kirk writes in his essay 'Machines of Loving Grace' in the book *Imagine Nation: The American Counterculture of the 1960s and 70s*, perhaps the most striking aspect of *The Whole Earth Catalog* was what it suggested about the idealistic abstractions of the counterculture giving way to physical manifestation:

> *Brand's creation perfectly captured the post-Vietnam counterculture movement of the mid-1960s with its emphasis on lifestyle and pragmatic activism over utopian idealism and politics.* [The Whole Earth Catalog] *marketed real products, not just ideas, and*

47

> *the focus was always on theoretically feasible, if not*
> *always reasonable, solutions to real-world problems.*
> *For Brand and his colleagues,* Stop the 5-Gallon Flush,
> *a guide to stopping water waste with simple household*
> *technological fixes, was just as revolutionary a book as*
> Das Capital, *maybe even more so.* *

Although personal computers were still several years away when he first set up shop, Brand was quick to jump on their existence as a means by which to achieve technological nirvana. 'The success of *The Whole Earth Catalog* became a carrier wave for this enthusiast approach to technology rather than being a complainer about technology,' he says. 'This was at a time when E. F. Schumacher was putting out "small is beautiful" and all that. Now Fred Schumacher was anti-technology too; he didn't even like automatic roll-up windows in cars. In a sense we were saying that small *technology* is beautiful – let's take it one step further. Small is about empowering the individual.'

Steve Jobs was far from alone in being influenced by Stewart Brand. When *The Whole Earth Catalog* was published as a book in 1971, it became a national bestseller and inspired numerous follow-ups and copycats. As a young teenager when the first issue was released, Jobs was certainly on the junior end of the reader spectrum, being altogether too young to run off and join a commune as Brand was enthusiastically advocating. The excitement which undulated from every oversized page, however, absolutely registered with Jobs. 'That's a formative time,' Brand says of Jobs' teenage years. 'Steve was in the middle of it all growing up in Cupertino, so when his twig

---

* Jobs himself would later come up with his own version of this realisation, when he commented, 'maybe Thomas Edison did a lot more to improve the world than Karl Marx and [guru] Neem Karoli Baba put together'.

was softest and being bent, it was subject to the kind of stuff that I and others were putting out there. When he got into his twenties that was the direction his twig was going.' Certainly it is difficult to find a better summation of Apple's ethos than the one which appeared printed on the inside of every edition of *The Whole Earth Catalog*. The Brandian manifesto laid out the ideals of personal autonomy, freedom from traditional centralised forms of power, and man's ability to change his world for the better, which Apple would later champion – all underscored by a certain arrogance that would also become characteristic for the Cupertino company:

> *We are as gods and might as well get good at it. So far,*
> *remotely done power and glory – as via government,*
> *big business, formal education, church – has succeeded*
> *to the point where gross defects obscure actual gains.*
> *In response to this dilemma and to these gains a realm*
> *of intimate, personal power is developing – power of*
> *the individual to conduct his own education, find his*
> *own inspiration, shape his own environment, and share*
> *his adventure with whoever is interested. Tools that aid*
> *this process are sought and promoted by The WHOLE*
> *EARTH CATALOG.**

Years later, in Jobs' uncharacteristically candid 2005 commencement address for graduating students at Stanford University, he ruminated on the *Catalog*'s influence on his life,

---

* The opening sentence was borrowed from a similar sentiment expressed by the British social anthropologist Sir Edmund Leach in his 1968 book *A Runaway World?* 'Men have become like gods,' Leach wrote. 'Isn't it about time that we understood our divinity? Science offers us total mastery over our environment and over our destiny, yet instead of rejoicing we feel deeply afraid. Why should this be? How might these fears be resolved?'

while simultaneously introducing Brand's work to a whole new generation. 'This was ... before personal computers and desktop publishing, so it was all made with typewriters, scissors and Polaroid cameras,' Jobs said. 'It was sort of like Google in paperback form, thirty-five years before Google came along. It was idealistic and overflowing with neat tools and great notions.' Jobs was particularly impressed by a photograph which appeared on the back cover of the *Catalog*'s 1974 issue: an image of a picturesque country road, taken north-east of the Bay Area by a friend of Brand's. Jobs described it as 'the kind [of path] you might find yourself hitchhiking on if you were so adventurous'. Above it, in simple white-on-black letters, were the words 'Stay Hungry, Stay Foolish'.

Stewart Brand and *The Whole Earth Catalog* had another, altogether more immediate impact on Jobs' life – although Jobs did not realise it at the time. On 21 June 1971, Brand had decided to throw a somewhat premature 'Demise Party' for the *Catalog*, to mark what he thought might be the end of its run. The gathering took place at San Francisco's Palace of Arts and Sciences and drew a crowd of around 1,500 revellers. Brand wore a black monk's cassock, and presided over activities which included an indoor non-stop, non-scored game of volleyball, and a designated area filled with balloons from which people could inhale nitrous oxide. A band called the Golden Toad performed a soundtrack comprising everything from bluegrass to belly-dance music. When the party was reaching its apex, the evening's master of ceremonies, Scott Beach, stepped up to the microphone and informed the gathering that Brand had handed him a bag containing the sum of $20,000 in $100 bills. This was to be given to the party-goers, so long as they could unanimously agree on what should be done with it.

'Use this as a seed,' Beach said. '*The Whole Earth Catalog* ceases. The seeds have been planted already. Your consensus will decide what will be done with this money. There are microphones, there are causes, there are lots of possibilities.'

Brand would later recall the crowd's frenzied reaction: '*"Flush them down the toilet!"* *"No, don't!"* *"Give it to the Indians!"* *"Bangladesh!"* *"Our commune needs a pump or we'll all get hepatitis!"'* The debate continued until nine o'clock in the following morning, when a dozen remaining partiers agreed to give the money to a local dishwasher by the name of Fred Moore. Moore agreed that he would use the money to perform a public service to benefit everyone. He did just that when, several years later, he founded the Homebrew Computer Club.

Following high school, Jobs enrolled at Reed College, a private liberal-arts college in Portland, Oregon, which still carried the pungent smell of marijuana wafting on the breeze from the 1960s. With a relatively small student body of 1,200 and a reputation for attracting freethinkers, Jobs saw it as a chance to reinvent himself far away from home. 'I just wanted to be [perceived as] an orphan from Kentucky who had bummed around the country hopping freight trains for years,' he told one biographer. Once he had arrived at Reed, however, he found that he didn't fit in any better than he did on his home turf. He was the same psychologically intense young man, only shifted seven hundred miles down the road. '[Steve] was one of the freaks on campus,' Robert Friedland, one of Jobs' few friends at Reed, told biographer Michael Moritz. 'The thing that struck me was his intensity. Whatever he was interested in he would generally carry out to an irrational extreme. He wasn't a rapper. One of his numbers was to stare at the person he was talking to. He would stare into their fucking eyeballs, ask some question and would want a response without the other person averting their eyes.' This rather unnerving trait served Jobs well over the years, especially when dealing with computer programmers who generally didn't like to make eye contact to begin with. But it was one thing when he was using it to sell high-tech devices with all the charismatic conviction of a televangelist, another altogether, when he was trying to interact normally with his fellow students. It didn't help that the subjects he chose to discuss struck people as unusual. For a while Jobs seized upon the idea of leading a mucus-free lifestyle for health reasons after reading Arnold Ehret's 1922 book, *The Mucusless Diet Healing System*. This became his only topic of conversation. However a conversation might start, it would eventually find one way or another of winding around to the topic of glandular secretion before all was said and done. 'The whole world revolved around the elimination of mucus,' Friedland recalled.

It was at Reed where Jobs met Dan Kottke, a gentle-natured boy who had grown up in Pelham, New York. 'It was a kind of bedroom community for New York City,' Kottke recalls. 'I would not say we were rich kids, but it was pastoral. [My friends and I] spent a lot of time hanging around and getting high. I wasn't really into cars; I had other friends who were into making music – but I wasn't. Personally I was interested in reading, and I read widely. Throughout high school I was reading science fiction.' Unlike Jobs, who had specifically sought out Reed, Kottke had wound up there more or less by accident. 'I thought I had a good chance to get into Harvard, but I didn't,' he says. 'I wanted to live in Boston, because I had other friends there, but I didn't get admitted to Harvard and I decided I didn't want to go to Antioch. I didn't know what to do, so eventually the high-school guidance counsellor recommended Reed.' Kottke had no idea what it was that he was working towards. He played the piano and was studying as a music major, but saw no future as a paid musician. For a while he considered training to become a psychologist, but was put off when he discovered that he would have to carry out

Skinnerian behaviourism studies involving white rats.

The discussion of electronics did not enter into Jobs' and Kottke's friendship. Instead, what the two bonded over was reading. Despite years later damning Amazon's Kindle by stating that 'the whole conception is flawed at the top because people don't read anymore', Jobs was a voracious devourer of books throughout his childhood and adolescence. A sixth-grade report noted that he was 'an excellent reader' and this hobby continued well into his teenage years. The book that he and Kottke spent most time discussing was *Be Here Now*, a seminal 1971 volume concerning spirituality, yoga and meditation by the Western-born yogi and spiritual teacher Baba Ram Dass – born Richard Alpert. Just as striking as the subject matter was the book's somewhat unusual price: $3.33. For a generation that included Jobs and Kottke, *Be Here Now* was a key ingredient in the psychedelic experience. 'My friends and I were all taking psychedelics at high school, but we didn't have a conceptual framework,' Kottke says. 'That book is very much about psychedelics, as relates to consciousness. Once we started reading all this Eastern literature, psychedelics became much more interesting.' Other spiritual books read by Kottke and Jobs during this period included *Zen Mind, Beginner's Mind* by Shunryu Suzuki and *Autobiography of a Yogi* by Paramahansa Yogananda. Almost forty years later, the latter would become the only book Jobs downloaded to his iPad 2. He and Kottke also volunteered at a 220-acre apple farm-turned-commune, which was located forty miles south-west of Portland. Rechristened the All One Farm, it was run by Jobs' friend Robert Friedland. Before long, however, Jobs became disillusioned when he saw how Friedland was profiting from the work that people like himself were doing for free. 'It started to get very materialistic,' Jobs said. Years later Friedland became a billionaire copper and gold mining executive.

Additionally, Jobs' tenure at Reed saw him embrace the

various regimented diets he would continue to experiment with throughout the rest of his life. One such diet had him and Kottke eat nothing but fresh fruit, while another allowed only almonds and dates. One decision that the pair made was to become vegetarian. Jobs had read the bestselling book *Diet for a Small Planet* by Frances Moore Lappé, in which the author argued that eschewing grain-fed meat demonstrated the individual's power to create a saner world. Kottke, meanwhile, picked up a copy of the *Vegetarian Times* and was horrified by its photographs showing slaughterhouse conditions. When he went home for the holidays and told his mother that he could no longer eat her cooking, Mrs Kottke burst into tears.

□ □ □

Much has been made of Jobs' status as a college dropout. While it was certainly unusual in such a high-flying Silicon Valley executive, in fact it was par for the course at Reed, where more than a third of pupils never made it to graduation. 'After six months, I couldn't see the value in it,' Jobs later recounted in his 2005 Commencement address at Stanford University. 'I had no idea what I wanted to do with my life and no idea how college was going to help me figure it out. And here I was spending all of the money my parents had saved their entire life.' Remarkably, Jobs was allowed to hang around the campus, crashing in friends' dorm rooms, even though he was no longer paying tuition. He was also able to audit the classes that interested him. He rented an unheated garage apartment for $20 a month and subsisted on free meals from the local Hare Krishna temple, along with ones prepared by Kottke's girlfriend Elizabeth Holmes. Even at this point Jobs refused to compromise on his diet.

In February 1974 Jobs moved home and scanned the jobs pages of his local *San Jose Mercury* looking for work. As biographer Walter Isaacson points out, there was hardly a shortage of possibilities. At peak times that decade, the *Mercury* would be fit to bursting with up to sixty pages of 'help wanted' ads. The one which caught Jobs' eye was for the video-game manufacturer Atari. 'Have fun, make money,' it read, both of which sounded good to Jobs. Atari was another local success story, whose trajectory from high-tech start-up to big business hit all the countercultural checkpoints. Twelve years older than Jobs, founder Nolan Bushnell had been an electrical-engineering graduate at the University of Utah, a pioneering centre for computer graphics. While he was there he came across an early digital computer game called Spacewar, which had been developed by hackers at MIT in the early sixties. Spacewar was a two-player game pitting rival two-dimensional spaceships against each other. Inspired by their underground effort, Bushnell created his own version, Computer Space, which he attempted to market commercially. It proved a massive flop, however, and Bushnell went back to the drawing board. What re-emerged was Pong, an ingeniously addictive game in which disembodied paddles batted a square ball back and forth across the screen like an existential game of tennis set in the cold outer reaches of space. Pong was a smash hit and within six months Nolan's tiny start-up company had transformed into a marketplace leader in the amusements business. If computers were still viewed as the remit of a technological priesthood to the general public, then video games reflected the technological power-dreams of the Cold War in a manner that pinball tables and other electro-mechanical machines would never be able to.* Of course, Pong's

---

* For reasons such as this recontextualisation of technology from tool of the military-industrial-academic complex to consumer product, early video games such as Pong are often confused with developments in computing from the time. In fact, the original game of Pong featured no line of program code. It did not use a microprocessor, or a custom integrated circuit, but, rather, a digital logic design made from components familiar to a television engineer like Al Alcorn, who understood the different ways in which pulse waveforms could be both generated and manipulated.

overnight success immediately meant that rival companies piled in to join in the gold rush. Within a year of its release, more than fifteen different companies were competing for the coin-operated video arcade market that had previously been Atari's alone. Most of these modelled their output so clearly on Pong that they were little more than carbon copies. Ramtek's Clean Sweep, for example, asked users to clear a pattern of dots from a screen by hitting them with a ball, bounced off a paddle controlled by the player. Bushnell might have been annoyed, but he wasn't worried. 'I felt we could out-innovate them,' he said. Flush with cash, Atari began bringing in the top creative programmers and created a utopian environment in which they could work. Egalitarian to a fault, the company represented Silicon Valley countercultural libertarianism in all its geeky glory. 'We treated programmers like mini-gods,' Bushnell recalled. 'We gave the best ones private offices. Then we put a hot tub in the engineering building. We hired the best-looking secretaries we could find for that department. Atari beer busts – the all-night parties we'd throw to celebrate revenue goals – had already become legend. Were there planning sessions where we smoked pot? Absolutely. Did we have some incredibly crazy parties? Probably true. Was the company probably the hardest-working company in the world? Probably true as well.' To make sure that Atari didn't deviate from its freethinking course, he drafted a two-page company manifesto, which assured employees that it would 'maintain a social atmosphere where we can be friends and comrades apart from the organizational hierarchy'. In particular there would be no discrimination in terms of 'the short hairs against the long hairs or the long hairs against the short hairs'.

In short, it sounded perfect for Steve Jobs. Despite his raggedy appearance he was given a $5-an-hour technician's job by chief engineer Al Alcorn – even if Jobs' distinct lack of personal hygiene at the time hardly made it the most popular decision with some of the company's other employees. 'The engineers didn't like him,' Al Alcorn would later recall. 'He smelled funny.'

□ □ □

Jobs had not been Atari's fortieth employee for long when he announced that he planned to travel to India on a spiritual journey. The only problem, he explained to Alcorn, was that he didn't have enough money. Would Atari be interested in paying his air fare? 'Bullshit,' Alcorn responded. 'I'm not giving you any money to go see [a] guru.' Eventually a compromise was reached whereby Jobs would first fly to Europe to sort out some technical issues with kits that Atari was shipping to Munich. He would have to pay his own way from there, but at least he would be on his way. Due to the surge of interest in Eastern spiritualism among US students, India was a popular destination for those looking to tune in, turn on and drop out. In particular Jobs had been excited by the stories told to him by Robert Friedland, who described an 'electric-charged atmosphere of love'. That was enough to sell him and Dan Kottke on the trip. 'The books we were reading were filled with miracle stories of holy men, so we were just wondering what was going on,' Kottke says. But India proved a disappointment. In the same way that both were too young to have gone to Woodstock, so Jobs and Kottke were a couple of years late to have been part of the buzzy furore caused by *Be Here Now*. By the time they arrived, the gurus had mostly moved on. Ironically, many had travelled to America, seeing it as a growth market. As a result, Kottke says, 'We didn't have any miracle experience or profound realisations ... We were glad we had gone to India, but I think Steve and I both felt that the real action was happening in America.'

Kottke did see sides of his friend he had never witnessed before, however. 'You would buy your food on the street from vendors,' he recalls. 'Steve had a very keen sense of what the right price was. He would look to see what other people were paying and would want that same price. But really we didn't fit in and it was obvious we were foreigners, so the price was higher for us. That was just the way it is. Most people who lived there were very, very poor ... There were several times when Steve got into disputes with these people because he wanted the local price. I could never understand what his problem was. From my point of view it was just pennies – just pay them what they want. [But] that quality of wanting to get the cheapest price possible, you could see how that developed later in life for him.'

□ □ □

When Jobs returned to the United States he was ready to work. He went back to Atari and was able to walk straight into his old job, despite looking – if possible – even more bedraggled than when he had left. This time he was offered the chance to design the circuitry for a new Atari arcade game called Breakout. Despite being a passable engineer, this was far beyond Jobs' level of expertise. He called in Steve Wozniak, who Jobs knew would practically salivate at just this kind of problem. Although Wozniak was employed by Hewlett-Packard, he was no stranger to the Atari offices. Prior to going to India, Jobs had been so successful in rubbing his Atari colleagues up the wrong way with his insulting comments that Al Alcorn arranged for him to work a nocturnal schedule only. The official reason given was that Jobs' open-toe sandals should not, for health and safety reasons, be allowed in an area where there might be heavy lifting going on.

This suited Jobs down to the ground. Once Wozniak had got off work and eaten his supper each day, Jobs could invite him over to keep him company while he worked. In exchange for free games on Gran Trak 10 – Atari's first driving arcade game, which had fast become the company's hottest property behind Pong – Wozniak would help Jobs with any tricky technical problems he encountered.* Breakout was a chance for him to do that with a bit more legitimacy, and to be paid for it! As a game, Breakout was not remarkable. It was, essentially, a single-player version of Pong, only with smashable bricks at the other side of the screen rather than an opposing player who could volley back the 'ball'. It had been conceived in a typically Atari way, however: in a brainstorming session held at an ocean resort, during which any idea that an employee could dream up would be considered by the group.

As much as Bushnell liked the suggestion, he was also worried about the number of integrated circuits it would take to realise the game. Setting Jobs a challenge, he offered him a bonus for every microchip he was able to cull from the design. Anticipating that it would take dozens of integrated circuits, Bushnell was pleasantly surprised when Wozniak delivered a remarkably compact, and typically innovative, design. Bushnell paid Jobs several thousand dollars' worth of bonuses, in addition to the $700 he had promised him for the work. (Jobs wrote Wozniak a cheque for $350 and never mentioned the rest of the money. Years later, when Wozniak read about this for the first time in an Atari history book, while flying to a promotional event for Apple, he started crying. 'He realised Steve hadn't told him the truth about how much they were getting,' says Apple engineer Andy Hertzfeld, who first alerted Wozniak to the fact.) Breakout wound up becoming the most popular arcade game of 1976, and the following year was included on the Video Pinball home games console.

At the same time as this was going on Jobs had begun attending meetings at the Homebrew Computer Club with Wozniak. If Wozniak was par for the course at Homebrew,

---

* Despite its vast popularity, Gran Track 10 very nearly put Atari out of business. Due to an accounting error, the arcade machine was issued horribly underpriced, meaning that the company lost money on every unit sold.

then Jobs was a standout. It was not because his genius was immediately apparent to everybody present, and nor was it for his tendency towards dressing in hippie garb, which raised precisely no eyebrows in a place where, in the words of one Homebrew member, 'the average hair length was well past anybody's collar'. If there was a reason why Jobs stood apart from everybody else it was because – at least by the standards of those willing to attend such a club – he didn't seem particularly enamoured with computers. 'He was a fish out of water,' says Homebrew member and early Apple employee Randy Wiggington. 'We were in Silicon Valley, so we were surrounded by computers and technology, but I think that if Steve had been surrounded by a different set of friends he would have taken over and transformed some other industry.' The unanswerable question, of course, is whether Jobs would have been so successful had he been born and grown up somewhere else. Steve Jobs, like his on/off friend and rival Bill Gates, was born at a critical juncture in the personal-computer revolution. He was one month shy of turning twenty when *Popular Electronics* magazine ran its famous January 1975 cover story hailing the Altair, the 'World's First Minicomputer Kit to Rival Commercial Models ...' Much older and he would have been firmly entrenched in a career with a company such as IBM, and would have struggled to make the leap to capitalise entrepreneurially on the Altair. Much younger, and he would have been in no position to capitalise whatsoever.

Of course, although Jobs was undoubtedly fortunate in terms of when and where he was growing up, there were many children of 1955 with an interest in electronics present in Silicon Valley, only one of whom became *Steve Jobs*. Right from the beginning Jobs had a wild-eyed magnetism, which gave him a distinct advantage when coupled with his easy sales patter. 'Steve had this amazing charisma,' says Wiggington, in what is something of a popular refrain. 'Just being around him and talking to him made you want him to like you.' Jobs saw new value in his friend Wozniak when the older engineer mentioned that he was working on a computer, much like the Altair, that would allow people to play at home the kinds of games Atari was working on. Jobs was intrigued.

'What are you going to do with it?' he asked.

'I'm not sure,' Wozniak said. 'I'm probably going to Xerox the blueprints and hand them out to whoever wants them at Homebrew.'

Jobs was appalled. *Give them away?* Clearly Wozniak wasn't thinking straight. As something of an outsider to the group, Jobs never fully grasped the Homebrew ethic. Like academia it was a share-and-share-alike environment. Wozniak was so proud of the intricacy of the printed circuit-board layout he had come up with that his ambition extended only so far as a desire to see other members use his plans to build their own computers. After all, he was already working full-time for Hewlett-Packard, where he had all the access he wanted to a mainframe computer. While the Homebrew Club represented a fun evening activity, he failed to believe that anyone would shell out actual money for his computer design. Jobs couldn't understand it. He wasn't yet at the stage in life to be suffering from the inventor's dilemma. Instead he was suffering from *seller's dilemma*; the dilemma being that he needed something to sell. 'He had a tremendous drive to start a company ...' Wozniak would later recall. 'Steve wanted to have a successful product, go out and start selling it, and make some money. He also had excellent product ideas for the upcoming home personal computer.'

This ideological divide is not uncommon. Silicon Valley has long been defined by the innate tension between the technologist's urge to share information and the industrialist's incentive to profit. Of course, the fact that the Two Steves (and, more specifically, Steve Jobs) so quickly came up with an idea to turn the homespun idealism of the computer hobbyists into a viable business enterprise has led some to suggest that Apple effectively 'sold out' whatever countercultural cachet it had right from the start. This, in turn, gives rise to a fundamental myth about the counterculture that it is important to dispel: that it was ever unanimously opposed to capitalism. Simply put, this is not true. Certainly there were aspects of the counterculture that were staunchly anti-capitalist in their views. The New Leftists represented one perspective. Men like Jobs epitomised the opposite extreme. 'There were two main wings of the San Francisco counterculture,' explains John Perry Barlow. 'One of them was definitely Marxist and the other was largely apolitical and more of the view that whatever worked to empower and liberate was fine, and capitalism was not necessarily a tool of enslavement.'

During the early stages of the personal-computer revolution, in situations like the Homebrew Computer Club, these two stances could comfortably sit next to one another, in much the same way that the more psychedelic hippies and the New Leftists could, for the most part, cheerfully coexist without the former group accusing the latter of being politically overwrought, and the latter countering that the former were nothing more than frivolous timewasters. But these two differences of philosophy, both within the counterculture and among the techno-geeks, do illustrate a fundamental schism. For the hobbyist-hackers, the aim was not an anarchic one aimed at bringing down an existing symbolic order that it saw as innately wrong, so much as it was about garnering its approval and thus gaining access to its technology. What they were searching for was the creation of a new Jeffersonian democracy based not upon equal distribution of land, but on equal access to information. In other words, it wasn't that technology was bad, computers were evil, or that business was wrong – it was that they wanted a piece of it and weren't currently getting it. The New Leftists took a different, more radical view. With concepts like Lee Felsenstein's Community Memory project the idea was most assuredly to overthrow a system, and to distribute the tools of that hierarchy among the people it exploited and excluded.

From these disparate (although, at the time, temporarily convergent) goals, one can get a sense of the lines along which the computer industry has continued to develop. The hobbyist-hackers were after technologically achievable social status; the New Leftists were looking to effect social change. Thus, although companies such as Apple have continued to wave their world-changing ambitions over the years, no ideological problem exists with the notion of profiting from that same new world order. In Jobs' mind there was nothing to sell out of – apart from maybe computers, of which more could be ordered in. Not lost on anyone was the fact that the voices of the counterculture, musicians like Jobs' beloved Dylan, were churning out the soundtrack to a revolution while being paid multi-million-dollar sums by major record labels. For the most part, the hippies were actually *progressive* capitalists; fiercely espousing the free-market libertarian views that would become the prevailing ideology in the Western world as time went on. While the long-hairs found fault with corporate America, and laughed at the buttoned-down unhipness inherent in the repressed figure of the grey-flannel-suit-wearing businessman, their mythos celebrated the figure of the self-made, alternative entrepreneurs: the dope dealers and the underground artists, through to violent criminals like the titular bank robbers in *Bonnie and Clyde*. As Stewart Brand explains, 'As they followed the mantra "turn on, tune in and drop out", college students of the sixties also dropped academia's traditional disdain for business. "Do your own thing" easily translated into "Start your own business".' In this manner, the immaculate creation of Apple is portrayed not so much as an example of unbridled capitalism, but as a point of *rebellion*; an extension of the hippie DIY ethos.

Whether the effort to whitewash the necessary evil of making money was part of Jobs' initial pitch to Wozniak isn't known, but it is fascinating to note how almost every telling of Apple's early days – including the one put forth by Walter Isaacson in his authorised Steve Jobs biography – frames the creation of Apple not as a capitalist act, but, rather, as a 'fun adventure'. 'This was enticing to Wozniak,' writes Isaacson, 'even more than any prospect of getting rich.'*

Picking a good name for a company does not guarantee success, but picking a bad one can be ruinous. Today the branding of companies is an entire industry, with consultants, seminars and books helping to decide what name defines your company above all others? Years later Steve Jobs would be as guilty as anyone of going down that route.† In the middle of 1975, however, with barely two cents to rub together between them, Jobs and Wozniak didn't have that option. Riding around in Wozniak's beaten-up car, the pair shot names back and forth to try and

---

* An interesting notion to contemplate is how much the trend from New Left ideology to the altogether softer psychedelia of the aesthetic hippies owed to the changing situation in Vietnam. Men like Lee Felsenstein actively campaigned against it. Jobs was a thirteen-year-old freshman student at Homestead High School. Much of the official discourse surrounding the Vietnam War concerned how best to end it. In that year's presidential election, Richard Nixon campaigned on a promise to end the draft. There was little chance that Jobs' contemporaries would be conscripted, and this proved to be the case. One month after Jobs turned eighteen, the last of America's combat troops were removed from Vietnam, with President Nixon declaring that 'the day we have all worked and prayed for has finally come'. (Although the last men conscripted were those born in 1952, the Selective Service had, in fact, assigned tentative draft priority numbers for all men born in 1954, 1955 and 1956 – which would have included Jobs – although these were never acted upon.)

† Guilty is perhaps the wrong word, given that iconic brand names like the iPhone and iPod have successfully entered the public consciousness in the way that few products – regardless of quality – ever manage to.

come up with the right one. *How about Matrix Electronics? Maybe something simpler like Personal Computer, Inc.? Does Execuistek sound too sinister to you?*

'What about Apple Computer?' Jobs said suddenly.

Wozniak had nothing better. *Apple it was!* As names thought up by two college dropouts bouncing along the highway at sixty miles an hour go, Apple was a pretty good one. The image was fresh in Jobs' mind because he was just back from another trip pruning apple trees at the All One Farm in Oregon, but it is difficult to think of a title more ripe with symbolism. 'Do not eat the apple from the tree of knowledge,' God tells Adam in the Book of Genesis. 'You must not touch it, or you will die.' The pronouncement comes as a warning to the finite mind of man, which is supposedly capable only of primitive thought and largely lacking in rational insight. The reality, of course, is that the apple does not contain properties that make it injurious to health, but, rather that biting into it becomes the catalyst for revealing a greater, previously hidden truth, which ultimately frees man's mind and wakes him up to his own nakedness. The fact that it was the same name that the Beatles had chosen for their music company was just an added bonus.* Writing in his cyberdelic book *Chaos and Cyberspace* years later, Dr Timothy Leary notes, well-meaning but with a degree of inaccuracy, how:

> *The Personal Computer was invented by two bearded,*
> *long-haired guys, St. Stephen the Greater and St.*
> *Steven the Lesser. And to complete the biblical*
> *metaphor, the infant prodigy was named after the Fruit*
> *of the Tree of Knowledge: the Apple! The controlled*

---

* While to a couple of long-haired Beatles fans on the other side of the world this might have seemed a nice tie-in, it proved to be the start of a long and winding legal battle that would only be settled once and for all in 2006.

substance with which Eve committed the first original sin: Thinking for Herself!

For Jobs, Wozniak and the whole techno-utopian movement this is what the revolution was all about. If computers were towering, sterile mainframes, kept utterly apart from nature, then the Two Steves' new company would be just the opposite: small and inviting, fusing decentralisation with the ecological pastoralism of *The Whole Earth Catalog.* 'To a marketer Apple was an odd name,' Wozniak later admitted. 'It came from the days when you picked an interesting, fun name for a company. You do that when you're on a hobby basis ... [Conventional wisdom would suggest that] we had to have a name that suggested technology, number crunching, calculations, databases. We took the attitude that Apple is a good name. Our computer would be friendly. Everything an apple represents: healthy, personal, in the home.' All that plus it would put them ahead of Atari in the phone directory.

◻ ◻ ◻

Of all the demonstrations carried out at Homebrew, none would have the long-term impact of the showing which occurred one Thursday in April 1976. Standing up in front of the group, and having manned the only power outlet in SLAC, the Two Steves showed off the printed circuit boards for the Apple I. Wozniak did most of the talking, held up the board, chatted through its features and then sat down to a reaction that said ... not very much at all. It was somewhat disappointing to say the least. That Wozniak was working on a computer was not news to anyone who had spoken with him at Homebrew meetings. He had first brought the boards in the previous autumn after he had completed one and was busy working on a second. As business plans go – and for all Jobs' posturing – it hardly looked set to light the world on fire. The idea was to build computer boards for $25 and then sell them on for twice that. If they sold enough they might even make back the $1,300 they had already spent having Wozniak's handcrafted circuitry turned into a professional printed design that could be manufactured for sale. This had been pieced together through the selling of Jobs' red and white Volkswagen van and Wozniak's HP calculator, both of which appear as the sacrificial acts present in the Apple creation myth. But back to the demonstration. Even Lee Felsenstein, whose task at Homebrew could just as often be eliciting an audible reaction from a nervous young hacker as it was quietening down an overly rowdy group, appeared far from blown away. 'I wasn't terribly impressed by it, because it was a forty-character per line display, upper case only,' he says. 'I was a bit ho-hum in that sense. At the time I investigated the architecture of the display, because that's what I specialised in ... It turns out that it was using shift-register memory, which was by then an obsolete technology, which terminals had used starting in the mid-sixties. They were very clumsy ... You had to clock the data through them and wait until it came out the other end.'

If the news appeared grim, however, then it was worth remembering that it is always darkest before dawn. A local entrepreneur named Paul Terrell had opened the first of what would become a chain of computer retail stores. The Byte Shop had a ready and willing base of customers waiting in the wings, but before it could do anything it needed something to sell them. Terrell informed Jobs that if he could produce fifty Apple Is then he would pay $500 per machine: a total order worth $25,000. Of course, there was the small issue of just what comprised a computer. When Jobs took in the first batch for delivery Terrell looked at them in astonishment. There was no outer case, no keyboard, no power supply and no monitor. To a Homebrew hobbyist used to using their imagination to project what computers *might one day be able to do*, a printed circuit board was all Wozniak considered necessary to get someone up and running. Once this misunderstanding was negotiated, the Apple I business continued. By the end of 1976, 150 of the machines had been shipped out. With its mark-up, the Byte Shop was selling them for the somewhat devilish price of $666. Business, though, was beginning to wind down. 'With the Apple I nobody saw the big deal,' says Randy Wiggington. 'Honestly, the display was so limited. The whole computer was so limited really. People thought it was neat, they thought it was cool, but as far as being a big deal? I don't know anyone who thought it was revolutionary at the time.'

At one point Jobs and Wozniak took it into the offices at Atari to show off. 'I thought it was interesting, but I didn't consider it a consumer product,' says Mike Albaugh, an employee at Atari, who went on to enjoy a long career with the company. According to Albaugh, the problem with the Apple I was that it was neither fish nor fowl. 'I picture a continuum of people who might be interested in computers,' he says. 'At one end are the people who, nowadays, might be playing with Arduinos and, at the other, people who have their iPads.* At the time, I just felt that the Apple I wasn't easy enough for the equivalent of the iPad users, but wasn't cheap enough for the Arduino users of 1976. I mean, $666 – and you still had to add a power supply, keyboard and a monitor? That was one heck of a lot of money.'

If people like Albaugh were bemused, however, then Wozniak's father was livid; angry that his son's hard work was being exploited by someone who, in his eyes, had done nothing to earn anywhere close to 50 per cent of the Apple profits. 'You didn't do shit,' he reportedly told Jobs. For his part, Jobs had bigger problems to deal with – not least the fact that Wozniak

---

* The Arduino is a popular open-source, single-board microcontroller.

had started to lose interest in the Apple I business. At first, Jobs was annoyed. He badly wanted to get a product out there into the ether and Wozniak was impeding this process. But when Wozniak explained what he was thinking, Jobs softened. Wozniak was proposing the building of a new computer; one that would improve on the Apple I in every conceivable way. If anyone had doubts about Apple as a company they certainly wouldn't after they saw what Wozniak had up his sleeve. In 1976, the year that *The Enforcer*, a *Dirty Harry* follow-up, was making big bucks at box offices the world over, this would be Steve Wozniak's very own sequel to the Apple I.

What better name for it than the Apple II?

3

THE BiG GAY ELECTRiC KOOL-AiD ACiD APPLE

Norman Mailer: 'They're going to do it next.
People who work with computers see an extraordinary value in what the computer can
accomplish ... A computer will one day be
able to tell you how Don Juan made love.'

Dick Cavett: 'They can work into the past as
well?'

Norman Mailer,
interviewed on *The Dick Cavett Show* (1970)

In January 1977, the Two Steves tramped up a flight of stairs in Palo Alto. One Steve carried a JVC portable television and a cassette recorder, while the other tightly clutched a medium-sized wooden box, almost bursting with circuit boards and containing a rat's nest of tangled wires. Taken in its totality, this machine was the prototype of the Apple II: the world's first sixteen-colour personal computer. 'The Apple II came out of trying to improve the Apple I,' Wozniak would later say. 'It was faster, it had colour, it had high-resolution graphics, it had mixed modes on the screen with text. A lot of neat features made it look like this might be a nifty product.' A lot had happened in a short time span. Seeking investors, Jobs had gone back to his former boss at Atari, Nolan Bushnell, who in turn introduced him to Don Valentine, a legendary Silicon Valley venture capitalist. 'Steve was twenty, un-degreed, some people said unwashed, and he looked like Ho Chi Minh,' Valentine said of his first meeting with Jobs. While he eventually wound up investing in Apple, Valentine's first reaction to Jobs' arrival was one of bewilderment. 'Why did you send me this renegade from the human race?' he asked Bushnell. Valentine, however, was impressed enough to crack open his Rolodex and point the Two Steves in the direction of Mike Markkula, a semi-retired Intel employee who had made a fortune by investing his Intel money in the oil industry just at the moment that the energy crisis of the 1970s was picking up steam. Markkula agreed to provide Apple with a line of credit in exchange for a one-third stake in the company. That 3 January, while Apple was still operating out of Jobs' parents' garage on 11161 Crist Drive in Los Altos, the enterprise became incorporated.

To people outside Silicon Valley the notion of two twenty-something college dropouts successfully starting up a computer company seems highly unlikely. But while the Two Steves were undoubtedly young, they were no younger than many of their contemporaries – and their company, while small, was not remarkably so. Despite the homespun underdog status that Apple would cultivate over the coming years, its birth was, in fact, more representative of the rule than the exception. Of the roughly 3,000 electronics-manufacturing firms that existed in Silicon Valley at the tail end of the 1970s, 70 per cent employed less than ten people. Furthermore, the late seventies saw the largest wave of new start-ups of any point in the area's history. January 1977 was an important juncture in Apple's history for a second reason. Three months before the Apple II computer went on sale to the public, this month marked the genesis of the *brand* of Apple: an area in which the company would excel more than any other similar business in history. The people entrusted with this momentous task were employees of Regis McKenna, Inc., a medium-sized agency in the Bay Area which specialised in public relations and advertising for high-tech businesses, at a time when few such agencies were doing so. Regis himself had first come to Silicon Valley in the early part of the decade to work for one of the big early semiconductor companies. He quickly realised that marketing was an area that most tech companies needed but didn't really understand. He left to set up shop on his own and, within several years, had a nice little business ticking along. 'Regis knew everybody in the valley,' says Richard Melmon, today a successful venture capitalist who served as a partner in the McKenna Group during his career. Although Regis would later speak about Apple in reverential tones as one of his biggest clients, Jobs had to fight long and hard even to get a meeting in the early days. He harassed McKenna account executive Frank Burge almost to the point of distraction, routinely bombarding his office with three or four telephone calls a day. Eventually he got his invitation to come in. The reason for Jobs' persistence was an advertisement he had seen which the agency had created. It was for Intel's SDK-80 microcomputer, a relatively simple machine designed as an entry point for the more powerful MCS-80. Rather than marketing the SDK-80 – which came in kit form, with an assembly time of approximately six hours – as merely an engineer's tool, Regis McKenna's creative department had instead chosen to promote it as something akin to a child's science-project kit. This unusual bit of thinking appealed to Jobs, and suggested that Regis McKenna was a company he needed to latch on to. To use his parlance, *they might be hip to his way of thinking.*

The Two Steves arrived at the front desk and were shown into the boardroom. The short walk from reception to meeting room allowed for plenty of raised eyebrows; barely any of them related to the equipment that Wozniak and Jobs were carrying. While Wozniak wore a business suit for the meeting, Jobs had opted for his then usual wardrobe of sandals and jeans with the knees torn out. 'I think it's safe to say that anyone who met Steve for the first time would come away with the impression that he was somebody who was not only counterculture, but who enjoyed throwing that fact in the face of corporate America,' recalls one Regis McKenna employee. Bill Kelley, an account executive, agreed with the statement: 'Most computer hobbyists looked like your typical cartoon of an engineer, with the skinny tie, the white, short-sleeved shirt, and the pocket protector. Steve was kind of iconoclastic. His hair was relatively groomed, but he had this scraggly beard, which wasn't so much a beard as it was various hairs coming out of his face. He looked almost more like a street person than someone you'd want to do business with. He didn't present well from a sartorial standpoint, let's put it that way.'

The meeting, in which Regis himself was described by one person present as 'grudgingly in attendance', began with Wozniak connecting up the bare-bones Apple II, bringing it to flickering life on the table like an 8-bit Frankenstein's monster. To demonstrate the machine's extraordinary abilities he loaded in a program called *Life*, a computerised version of a cellular automaton simulator dreamed up by Cambridge-educated mathematician John Horton Conway several years earlier. On balance, this was a mistake. What the Two Steves had failed to realise – what techies and geeks all over Silicon Valley who ran the simulation to drum up interest in personal computers failed to realise – was that regular people outside the microcomputer bubble didn't understand the significance of *Life*.

Regular people saw personal computers as potential ways to balance their chequebooks, to keep track of birthdays and record collection, and to catalogue recipes. Where Jobs and Wozniak saw the miracle of existence – and all in glorious colour – the other people in the room saw only a bewildering series of dots creeping slowly around a flickering screen. 'It was interesting to watch, but it wasn't really clear to us what we were seeing,' Kelley recalls charitably, 'but the Two Steves were obviously very excited about it.'

For the Regis McKenna staffers it reinforced every belief they already held about these kinds of hobbyist computers. It was a matter of imagination, and of emphasis. The Regis McKenna staffers in the room looked at personal computers and asked what they could *do*. The Two Steves looked at personal computers and asked what they *could* do. Jobs and Wozniak did, after all, come from the generation of the road trip: the long cross-country voyages with no destination in particular, and no purpose laid out aside from the journey itself. To put a spin on it better reflecting the views of media maven Marshall McLuhan, *the medium itself was the message*. Thinking more practically, as Jobs and Wozniak figured, if computers were able adequately to simulate life, that most profound of abstractions, then surely everything else – typing up letters, running your accounts, reminding you that your tenth wedding anniversary was coming up – was simply a matter of scaling back ambition. Jobs was the first to speak. Sensing the general bafflement, he tried to convey in words the excitement of what was appearing on the screen. Although he hadn't yet developed into the superb orator he would become in later life, Jobs was already a compelling speaker: charismatic and concise. In a clear, well-rehearsed patter that he obviously believed in absolutely, Jobs spoke about the enormous potential of personal computers over the coming years. His dream was that one of these strange boxes would one day appear in every home in America and, later, the world. He wanted the Apple II to be to computing what Kleenex is to tissues.

The tide began to turn. People in the room were impressed, none more so than Bill Kelley. 'Steve was a man on a mission,' he remembers. 'He had that missionary zeal to him. He was evangelising for the concept of personal computing.' The meeting appeared to be back on course and probably would have stayed that way had Wozniak not seized the moment to unveil what he felt was his master stroke. Reaching into his shoulder bag, he pulled out a hefty fourteen-page treatise, which was half technical jargon and half manifesto. He leant across the table and handed it to McKenna's public-relations rep. Finally seeing an opportunity to jump in, Regis decided that this was the perfect time to point out the wonders that a good bit of public-relations tidying-up could achieve.

'We'll be able to help you edit that down a bit,' he said, indicating Wozniak's pamphlet.

He barely got the sentence out. Wozniak leapt to his feet, snatched the paper back and, clutching it to his chest, yelled, 'I won't have any of you putting your grubby marketing hands on my work.'

As far as McKenna was concerned, that was the end of the meeting. Civil conversation generally ends the moment words like 'grubby marketing hands' are bandied about. After the Two Steves had departed, the group sat around to discuss the fallout. 'Regis was a traditional business guy,' says one employee of the agency. 'He was so outraged by the guys' lack of professionalism that he sort of threw up his hands and said, "Kelley, you handle this. It's beyond me." I think it was all pretty shocking for the older members of staff. Although I think they clearly knew there was a goldmine there as well.'

□ □ □

New technology is always on a deadline. There is the ever-present fear of being pipped to the post by a competitor with a faster, smaller, cheaper, *better* version of whatever it is that you're working on. Beyond that there are investors to be paid back, bosses to be placated, or simply profits to be made. But deadlines can, more often than not, change. Doing so wreaks havoc with marketing departments, upsets magazine editors waiting on that all-important exclusive demo, and deflates customers, but at the end of the day it is always better to release a polished final product than to rush something out of the door that is clearly unfinished. In this case, however, the deadline was already decided – and if the Two Steves wanted to take advantage of it they would need to make sure that they met it. The deadline on this occasion was an event being promoted from 16 to 17 April 1977, which promised to be Silicon Valley's first personal-computer expo. Several rival companies, including Commodore, had already announced that they would be debuting new computers at the convention, and if Apple missed out it would lose a valuable opportunity.

The man behind this particular venture was a unique character named Jim Warren. Like a number of people who wound up making money early on in the personal-computer industry, Warren was a libertarian hobbyist first and a businessman a distant second. He had grown up in San Antonio, Texas, but moved to California in the mid-1960s on the advice of a friend. He quickly found himself thrust into a world gone mad. The Vietnam draft was going on, the Pill had recently been introduced and the Free Speech Movement was in full swing. Everywhere he looked there were psychedelics, upstart communes, great music, and unattached, no-strings promiscuity. One day the door of his rented La Honda home burst open and in walked Neal Cassady, the seminal Beat-era figure who allegedly formed the basis for every literary countercultural protagonist from Dean Moriarty in Jack Kerouac's *On the Road* to Randle McMurphy in Ken Kesey's *One Flew Over the Cuckoo's Nest*. Warren could only watch, dumbfounded, as Cassady marched through his house, followed by a group of hangers-on, repeating the seemingly nonsensical phrase: 'Got the mash, where's my stash?'

Warren had been a teetotaller up until 1966, but soon begun smoking joints and experimenting with LSD. At the same time he managed to pick up a teaching post, running the maths department at a local Catholic girls' college on the San Francisco peninsula. He split his leisure time between pursuing the kind of electronic hobbyism practised at the Homebrew Club and hosting some of the greatest naked house parties in the Bay Area. 'I lived just up the road from the best clothing-optional beach in the nation,' he recalls. 'I started going to it and met a bunch of people down there, and invited them back to my place. We ended up having this huge party; starting out clothed and eventually without clothes. There was no sex involved, just naked people dancing, smoking dope and having a lot of fun.' Word got out, and before long more than a hundred revellers – ranging from nigh unemployable hippies to IBM engineers – were turning up at Warren's doorstep on a regular basis. Eventually this aspect of his life collided awkwardly with his day job, courtesy of a front-page *San Francisco Chronicle* news story and an awkward telephone call which arrived during the Christmas vacation.

'Professor Warren, it has come to our attention that people are claiming you are hosting nude parties at your house. Is this true?' the college dean asked.

'Yep,' Warren admitted. There was a long pause.

'Well, what you do in your own private life is your business, of course, but in this instance one can't help but think that your private life has become rather ... public,' said the dean, speaking carefully.

Warren replied that perhaps it had.

'I think you'll have to agree,' said the dean, 'that at the very least, this is all rather incompatible with the philosophy of a Catholic girls' school.'

Warren asked whether the dean wanted him to resign. There was a sigh of relief on the other end of the phone.

'Oh, we would very much appreciate that, Professor Warren.'

Warren quit in the middle of an academic year and moved on to do something else. For a while he bummed around in various academic research jobs before his funding was cut. Then he spent a period editing the monthly programming newsletter, *Dr. Dobb's Journal*. The night before he was due to attend an early computer convention in Atlantic City, Warren sat down with a friend of his, Bob Reiling, who edited the Homebrew Club's (almost) monthly newsletter.

'What the hell are they running it in Atlantic City for?' he complained. 'There's nothing there. Everyone knows that the centre of micro-computing is right here in Silicon Valley.'

There was a pause. Both men looked at each other, each having had the same idea at the exact same instant.

'We should do our own trade show,' they said.

Flicking through a calendar they decided on a date. Then they phoned up Stanford University, where Warren had been a student, and enquired whether the Memorial Auditorium was free for the weekend they had selected. It was. Warren sat down at a compositor and cobbled together an 8½x14-inch advertisement for what they decided to call the West Coast Computer Faire. 'I spelled it that way after the Renaissance Faire they held up in Marin County,' Warren recalls. 'I felt like this was a Renaissance in technology, so let's name it after the Renaissance Faire.' The very next day he set out to fly to New Jersey. In his luggage was a pile of freshly printed flyers, informing everyone of the time, date and venue of the as yet unrealised West Coast Computer Faire.

The Atlantic City convention was exactly as Warren expected. 'It was miserably hot,' he says. 'It was jam-packed, just a terrible time of the year in an awful cesspool of a city. But it was *fun*. It was exciting.' He had no problem whatsoever in distributing his leaflets to a host of interested parties. When he arrived back on the West Coast, he immediately called up Reiling and told him that they needed to put their plan into action immediately.

'It's going to be great,' he enthused. 'I've got a whole bunch of people who want to be exhibitors or to give talks. This is going to be really, really wonderful.'

Reiling was excited, too, although he was more of a realist than Warren.

'Let's get everything nailed down,' he suggested.

Warren agreed. As soon as the two men had talked he phoned Stanford to book the hall.

'Okay. We'll take the venue on the dates we talked about,' he said.

'I'm sorry,' said the woman in charge of booking. 'One of the fraternities has requested the use of the Memorial Auditorium on that evening, so I'm afraid that it's no longer available.'

Warren turned white. *Oh shit!* he thought.

They scrambled to find a backup plan. There was another option: the downtown San Francisco Civic Auditorium, today known as the Bill Graham Auditorium. Warren and Reiling paid a visit and discovered that it was exactly what they were searching for. 'It was just perfect,' Warren remembers. 'There was a big centre area where we could have exhibits, four large meeting halls each side of it and all sorts of small rooms on the second, third and fourth floors for talks and seminars.'

'We'll take it,' they said at the conclusion of the tour.

'Excellent, gentlemen,' the venue manager smiled. 'I'll have my secretary draw up a contract. We charge a fee of $14,000 per day for exhibition days, and $3,000 for each day you're setting up or taking things down. Now, how will you be paying?'

This was the Computer Faire's second crisis. The major benefit of the Stanford Memorial Auditorium, aside from its location, was that it was unlikely to charge more than the most nominal of fees for venue hire. Almost $50,000 to rent out the Civic Auditorium was about $45,000 more than Warren or Reiling had to invest in the entire world. 'Hell, I was making $350 a month,' Warren says. The manager told them to take some time to think about it, but warned that the building was likely to get booked up before too long. Warren and Reiling left dejectedly. On the drive home they decided to stop at Pete's Harbour, an open-air café by the bayside marina, to talk things over. As they waited for their drinks to arrive, Warren pulled out a paper napkin and began scribbling rough calculations on it. When he began adding them up his face broadened into a smile. 'Up until that point I just wanted this thing to happen so that a whole bunch of people who were doing really fun things in this environment could talk to each other,' he recalls. 'I didn't give a shit about making money. It didn't even occur to me that it was possible. But then we were forced to make at least enough to cover the costs of the Civic Auditorium, and for the first time I looked up at Bob and said, "My God! Not only [can] we afford the cost of the building, we might be able to make some money on this".' He booked the venue as soon as he got home.

When word got out about the West Coast Computer Faire, Warren was inundated with phone calls from interested parties. 'Neither Bob or myself had ever run a convention before,' he says. 'We didn't know what the shit we were doing. I ran it out of my house, with my bedroom as an office. I hired a girl who was a neighbour as a secretary, then hired more neighbours to do other jobs as we got closer to the time.' One of the phone calls Warren received was from Steve Jobs, who was keen to use the event to launch the Apple II.

'We want the best location possible in the hall,' Jobs said.

Warren told Jobs how much the space would cost him, and Jobs promised to send him a cheque in the post immediately.

Jerry Manock was thirty-three years old, a graduate of Stanford University, where he had earned a master's degree in mechanical engineering and product design. A short, firmly built man with a full, bushy beard, he had spent much of his adult working life as a technical nomad – jumping from one pursuit to another in an effort to find spiritual fulfilment. For a while he worked at Hewlett-Packard, where he specialised in designing electronic measuring equipment for testing microwaves, before he quit and travelled around Europe for nine months with his wife. When he returned, he worked for a company in Palo Alto that made electronic aids for disabled people, prior to packing that in to start up his own design outfit, Manock Comprehensive Design.* When he got a phone call from Steve Jobs, telling him that he had a major project to discuss, Manock thought it might be just the opportunity he was looking for. He was close to broke and beginning to worry that his new venture was more about building his ego than it was about creating a viable business. He decided to throw the entire weight of Manock Comprehensive Design behind it. This wasn't altogether difficult to do: Manock was, at that point, the owner and sole employee.

Rather than inviting Manock to visit him at his office, Jobs suggested that they meet at the next Homebrew Club meeting to talk about specifics. Manock arrived at the Stanford Linear

---

* The name Manock Comprehensive Design was borrowed from one of Jerry Manock's heroes, futurist Buckminster Fuller, who referred to himself as a comprehensive anticipatory design scientist. 'That was too long for business cards so I shortened it to "comprehensive",' says Manock.

Accelerator Center and found himself one of a group of several people standing in a semicircle around Steve, discussing various aspects of the Apple II project. Rather than have a single conversation at a time, Jobs would, to borrow a term synonymous with computing in the 1970s, 'timeshare' conversations – limiting each person to one minute of talking before jumping to the next in line. When he finished with the last person he would then come back to the first and, without skipping a beat, continue exactly where he had left off with them. Manock was impressed. What Jobs wanted, he explained, was the casework to contain the Apple II circuitry; something that would appear beautifully machined, as if it had rolled off a conveyor belt at IBM, as opposed to having been bodged together in his parents' garage. The only catch, Jobs said, was that he needed it completed in two months' time. Despite this, Manock felt rather flattered that he had been approached. *He must like my work,* he thought. Later on he discovered that all the major design firms had turned Jobs down. (The only person who had been more than happy to take a crack at the design – an old friend of Jobs' from his Atari days – had sketched a radically shaped Apple II and suggested that it be coloured a deep shade of purple. Jobs declined to take him up on the offer.)* 'I didn't know enough to say, "No, it can't be done",' Manock says. 'Steve spoke with a lot of other people in Silicon Valley who said it was absolutely impossible to get it done in that short space of time. But I needed the work, so I jumped at the opportunity.'

As a designer, Manock's real strength was his ability to conceive aesthetics and 'hardcore engineering' in equal measure.

---

* A similar incident occurred years later when Jobs and employee Bud Tribble met with an award-winning British design firm regarding a prototype they had come up with for Jobs' NeXT Computer. After a lengthy build-up, the team revealed what they had been working on: a computer shaped like a human head. Jobs – who had flown thirteen hours from San Francisco to London to attend the meeting – was appalled.

It was important, he felt, to have a 'nice balance between manufacturability and human factors'. This was not necessarily as commonplace as it sounds. A lot of industrial designers worked strictly from the outside in; meaning that they were constantly butting heads with the engineers who were then forced to compress complex technological innards into spaces where no such considerations had been made. In the same way that a programmer must dream up a beautifully crafted line of code which is as elegantly minimalist as possible, Manock's job was about compacting a hefty bit of hardware into as small an area as he could without losing functionality. A personal computer may have been considerably smaller than a mainframe, but exactly *how* small could one be? In 1977, these remained hypothetical questions. When he got home Manock took out a pen and paper and began working it out. The Apple II would, he knew, have to include a built-in keyboard. Allowing for this and adding on the thickness of two reaction-injection-moulded outer walls, as well as a bit of clearance, he could ascertain the minimum possible width of the machine. Height was simply a matter of finding out the combined size of the power supply and motherboard. Naturally the machine had to have a flat surface, so as to allow for peripherals like the monitor and, later, the disk drive to be placed on top without falling off. (One neat detail Manock added was a slight recess built into the top cover so that any liquid spilled on to the unit would pool and could be mopped up before it frazzled the electronics below.) Depth was a little more flexible; just calculating the length of power supply when positioned at the rear of the keyboard.

Once this was done Manock flipped the problem on its head and began designing it a second time, now with the eye of an aesthete. Some ideas – such as the concept of robust side handgrips to make the machine portable – were ditched, only to reappear on later Apple machines. Within several days, Manock had created a cardboard model of the Apple II, which he took into the offices of Regis McKenna to show off to the team working on the account. The consensus was positive, although Bill Kelley had one piece of advice. 'Knowing that this computer was going to be very popular, I suggested that instead of taking a flat fee for the design he should just ask for a 25¢ per unit royalty,' Kelley remembers. 'I said, "You'll be a lot happier that way". Jerry said, "I know that's probably true, but I need the money now."' Manock was, in fact, making $20 per hour as a design consultant for Apple. Sometime later, when the Apple II was selling like injection-moulded hotcakes, he retroactively brought up the idea of royalties.

'Gee, Steve, wouldn't it be nice to have a dollar royalty on the objects that have been shipped,' he said in as casual a manner as possible.

Jobs fixed him with a stern gaze and replied, 'You're good. But if you knew how many units we were expecting to sell over the next two to three years, you're not *that* good.'

To his credit Manock retains a good sense of humour about the arrangement and even lets on that he missed out on the chance to become Apple employee number five or six (as per Silicon Valley convention, many employees were part paid in stock options) because he wanted to keep his independent contractor rate going. 'A very shrewd financial move on my part,' he laughs. 'But at the time I didn't know whether this company where the president wore Birkenstock sandals, frayed jeans and a flannel shirt with holes in it would make anything, or even survive. I wanted my $20, guaranteed.'*

---

* Unbelievably Jerry Manock made another error of judgement when he eventually joined Apple as an official employee. Still wanting to be able to work for other companies part-time, Manock negotiated a contract that would see him only working a half-week at Apple. 'I realised after six months that I was working seventy-five hours a week for half a normal salary,' he says. 'It took me until 1979 to figure out I needed to join them full-time on a full salary.' Manock became Apple's first corporate manager of product design and the 'father' of its industrial-design group.

☐ ☐ ☐

Design work on the Apple logo, meanwhile, was handed over to Rob Janoff, a 29-year-old junior art designer at Regis McKenna, with a mop of dark, curly hair and a thick moustache. 'To geeks, the idea of using a computer wasn't threatening, but to everyone else – me included – it was terrifying,' he says. 'I was a creative, visually minded guy. Not a math guy. Not a science guy. The whole deal with having to work on something involving a computer was pretty daunting. I projected that on to everyone else, that it would be a scary idea to a lot of people to have a computer in their home. I said we should make this thing as non-threatening as possible.'

Apple already had a logo in existence, designed by a middle-aged man named Ron Wayne, who had briefly invested in Apple before panicking and pulling his money out – and losing an estimated $2.6 billion in the process. The image Wayne had come up with was drawn in the style of a Victorian woodcut, and depicted Sir Isaac Newton sitting beneath a tree, a solitary apple hanging precariously over his head. Printed around the border was a quotation from William Wordsworth's *The Prelude*: 'a mind forever voyaging through strange seas of thought, alone'. It was an attractive picture, despite its oddly glum corporate slogan (did happy families buying their first personal computer *really* want to be doomed to a life of never-ending journeys and empty oceans of contemplation?), but was also too parochial and difficult to reproduce.

Jobs gave Janoff two directions: don't make the logo cute, and find some way of incorporating the idea of the Apple II's sixteen-colour display. With this brief, Janoff sat down at his desk and began sketching out the shape of an apple on a pad of A2 paper. Before long the sheet was covered in doodles – some of them realistic, fine-art reproductions, complete with shading, line depth and shadows; others basic cartoons. The one that stood out most obviously was a simple stencil outline of an apple; a single leaf hovering, unattached by a stalk, just above the main body. Something was missing, however. Janoff considered for a moment, then took an eraser and rubbed out a section of the image on the top right hand of the image. Instead of redrawing a curved line, he sketched a deep gouge out of one side of the apple, as if someone had taken a single bite and placed it back down. Janoff felt it lent the logo a bit of character. More importantly, it meant that if the picture was shrunk down or enlarged it would retain its sense of scale. After all, what looked like an apple on a one-sheet poster might closer resemble a cherry when reduced to business-card size – and at the rate that upstart computer companies were emerging, could 'Cherry Computers' really be all that far away? At that moment Bill Kelley walked past and glanced over Janoff's shoulder to see what he was working on.

'You know there's a computer term called a byte, right?' Kelley said, half joking.

Janoff didn't, but as soon as he heard that there was everything clicked. *This was it. This was the logo.* The next task, of course, was to find a graphical way of demonstrating the Apple II's colour display. Without thinking, he picked up his pencil again and drew five horizontal lines, one below the other, through the image of the apple, splitting it into six equal segments. Reaching for a set of felt-tip pens he filled each segment in a different colour. The effect was striking: it looked like an apple on acid. 'I had a big hippie influence myself, having grown up in North California in the late sixties,' he says. 'I'd dabbled in all the psychedelic stuff, as well as the music and the visuals. To me, it was more interesting to draw it like that than, say, a *red* apple. The idea was to make it appealing, and to differentiate it from everything else that was out there.'

It was certainly different. Computer advertisements of the era

– most of them business-to-business – were almost unanimously dour, as if trace levels of humour might contaminate the pressured, sterile environments in which high-tech equipment needed to be kept. A few consumer companies had started making adverts targeted at the hobbyist market, but these tended to aim for a level of *Star Trek* futurism, with chunky, technical fonts that resembled the machine-readable numbers on a cheque. This wouldn't do at all. 'Steve specifically said he didn't want something that was too *meep*,' Janoff remembers. '*Meep* meant the popularisation of technology, where the computer or robot says "meep". Steve was smart enough to see that a computer had to tell people that it was accessible. A highly technological typeface would imply that it was not human, not friendly. He wanted to do the opposite.' For the lettering Apple wound up opting for a modified version of Motter Tektura, an Austrian typeface designed three years earlier, which appeared modern but not overly decorative. (It would later be adopted for the corporate logo of Reebok.) At first, the initial cap of Apple was made lower case and, through a bit of design magic, this could be tweaked in such a way that it would echo the shape of the bite. 'I probably wouldn't use it today, but what can you do?' chuckles Janoff. 'The popular mantra among designers back then was "anything but Helvetica".'

As one of the most recognisable corporate logos in the world today, the Apple emblem has received its fair share of scrutiny over the years. 'One of the deep mysteries to me is our logo,' Jean-Louis Gassée, later the president of Apple products, has said. 'The symbol of lust and knowledge, bitten into, all crossed with the colours of the rainbow in the wrong order. You couldn't dream of a more appropriate logo: lust, knowledge, hope and anarchy.' For the religious-minded (which didn't include designer Rob Janoff) the logo has biblical overtones, recalling Adam and Eve in the Garden of Eden. The bite thus represents the acquisition of knowledge, but simultaneously the fall from grace. For the suspicious, the question becomes, *why only one bite?* For the engineer, the logo is an undisguised reference to Alan Turing, the father of computer science and formaliser of the algorithm concept, who committed suicide in the summer of 1954 by taking a bite out of an apple laced with cyanide. Turing's death – coming two years after he was prosecuted as a homosexual, and was faced with enforced chemical castration if he wanted to avoid prison – was supposedly meant as a re-enactment of the Snow White story. Jobs, for his part, was more preoccupied with the colours chosen for the logo, and kept vetoing Janoff's increasingly frustrated suggestions to the point where a ticked-off Bill Kelley was forced to drive down to meet Jobs with a colour swatch book, and to wait while the Apple co-founder leafed through the pages deciding exactly which shade of green best summed up his particular microprocessor.

The colours, too, have prompted much speculation over the years. It is interesting to note that, for one thing, unlike most rainbow flags, the Apple logo had six colours, rather than seven. One reading suggested that the colour bands marked Apple out as the world's first pro-gay computer company, utilising the LGBT (lesbian, gay, bisexual and transgender) flag that had been designed by artist Gilbert Baker in San Francisco. This is an intriguing guess, but unfortunately holds no water. The banner in question – often called the Gay Pride flag – would not come into existence until the following year. Another fascinating supposition, given Jobs' interest in Eastern spirituality, is that the bands of colour are a reference to the flag designed by the Indian spiritual teacher Meher Baba, in which the colours represent the seven planes of involution, in addition to the seven kinds of sanskaras. 'The colours in the flag signify man's rise from the grossest of impressions of lust and anger – symbolized by red –

to the culmination in the highest state of spirituality and oneness with God – symbolized by sky blue,' Baba stated. The groundwork for another, altogether more revolutionary interpretation of the visual motif was laid out in a study entitled *L'Étoffe du diable*, by Michel Pastoureau, a historian of Western symbolism. Setting out to examine the history of striped patterns from the Middle Ages up to the present day, Pastoureau noted how after 1775 stripes became emblematic of the American Revolution, at which point they became shorthand for a particular United States brand of freedom. Writes Pastoureau:

> *... it may well be that the American revolutionaries chose a striped cloth, symbol of slavery (around 1770, the striped uniform is already the dress of prisoners in the penal colonies of Pennsylvania and Maryland), to express the idea of the slave breaking his shackles and, thereby, reversed the code of the stripes. Once a sign of deprivation of liberty, it becomes, with the American revolution, a sign of freedom gained.*

In fact, the notion of stripes equalling 'freedom gained' through computing had already been seized upon by IBM, whose Paul Rand-designed logo featured the name of the company divided up into stripes. Apparently by accident, Janoff was therefore responsible for Apple's first blast of the trumpet against its soon-to-be sworn enemy, IBM. In the book *Visual Identities*, Jean-Marie Floch asks whether:

> *We might ... find, in the counter-use made by Apple, a form of historical irony: in two centuries of American culture, stripes change from being a symbol of slavery to being a symbol of freedom and back again to being*

*a symbol of slavery – at least within the Apple counter-
culture. The Apple logo ... would then be to the IBM
logo what (for the audience at Woodstock in 1969) the
'Stars and Stripes' played by Jimi Hendrix was to the
official national anthem.*

In a highly un-Jobsian move, Baba specified the placement of
only two colours (red at the bottom and sky blue at the top) and
left the rest of the choices to the flag maker. For those at Regis
McKenna, the colours themselves were of little importance.
What concerned them was the cost of reproducing the individual
colour bands for publicity materials, which had to be done to an
extremely high quality to prevent the colours bleeding into one
another. This proved such a worry that one member of the team
actually tried to *un-sell* Jobs on the idea, warning him that the
prohibitive cost of producing stationary might cause Apple to
go out of business before it had even really started. Jobs failed
to listen.

The Apple logo today remains the most consistent aspect
of the company's corporate image; existing in a slightly more
symmetrical version minus the colour bands, which vanished
upon Steve Jobs' return to Apple in 1997. Bill Kelley calls it
'a really nice, tidy, first-class logo design'. Rob Janoff – who
dreamed up the design while an employee of Regis McKenna
and has thus never received any royalties for one of the most
iconic images of the last century – entered the logo into a few
competitions the year that he designed it. He didn't win anything,
or even place significantly. A picture of an apple for a company
called Apple? It was hardly very memorable.

❏ ❏ ❏

By this time Apple had its first president – and his name wasn't
Steve. Mike Scott was a friend of Mike Markkula's, who had

known the angel investor since their days together at Fairchild Semiconductor. Examining when Apple became a fully-fledged commercial entity is not as straightforward as is the case with the majority of companies outside Silicon Valley. For one thing, although logos were being designed and products manufactured, there were absolutely no profits being made; just a big pile of start-up cash being burned through all too rapidly. 'We'd already stopped Apple I production so there was no cash coming in,' Scott recalled years later. 'We had to get the Apple II out ... or we wouldn't have been in business. So Markkula put up a quarter million dollars and that's the cash we used to get started.' In few other lines of business is there the same two-way membranous divide between hobbyist fun and bottom-line-driven business that exists in high-tech. It would be a rare accountant who would spend his off-time balancing books for no reason other than for the sheer pleasure of it. Similarly, few lawyers would voluntarily take a book of obscure by-laws on holiday with which to unwind by the pool. As the Homebrew hobbyists had effectively demonstrated, however, there were plenty of computer geeks who would happily invest countless hours in a workmanlike activity for no reason other than the joy of tinkering away with electronics. It's of little wonder that when the money did eventually start to roll in – and money flows in Silicon Valley like few other places on earth – the people on the receiving end of it would think that they had found a hitherto undiscovered loophole in the idea of labour-for-wages. 'In Santa Clara, the people to your left and right are almost universally talking about semiconductors, operating systems, networking typologies, interfaces, high technology start-ups and high technology rocket drops,' one journalist commented on the sheer pervasiveness of technology love in the valley. 'It's constant. It's everywhere. In the malls, at church (if you have time to go and haven't given up that East Coast custom), in the newspapers, on the television, in the bank queue. It is a twenty-four-hours-a-day, seven-days-a-week activity.'

And while the computer hobbyists treated their leisure activity as serious business, serious business in Silicon Valley behaved as though it were, in fact, a fun hobby. Mike Scott himself had come up through the ranks at Fairchild, and had learned much of his management style there. Fairchild, along with Hewlett-Packard, did more than any other company to establish the prevailing corporate culture in the Santa Clara region. As its commercials of the time stated with some accuracy, 'We started it all'. Rather than hiring experienced managers, Fairchild instead opted to give the jobs to young engineering graduates; cultivating an atmosphere of youthful enthusiasm and providing many with their first big break in the business. Disenfranchised long-hairs, eager to prove the disdain the Establishment felt for the country's youth, presented themselves to such companies in all their unwashed, dope-smoking glory and found ... that they were accepted. Not only were they accepted, but they were given high-paid jobs, with plenty of responsibility and their own secretary. *Maybe this big business thing wasn't so bad*, they thought to themselves.

Although Silicon Valley companies would come to represent the next level of streamlined capitalism after Fordism, they managed to do a reasonable job of disguising themselves as little havens of socialism: communes which also just happened to turn a profit. Companies cultivated a familial attitude by providing their employees with stock options; giving workers a vested interest in their employer's success in a place where people change jobs as people elsewhere in the world change underwear. At the same time, these companies stripped away the traditional hierarchies of corporate life. Reserved parking spaces were done away with, executive dining rooms became a thing of the past, and – in the case of Intel – everyone from the lowest-paid service-department technician to the highest-earning departmental manager, who might drive to and from work in a polished gold convertible, worked in an open-plan office from the same second-hand metal desks.

This eagerness to do business differently from the way it was done back East was part of the same nose-thumbing that fuelled the hippie dream. No one demonstrated this idea better than Robert Noyce, whose co-founding of both Fairchild and Intel, along with his invention of the integrated circuit or microchip, earned him the nickname 'The Mayor of Silicon Valley'. Despite Noyce's high level of success he appeared never to outgrow the notion of raising his middle finger at the perceived Establishment. When he divorced and remarried in the mid-1970s, he was blackballed by the elite from joining the upmarket Los Altos Country Club because, supposedly, its management disapproved of his new wife. Muttering 'to hell with them', Noyce commissioned an exact replica of the club's facilities to be built in his own grounds, within easy sight of the Los Altos clubhouse.

Not everyone could be Robert Noyce, of course, but the new generation of Silicon Valley managers were all keen to emulate the kind of relaxed, 'work hard, play hard' ethic that he personified. Joining Apple, Mike Scott irreverently gave himself the employee number seven – despite joining the company fifth – so that he would be recognised on all official paperwork as 'employee 007'. Scott's authoritarian style was, in the words of one former Apple employee, 'management by walking around'. This technique had been popularised by Bill Hewlett and Dave Packard, who suggested that managers spend a good part of each day roaming about and engaging employees of all levels in unplanned conversations. At noon Scott would grab the three or four people nearest to him – regardless of their department – and take them out to lunch, for which he would pick up the tab. It was an excellent way of keeping employees happy. It also proved a useful method by which to keep inside tabs on what was going on beneath the company's surface.

□ □ □

The next job for the small team at Regis McKenna was to create the two-page ad that Jobs had asked for. The right-hand page was copy-driven, listing the comparatively impressive specifications of the Apple II, beneath the headline 'You've just run out of excuses for not owning a personal computer'. The left-hand page was a colour photograph, demonstrating the seamlessness with which the Apple II could be slotted into the average family home. The picture showed a man – wearing a dark turtleneck similar to the kind that Steve Jobs would adopt as a uniform twenty years later – tapping away earnestly at his computer, gazing intently at the monitor which showed a complex-looking graph. In the background his wife chopped vegetables on the kitchen counter and risked a glance over her shoulder to look adoringly at her husband. The ad ran in a summer issue of *Byte* magazine. 'Within just a few weeks Steve received a letter from a woman in Oregon complaining that the ad was sexist – which it very clearly was,' says Kelley. 'The woman was obviously subordinate to the man and, worse, was left out from partaking in the wonderful advance of technology. Steve took that lesson to heart.' Because Apple didn't have any money left for a second campaign, Kelley begged a photographer friend to pitch in for free and borrowed a girl from the Regis McKenna art department to stand in as a model. This time they were sure not to make any mistakes. The art-department girl was given an Apple II to sit in front of, showing a complex intermodulation curve on the monitor, while Kelley himself posed as a decidedly browbeaten male, staring intensely at a decidedly simpler, low-resolution display.

The image was added as a replacement inset, and there were no more complaints from women in Oregon.

Where there *were* complaints was from Jobs himself. Having taken over completely dealing with Regis McKenna from Apple's end (Wozniak really had no interest in marketing), the 22-year-old co-founder began exerting his exacting influence. It started well. Early on, the copywriter for the account came up with the line 'You don't have to know a Ram from a Rom to enjoy Apple II'. Jobs loved it. 'That's exactly the tonality we want,' he said, ecstatic. But his enthusiasm quickly cooled as the pursuit of perfection took over. As he changed his mind over the tonality and amount of factual information they should be conveying to the public, draft after draft of the 'Introducing Apple II' brochures was rejected – even to the point where it seemed like there was no way they would possibly be ready on time. Finally, on the fifteenth revision, Jobs caved and the design went off to the printers, one day before the deadline. Kelley had to drive to the printers on a Saturday morning to pick up the practically still wet brochures. Then he drove the thirty-minute journey from the print shop to downtown San Francisco. His destination: the West Coast Computer Faire.

Crowds began amassing early on the first morning of the Faire. By eight o'clock, one hour before the doors would officially open, the line of people waiting to get in already reached three-quarters of the way around the block. By midday, 8,000 computer geeks had been admitted, and a temporary crisis in which badges and programmes nearly ran out had been narrowly averted. A little later Jim Warren, zipping around the auditorium on roller skates, breathlessly told a reporter that he estimated that 10,000 people had already made their way through the ticket booths. (The official attendance, in fact, was 12,657.) 'But,' Warren added, '10,000 as a number is very, very different than 10,000 walking, talking computer freaks – that's for sure. This was a mob scene.' A person, especially an outsider, walking into the main exhibit hall for the first time would be overwhelmed by a cacophony of sounds and images; a number of music and speech synthesisers jockeyed for attention with computer demos that had been thrown up on big screen Advent projection-TVs. Throughout the weekend, 'fascinating banquet speakers' delivered sermons to rapt congregations on subjects such as 'Robots You Can Build For Fun & Profit' and 'The 1940s: The FIRST Personal Computing Era'. 'I think 95 per cent of the people who have been here have been really excited over it,' Warren said. In all, 175 exhibitors had turned up, each desperate to prove that their product was the absolute last word in personal computing. Every booth with something to sell had an average wait time of at least fifteen minutes, before customers had the opportunity to ask a question or two, and part with their money.

As Steve Jobs had hoped, Apple was at the centre of it all. Their stall occupied the four booths directly facing the main entrance, so that the Apple II was the first computer that people saw when they set foot in the Civic Auditorium. While the Two Steves' competitors had plumped for little more than flyers and poster-board signs to advertise their products, Apple had opted for the more attention-grabbing, if expensive, approach of having Rob Janoff's logo imprinted on a large pane of backlit Plexiglas for the event. This new streak of upscale professionalism extended to Jobs and Wozniak as well, who, upon the advice of Mike Markkula, had visited a local tailor to be fitted out in three-piece suits.

The Apple team had arranged twelve Apple II computers in a row on the tables, which had been especially draped with black velvet (this *was* the seventies!). On the floor were piles of additional boxes, to give the impression that the visible stock represented the merest tip of a high-volume iceberg. Of course, this wasn't exactly true. The boxes were largely empty, and each Apple II on display had been individually hammered together, sanded and machined separately. One computer's lid wouldn't fit on to another's body. 'But nobody knew that,' says Jerry Manock. Personal computers, at least within the geeky confines of Silicon Valley's hobbyists, had reached a critical mass in terms of popularity. And Apple, which received three hundred orders that weekend alone, was one of the big winners.

□ □ □

Everyone inside Apple was keen to spread the word about the Apple II. The West Coast Computer Faire had been a roaring success, but Mike Markkula hadn't invested $250,000 of his own money to be content selling computers to hobbyists. That was, after all, like preaching to the converted. Markkula spoke with Richard Melmon, the general manager at Regis McKenna, about his quandary.

'I want to really make an impact on the public, but we don't have enough money to do a big television campaign,' he said. 'What can we do?'

Melmon thought about it and suggested a radio campaign. With radio, he explained, the overheads would be far lower. Forget about cameras, lighting and expensive special effects – all you needed was an entertaining – ideally funny – script, and a voice that people immediately recognised and trusted.

'Who would you suggest?' asked Markkula.

'How about Dick Cavett?' Melmon said.

Cavett was an American talk-show host, considered by many to be the thinking man's Johnny Carson. Apple approached the television star through his agent, who dutifully passed the offer along. Cavett had previously done commercials for Buick cars and Kellogg's breakfast cereal. While the request to be the mouthpiece for a personal computer was somewhat unusual, he had learned to roll with the punches in his career. After all, he'd been out for dinner with the notoriously mercurial Marlon Brando the night the actor had knocked out a paparazzo in a New York restaurant, breaking his jaw. He'd been presenting *The Dick Cavett Show* the time Norman Mailer got belligerently drunk and head-butted Gore Vidal in the green room. When Mailer came out to do his interview later that particular evening, he was in no good humour.

'Why don't you look at your question sheet and ask a question?' he snapped.

'Why don't you fold it five ways and put it where the Moon don't shine!' Cavett responded.

Mailer, taken aback, asked the talk-show host if he'd come up with that line himself.

'I have to tell you a quote from Tolstoy?' Cavett said, aghast.

If I can deal with that, I can deal with anything, Cavett figured. He was, however, still slightly apprehensive. 'I was a member of the uninformed,' he recalls. 'My image of a computer at the time was what I had seen in photographs and articles, where it occupied most of a 30x40-foot room, and looked as if it would practically take a railroad car to back one in. All of a sudden they were these small machines.' Even though he was familiar with the college-aged demographic that the Two Steves represented, the people at Regis McKenna still felt it necessary to give him due warning – particularly about Jobs' countercultural tendencies. 'They would almost always attach the phrase that one of them had lived in a tree for a time,' Cavett says. 'That was Steve Jobs. Apparently coming down from that tree he became quite successful.'

A dinner was arranged whereby Cavett, Jobs, a senior member of the Apple team (probably Markkula) and a staffer from Regis

McKenna could meet to discuss the campaign.

'What's it like to be Dick Cavett?' were the first words out of Jobs' mouth.

Later on that evening the public-relations executive from Regis McKenna took Cavett to one side and said, 'I have to apologise for Steve's opening question.' 'I didn't mind at all,' Cavett told me. 'I really liked him, and I'm sorry we lost touch in the years afterwards.'

The campaign with Dick Cavett was a great success. As Markkula had hoped, and Melmon had predicted, it made the Apple II appear a hip, youthful machine for the young, in-the-know generation; *computing's answer to the popular Nehru jacket*. As profits began to pour in, the commercial disappeared from radio and made its way to television screens. In the new advertisement, Cavett played a caricature of himself, subtly lampooning the preconceived ideas around computers in the home.

'I'm here with the average American homemaker with her own Apple personal computer,' he says, sitting next to an attractive blonde woman. 'Now, Jill, do you use your Apple for household budgeting?'

'Actually I'm working in gold futures,' she responds.

'You probably put a lot of recipes in there, eh?'

'Mmm-mmm. And you can do trend analyses, generate bar graphs—'

'Are you really a homemaker?' Cavett interrupts sharply.

'Well, of course,' Jill simpers.

Cavett concludes that Apple is the computer 'for all those pesky household chores'. Jill interjects with 'I also own a small steel mill' as Cavett throws one last exasperated look at the camera.

'It's embarrassing to look at now,' he recalls. 'The version of Dick Cavett that I played in that commercial – almost calling her honey, and suggesting that she would be better off going home and putting on an apron. But it was more hip than the other commercials out there.' Indeed, it would prove to be the start of a decades-long affair between Apple and screen advertising that was just that bit smarter than the other commercials in existence for similar products.

Not long afterwards, Regis McKenna came up with another stroke of genius. It was all well and good having celebrities endorse Apple products – thanks to Jobs' habit of sending out Apple computers to almost any famous person willing to have one, the company fast built up a list of Beautiful People only too happy to do so – but McKenna had a better idea up his sleeve. Silicon Valley may have been filled with antisocial geeks, introverted engineers and nerds with bad haircuts, but it was also populated by charismatic individuals like Robert Noyce. Jobs, too, fell into this category. He may not have been an engineer himself, but he spoke the language of engineers. At the same time he could speak the language of regular people – which is to say that he could speak. Many of the best computer scientists in Silicon Valley were capable of envisaging world-changing innovations without any trepidation whatsoever, and yet became entirely tongue-tied, or just plain terrified, at the thought of having to explain them to the layperson. *Why would anyone want an underpowered machine like a personal computer, with no discernible use beyond the hypothetical and a price tag a third the cost of a new car, taking up space in their living room?* 'Well, gee, um, when you phrase it like that ...' Not Steve Jobs. Right from the start McKenna realised that if he struck early enough he could turn his young client into the acceptable face of personal computing. It didn't hurt that, unlike many of his high-tech peers, Jobs was strikingly photogenic. It wasn't simply that he was attractive (although he was) but rather that, as is the case with those the camera truly loves, he had an *interesting* face. His features were bisected by a thin, angular nose. Those on the left side were softer, mischievous, beguiling. Those on the right were hard, with a surprising air of severity that could take one off-guard. If one half was for seduction, then the other – by contrast – was for cruelty.

Not on the day of Jobs' first television interview, however. On that day Jobs' entire face registered nervousness. Sitting in the studio of a local California TV station he shifted awkwardly in his seat and repeatedly ran his fingers through his hair. For the day he had opted to wear thin-framed glasses, a white shirt with tapered collar, a dark tie and a casual tan suit, which had the unfortunate – and unintended – effect of making him appear like a recent university graduate on a job interview. He looked distinctly uncomfortable, like someone who would rather be anywhere else in the world at that particular moment. Then suddenly he spied something which filled him with a sense of wonder.

'God, look at that! Look at that!' he said pointing excitedly. Jobs had caught a glimpse of himself on one of the myriad video monitors dotted around the studio. 'I'm on television!'

'Isn't that amazing?' said a bored service technician.

'Yes, it is,' Jobs replied earnestly. He was starting to relax.

'This is going to be on in New York, too,' another person noted helpfully.

Jobs froze.

'No, no,' he stuttered. 'Are you serious?'

They were. The sound recordist began fitting an earpiece into his ear. 'You see what that is? That's a talkback,' the man said. 'They're going to use this to speak to you.'

Jobs wasn't listening. His left eye had started to twitch.

'God!' he cursed, hissing through his teeth. He could feel his stomach beginning to play tricks on him. He asked someone to bring him a Styrofoam cup of water and drank this down in a loud gulp.

'You need to tell me where the restroom is,' Jobs said, speaking with some urgency. 'I'm deathly ill and just about ready to throw up at any moment.'

The woman who had brought him the water giggled.

'I'm *not* joking,' he exclaimed.

'It's across the hall,' someone offered.

Jobs looked around in quiet desperation. 'God!' he said again.

Whether Jobs ended up vomiting isn't known. What is known is that, as always, when the lights came on and the camera rolled, he more than delivered. As he had in the offices at Regis McKenna, Jobs spoke patiently, eloquently, enthusiastically about the utopian future promised by computers – and by one brand of computer in particular. Steve Jobs had got his wish: he had his own company. And Apple Computer, by way of return, had found its star.

If Steve Jobs had found his niche at Apple, then Christmas 1977 would see Steve Wozniak truly settle into his. Up until this point the Apple II had relied on cassette tapes for data storage, but these proved unfathomably slow. Aided by his former Homebrew minion Randy Wiggington – seventeen by this time – Wozniak sequestered himself away over the Christmas break and subsisted on takeouts from McDonald's while he pieced together the Disk II, Apple's first 5¼-inch floppy-disk drive. The pair worked together every day with the exception of 25 December (Wiggington took it off, Wozniak didn't). The finished disk drive was released the following year and proved almost obscenely profitable. For only $140 in components, a disk drive could be manufactured that would sell for upwards of $495. It was the finest piece of engineering Wozniak ever engaged in; proof positive that a two-man team could pull off something remarkable. 'That was so awesome, so much fun,' says Wiggington. 'It wasn't about money or fame, it was a nerdy accomplishment that we were proud of.' It was also the last time that something like this would happen at Apple. The company was growing quickly and a couple of hobbyist-hackers piecing together a project for the geeky fun of it was no longer on the agenda.

From this point on those little one- or two-person operations would come from the outside.

4

A TALE OF TWO WORD PROCESSORS

'You may have heard about me. In the computer business I'm known as the Oregon Hermit. According to rumour, I write personal computer programs in solitude, shunning food and sleep in endless fugues of work. I hang up on important callers in order to keep the next few programming ideas from evaporating, and I live on the end of a dirt road in the wilderness. I'm here to tell you these vicious rumours are true.'

Paul Lutus,
author of Apple Writer

If it appealed to customers, then programmers, too, liked the Apple II's relative ease of use, which made it a straightforward computer to design software and peripherals for. In addition to including an inbuilt programming language as standard, the machine featured plenty of expansion slots. It also helped that Apple was an approachable company. During office hours, its headquarters at 10260 Bandley Drive (Bandley One as it became known) became a hive of activity, welcoming whichever vendors, retailers, prospective employees or just plain enthusiasts wanted to pop in for a chat. One day Bruce Tognazzini, a typically hippie programmer (he had previously helped organise the Peace and Freedom Party) stopped by with two clever lines of code he had written, which added a String Array function to the Apple II BASIC language. He found himself in the role of salesman in front of Steve Jobs.

'What have you got for me?' Jobs asked.

Tognazzini showed how String Array function added considerable power for manipulating text within the Apple II, and Jobs cut him a cheque on the spot for the discovery. 'We hammered out an agreement for $50 for the two lines,' Tognazzini remembers. 'Steve swore me to secrecy because he was afraid that someone else would hear about how much he paid per line, and come in with a 2,000-line program, demanding $100,000.'

The Holy Grail that every programmer was chasing was the 'killer app': that one piece of software so indispensable that it could single-handedly persuade punters that a particular model of computer was worth shelling out the money for. Personal computers were the great democratiser. The reason for Apple's open-door policy was because, quite literally, no one knew where the next great idea was going to come from.

Or even what it was going to look like.

◻ ◻ ◻

When the average person in the late 1970s, not yet initiated into the world of high-tech, dared to imagine the stereotypical computer hacker – all missing teeth and poor personal hygiene – it was John Draper whom they found themselves shuddering at the thought of. With a cactus-like beard and an unkempt mane of hair, through which madcap, darting eyes could be seen glistening, Draper (better known to those in the business as Cap'n Crunch) was Silicon Valley's very own answer to the Wild Man of Borneo. Before winding up on the West Coast, his rootless childhood had been spent jumping from place to place to accommodate the career of his air-force-engineer father. 'I went to a different high school every year which kind of sucked,' he says. 'I had to make new friends each time.' In 1964 Draper enlisted as an air-force technician himself and spent four years working with radar and secure communications before being discharged. For a while he bounced around the Bay Area, working first as an engineer for National Semiconductor, then as a radio engineer for a local FM station, then at a small electronics company, Hugle International, where he started designing a cordless phone.

His life took an odd turn when he met a young blind man named Denny Teresi who showed him how the free whistle which came inside a box of Cap'n Crunch cereal was tuned to the exact frequency to enable it to control the long-distance calling switches of the AT&T telephone network. From this came Draper's introduction to the world of 'phone phreaks', a sub-subculture of hacker interested in exploring the vast outer reaches of the telephone network. He set about using his technical expertise to create a multi-frequency tone device, called an M-Fer, which could produce the necessary tones to gain access to different parts of the phone system. He discovered that he could not only make free international telephone calls, but also relay these calls all over the world. On one occasion he called himself from around the globe: first M-Fing into Tokyo, then west to India, then to Greece, then south to Pretoria, South Africa, then to South America, back to London, across the Atlantic into New York and finally to the payphone right next to him in California. He picked up. The total delay was twenty seconds. In 1971 Draper became a minor celebrity under his Cap'n Crunch pseudonym, after *Esquire* published an article about phone phreaking, written by Ron Rosenbaum and entitled 'Secrets of the Little Blue Box' ('a story so incredible it may even make you feel sorry for the phone company'). Steve Wozniak, then a student about to start college at UC Berkeley, read it after his mother left the magazine open for him on the kitchen table. From the opening paragraph he was intrigued. Phone phreaking tapped into the same hacker ethic that would later appeal to him at Homebrew. 'It wasn't just about free calls,' he noted in his memoir. 'One of the guys [in the article] said he was basically trying to do a good thing by finding flaws in the system and letting the phone company know what they were.' Wozniak dedicated himself to meeting Draper, eventually tracking him down through KKUP, a local Cupertino radio station. When Draper arrived at Wozniak's dorm room, its excited inhabitants – including Wozniak, his roommate and Steve Jobs – were taken aback. They had expected a world-class super engineer-cum-ladies' man (the *Esquire* article noted that he had a girlfriend) and *this guy* turns up? 'He smelled like he hadn't taken a shower in two weeks, which turned out to be true,' Wozniak later recalled.

'Are you Cap'n Crunch?' he asked doubtfully.

'I am he,' came the theatrical response.

For the next few hours Draper put on a one-man show: playing tricks with the communal dorm phone, making international calls, ringing several dial-a-joke numbers and topping it off with his famous round-the-world telephonic relay.* Meanwhile, Draper's knowledge of the phone system continued to expand. In addition to making and receiving free calls, he now learned that he could also patch into parts of the phone system that would allow him to eavesdrop on other people's conversations.

---

* The Two Steves wound up selling blue boxes of their own. Their guerrilla sales pitch involved knocking on the door of prospective customers (they decided to start with the male students at Berkeley) and asking, 'Is George there?' When they received the inevitable baffled response either Jobs or Wozniak would elaborate: 'You know, *George* – the blue-box guy. The guy who has the blue boxes to make free long-distance calls.' If this spiel received any sort of interest, the pair would offer to provide a further demonstration of their wares. If not, they would apologise and claim that they had somehow knocked on the wrong door.

One day some of his fellow phreaks were over at his home in Mountain View when they began listening in on a government line to what they believed to be a secret CIA discussion. At a certain point the code word 'Olympus' was mentioned. When this happened, none other than Richard Nixon appeared on the call. The phone phreaks could barely contain their excitement. One of them raced to get Draper, who was downstairs talking with friends. He arrived just in time to hear the end of the conversation. From here they came up with a plan. They would ring the number again and try and get through to the President of the United States. Hearts in mouths they dialled.

'9337,' said the person on the other end of the phone.

'Olympus please.'

'Yes, sir. One moment, sir.'

There was a pause of about sixty seconds, during which only static could be heard down the line. Then a voice remarkably like that of Richard Nixon came on.

'What is it?' the voice asked.

'We have a crisis here in Los Angeles.'

'What's the nature of this crisis?'

'We're out of toilet paper, sir.'

The group fell about laughing as the voice bellowed down the earpiece. 'To the best of my recollection, this was about four months before Nixon resigned because of the Watergate crisis,' Draper says. It was the phone phreaks' best ever prank. When Wozniak heard about it he was jealous for months.

Later that year Draper was on his way home from a FORTRAN programming class when he was apprehended by several men in dark suits. It transpired that they were federal agents and had been following him for some time. He was given a $1,000 fine and placed on five years' probation for wire fraud. Draper insisted that he had given up phone phreaking, having redirected his interests towards computers. He was now working at Call

Computer, a time-sharing computer-services company, and had joined the People's Computer Company, a group of hackers who met weekly in Menlo Park to talk tech and eat potluck dinners. This wouldn't be the end of his troubles with the law, however. In 1974 Draper was arrested again, this time for phone phreaking on an FBI-tapped payphone outside the People's Computer Company storefront. In October 1976 – as Jobs and Wozniak were celebrating six months since the founding of Apple Computer – he was sent to prison in Lompoc, California.

□ □ □

In 1975, the year the Vietnam War officially ended, the Suez Canal reopened after an eight-year closure and MITS released the Altair 8800, Paul Lutus decided that he had had enough. A 31-year-old with red hair and an appearance best described as that of a slightly eccentric maths teacher, Lutus decreed that he was fed up with the rat race, fed up with struggling to stay one step ahead of creditors, and fed up with slogging his guts out each day simply to earn a crust. So he dropped out. He resigned from Rockwell International, the NASA subcontractor for whom he had been working as a $20,000-a-year space-shuttle engineer, sold off all his belongings, and moved to a 12x16-foot log cabin he had built for himself on Eight Dollar Mountain, in the middle of rural Oregon, twenty-three miles from the nearest town. Being on the wrong side of thirty (an age that sixties activist Jerry Rubin warned us not to trust anyone above the age of) it was a hippie's midlife crisis.* What to do now? More than anything Lutus wanted to recapture the freedom he had felt as a youth,

---

* Jerry Rubin, incidentally, followed up on his 1960s anti-Establishment activism by becoming a successful entrepreneur in the 1980s. 'Wealth creation is the real American revolution,' he maintained, while going on a speaking tour with Abbie Hoffman entitled 'Yippie versus Yuppie'. One of Rubin's most fruitful investments? Buying shares early on in Apple Computer.

when he had hitchhiked across the country, taken an abortive crack at becoming a folksinger in New York's Greenwich Village, and dropped acid to try and mentally sort out the complexities of twentieth-century America. Oregon was his chance to start over again. It would be just him, his thoughts and his beloved equations. 'I was intent on having a dignified poor lifestyle, with a vegetable garden and a piece of property that I owned outright, in a part of the country where land prices weren't too high,' he says. At first it came close to the romantic image he had envisaged. He kept chickens, designed ham-radio equipment for his own amusement and offered his numeracy skills gratis to whoever wanted them in the local area. Occasionally NASA would drop by with a bit of lucrative freelance consulting work that he could carry out from home. Then the worst possible thing happened. He started to get bored. One evening in 1977, two years after moving to the wilderness, Lutus was flipping through a copy of *Scientific American* when he stumbled upon an advertisement for the Apple II. 'It occurred to me that if I could get my hands on a personal computer I would be able to do all these wonderful, exciting, creative pieces of mathematics,' he says. The next day he cycled to the closest payphone and placed an order. 'Even though I couldn't really afford it, I plonked down a couple of thousand dollars and bought myself one.' Suddenly his plan to live without any electricity, relying solely on batteries and kerosene lamps, appeared slightly short-sighted – as if he had mentally forgotten to carry the one in his calculations. To power the Apple II he was forced to run 1,200 feet of extension cable through the woods from the nearest electricity source up a steep hill to his cabin. The timing proved fortuitous. A magazine had approached him about writing an article explaining Einstein's theory of relativity and Lutus had agreed to do it. He immediately began compiling the dozens of notes he would need to write the story. Before long the floor of the cabin was littered with scraps of paper. 'Then I looked over and saw the Apple II and went, "wow, I could write a program that would take all of these notes that I had scattered around and organise them internally within the computer's memory",' he recalls. 'For the next month I spent half my time writing a program to organise my thoughts, and half the time actually writing the article to meet the deadline.'

The program was not intended for public consumption. It did the job that it was designed to do – allowing for the entering and manipulation of text – but in the most gear-grindingly unintuitive way possible. 'It was extremely primitive,' Lutus admits. 'It appeared on a 40-column-wide screen which didn't have upper- or lower-case display. One was obliged to use the shift key to enter upper-case characters, which appeared as black-on-white, while lower case was white-on-black. The user of the program was expected to remember what all of that meant when they were using it. It was all very silly.' It did, however, carry several nifty innovations. One function allowed the user to scan quickly through the document to find a particular word or phrase, which could then be amended or moved. Equally impressive was a feature which automatically stored in the computer memory whatever was being written, in the unfortunate event that the user accidentally hit the 'reset' button while working. This particular feature was accidentally tested by Lutus in dramatic fashion. One day after finishing the program, he decided to use the software to write up an accompanying instruction manual. The rain was lashing down outside his window, but Lutus was so wrapped up in his work that he didn't stop to consider that he was sitting in a cabin perched on a 400-foot ridge, in the middle of a storm. Suddenly there was a loud crash and a shower of sparks shot across the room. Lutus's Apple II went haywire. It transpired that one of the trees outside had been struck by lightning. Half of the diskettes which had been lying on the table next to him had been erased by static discharges, but as the

Apple II stuttered back to life he noticed that the work he had been doing was still on screen, the cursor resting next to the last word he had written. *Well that feature works*, he thought.

□ □ □

Steve Jobs didn't like John Draper. Everything about him – from his scabrous appearance to his invitations to weird physical-exercise sessions – made the young Apple co-founder wary about spending too much time with the so-called Mad Hacker. Nor did Draper easily relate to Jobs. 'He was really egocentric and kind of weird,' he recalls. Wozniak was another story, however. 'He was the person I had the best relationship with,' Draper says. Wozniak, too, had a soft spot for his offbeat fellow engineer. For a while he employed Draper at Apple, where the former phone phreak began work on a so-called Charlie Board, a plug-in printed circuit board capable of producing pitch-perfect dial tones, which turned the Apple II into an automatic phone dialler. Wozniak thought it was great, one of the best possible uses for the computer. He programmed it to telephone one of his friends mercilessly, again and again. The majority of Apple employees sided with Jobs, though. Draper was too weird, they said. He made them uncomfortable. His short stint as an Apple employee came close to ending when Draper broke down in hysterics at Mike Scott's lighting a cigarette indoors, a phobia he had had since he was a young child. It *did* end when it was discovered that he had used the Charlie Board to steal $50,000 worth of phone calls. He was off to prison again.

This time Draper was put on a day-release programme. With access to an Apple II during working hours he set about creating a word processor which he could use as a documentation tool for his beloved FORTH programming language. 'I only had a rough idea of how programming worked, but I have this uncanny ability of going in and picking apart other people's code,' he explained.

In this way, software was no different from hardware, in that one could open it up and peer at the innards to discover how a certain function worked. With Paul Lutus not having completed Apple Writer at this time, there was only one real word processor on the market, which had yet to be ported to the Apple II. It was called Electric Pencil, created the previous year by semi-retired hobbyist Michael Shrayer, who had moved from New York to California in search of a quieter life. But Shrayer had shown little interest in expanding his program beyond his initial version. Perhaps there was a business opportunity here, Draper figured. Coding in FORTH, he started work. During office hours he would have access to a computer, before returning to Alameda County Jail at night. 'It was an ideal programming situation,' he recalls. 'Nothing could have been better than that. At the end of the day, before I went back to jail, I would print out a complete listing of the code so that I had a hard copy when I went back in. Then I would look over the code carefully overnight and find the mistakes, without the use of a computer. I became a lot more familiar with the code that way.'

He would also write out new code in longhand on scraps of paper, ready to type in when he returned to work the following day. The wardens were interested and Draper enthusiastically talked them through what he was doing. 'Naturally, the info went in one ear and out the other, but the guards were happy that I was so focused on my work.' One day not long after, Draper was having a conversation with a friend named Steve Sawyer. Sawyer was impressed by how the word processor was coming along but wanted to know what he was going to call it. Draper wasn't entirely sure. He had been thinking along the lines of TextWriter.

'No way,' Sawyer said. 'You can come up with a better name than that.'

'What ideas do you have?' Draper asked.

Sawyer had recently watched the Dennis Hopper film *Easy*

Rider, about two freewheeling hippies who bike through the American South with the aim of achieving freedom. Now *there* was a story.

'How about EasyWriter?' he suggested.

Draper thought that was a great idea. EasyWriter it was. By this time he had been working on the program for several months and there was enough of it finished to show at the upcoming 1979 West Coast Computer Faire. Another of Draper's friends was the aptly named Matthew McIntosh. The two had first met at an Apple user-group meeting called the Apple Core Club, and McIntosh was thrilled at the chance to speak with the legendary phone phreak Cap'n Crunch. 'He's not a genius,' McIntosh would later say of his friend. 'He's not real intellectual. He's not educated. But he really knows about phones.' At the West Coast Computer Faire, McIntosh was in charge of running the San Francisco Apple Core Club booth and arranged for a demo of Draper's word processor. By sheer good luck, the booth immediately to the left was for devotees of the FORTH programming language. Although EasyWriter wasn't technically complete, it was suitably simple and ran blazingly fast on the Apple II. When people turned up at the Apple Core booth and asked which language it was written with, Draper could point next door to the people manning the FORTH table. The people in the FORTH Interest Group returned the favour by answering questions about the latest exciting applications to be written in FORTH by directing interested parties towards EasyWriter. Before long there was a crowd of people at the Apple Core booth, a mixture of publishers and would-be customers, all swamping Draper with technical questions. More than a few were waving money at him, hoping to buy a copy for themselves.*

---

* Jim Warren, still on roller skates but faring quite well financially at this point, cottoned on to what they were doing, and the following year made the newly christened Cap'n Software fork out half the money for the rental of the San Francisco Apple Core Club booth.

While Draper held court, McIntosh feverishly made copies of the program.

'How much should we sell it for?' McIntosh asked.

'I dunno,' Draper responded. 'When did *Easy Rider* come out?'

'Err ... 1969, I think.'

EasyWriter thus made it to the market, with the altogether revolutionary asking price of $69.

Paul Lutus had managed to get Apple's attention. Launching into the world of personal computing with his typical blinkered focus, he had reached out after buying an Apple II and begun communicating by post with members of the various Apple user groups that were springing up like weeds around the country. Like an embryonic news group, they traded news, gossip, programming tips – anything they could think of to further increase the pleasure of owning a personal computer. One of the people Lutus communicated with happened to know someone who actually worked at Apple, and casually mentioned his program to them. Not long afterwards Lutus received a letter, postmarked Cupertino, asking whether Apple might take a look at the text editor they had heard he was working on. He was elated. He saved the latest version on to a 5¼-inch floppy disk and cycled to his nearest post office to mail it. A few days later he had his response: Apple wanted to buy it. The terms on offer were simple in the extreme: $7,500 for a one-off, outright purchase of the program. Lutus mulled it over and decided to accept. '$7,500 was about three times more money than I had in the world at the time,' he says. It was late 1978 when he signed his contract and sent it back to Apple's headquarters on Bandley Drive. They immediately stripped out his inbuilt copyright notice and replaced it with one of their own. The finished product was titled Apple Writer. By the following year they were selling it.

Having signed away any claim that he had to his text editor, this might well have been it for Paul Lutus's involvement with the little computer company from Cupertino. As fate transpired, it wasn't. When Apple Writer was released in 1979, priced at $75, it began selling. And selling. And selling. Lutus may have thought the interface was incredibly basic, but it did a job that very few computer programs at the time were capable of carrying out. Before long Steve Jobs was demanding the programmers he had on staff begin work on a second version; one that would go beyond the simple 'meat and potatoes' of the original to ensure that Apple kept its rightful place at the forefront of the nascent word-processor market. Unfortunately no one at Apple could understand the way Lutus had coded Apple Writer. Rather than use any of the other popular programming languages available at the time, he had written it in computer assembly language, essentially the native tongue of the computer. If BASIC programming was akin to putting together an Airfix kit from pre-formed plastic parts, then assembly-level coding was like building a scale model of a skyscraper out of matchsticks. One line in BASIC could easily transform into five lines of assembly language, which appeared utterly incomprehensible to someone unfamiliar with it:

```
$0946:  A5 FF    NEXT  LDA *DEFINE
$0948:  F0 01    BEQ END
$094A   60       RTS
$094B   A9 60    END  LDA *RTS
$094D   4C 83 0F JMP *PSHCOD
```

What Lutus liked about assembly language was the mathematical certainty of it all. Each of the non-numeric symbols and letters was simply shorthand for yet more numbers. For a maths whiz it was like filling in a sprawling cryptic crossword. If you had done it correctly then everything made perfect, logical sense –

and the finished product would run like the wind. By the time Apple grudgingly came back to him, Lutus had rewritten the entire word-processor program from top to bottom. When he told them they were thrilled. A contract would be in the post that day, they said. 'Not this time,' Lutus responded. 'I want to start over with you. This time I want royalties, I'm not looking to sell the program outright.' Apple was surprisingly receptive to this arrangement. Although the company was making money hand over fist, almost every penny of it was tied up in various projects. Lutus was told to go away and draft a deal that he was happy with. He did so, giving himself 25 per cent royalties (meaning that he would receive a quarter of every copy sold). In doing so he claims that he was wholly unaware that this was five times higher than the industry standard. Nonetheless he sent it off to Apple, who looked over it – and signed.

When Apple Writer II hit the shelves in 1980, suddenly the royalties began pouring in. In Lutus's words it was just 'a fantastic, ridiculous amount of money'. For the next five years the program remained a mainstay on Apple software charts around the world. When the company realised how much they were shelling out each month they hired lawyers to scrutinise every inch of Lutus's agreement. The lawyers were unable to find anything approaching a flaw. Lutus, it seemed, was as thorough with his contracts as he was with his coding. He had several meetings with Steve Jobs at this time, most of them unpleasant. Lutus had previously assumed that Wozniak, being the designer of the Apple II, would also be the person running the show. 'It was a bit disconcerting to discover that Steve Jobs, who I regarded as a glorified salesman, was actually in charge of things,' he says. 'It was perfectly obvious to an outsider who was in charge and who wasn't.'

❏ ❏ ❏

Eventually Apple found its killer app – and, as it happened, it wasn't a word processor at all. It was a spreadsheet program called VisiCalc, built around the metaphor of the blackboard and useful for totting up and calculating finances. It also was not created by a hippie-hobbyist like Paul Lutus or an ex-phone phreak like John Draper, but by a man named Dan Bricklin who had graduated from Harvard Business School. Bricklin was the start of a new breed of programmer: the self-assured, professional MBA who not only understood how computers worked, but also saw what they might mean in the economic long term. When VisiCalc hit the shelves in 1979, selling for the sum of $100, it immediately proved a big hit. Its unparalleled success took Apple one step closer to cracking the big-business market. From this point on there would be no need for Steve Jobs to see all and sundry who stepped through his doors at Bandley Drive.

Meanwhile the likes of Paul Lutus and John Draper carried on as best they could, although following their breakthrough moments of software transcendence, their lives would take divergent paths.

Draper is among the more tragic casualties of the early personal-computer age. After EasyWriter appeared on the Apple II he was contracted by IBM to port it over to the PC they were in the process of developing, a deal that wound up making him fairly wealthy. For a time he drove around in a Mercedes 240D and took expensive holidays in the Galapagos Islands, where he spent his days watching the tortoises and exotic birds. Eventually his money was all frittered away and the work dried up. For a time he was homeless. Later on he said 'screw computing', in the same way he had said 'screw phone phreaking' before that, and threw himself headlong into the rave-music scene, until a tumble in Istanbul aggravated an old back injury. Since he hadn't any health insurance, a Bay Area therapist agreed to help him on the condition that he carried out $10,000 worth of website design work for the practice. Today Draper lives frugally; his meals, at least at one point, supplied to him by an employee of Whole Foods who left food past its sell-by date next to a skip near his home at prearranged times.

Paul Lutus fared far better. He stayed in rural Oregon and kept his bicycle, but in 1983 moved to a large house twenty miles down the road, set on thirty-two acres of land, which he paid for in cash. He continued to be happiest in his own company, and starting in 1988 spent several years sailing around the world solo on a 31-foot sloop. Nowadays he mostly involves himself in working on a Java-based source-code editor for the Internet, called Arachnophilia, which he makes available for free download via his website.

Lutus never returned to his space-shuttle work at NASA, although as the money continued to roll in from Apple Writer – eventually peaking at $6 million in royalties – he did give a passing nod to his former career as a rocket scientist, battling vainly against the forces of gravity.

He bought himself an aeroplane.

A SUPERiOR OPERATiNG SYSTEM

'For both the anti-sixties bombast of
Newt Gingrich and from cultural studies'
celebration of difference, transgression, and
the carnivalesque, a curious consensus emerges:
business and hip are irreconcilable enemies,
the two antithetical poles of American mass
culture ... The historical meaning of hip seems
to be fixed: it is a set of liberating practices
fundamentally at odds with the dominant
impulses of postwar American society.'

Thomas Frank,
The Conquest of Cool (1997)

Today Mitch Kapor is one of Silicon Valley's premier venture capitalists; a man who founded the Lotus Development Corporation and sat on the board as the first chair of the Mozilla Corporation. He is also among the more colourful figures in Silicon Valley – a Hawaiian shirt-wearing multimillionaire with a passion for Transcendental Meditation (from whose yogic lotus position he extracted the name of his former software company) and an earnest belief in the cyber-libertarian ideals of personal computing. In the late 1970s, however, Kapor was just another East Coast wannabe hippie with a magnetic pull towards California. He had first got the itch to move out to San Francisco after attending the Woodstock festival, only to find that his girlfriend at the time wouldn't allow it. Instead he made do with reading *The Whole Earth Catalog* and *Rolling Stone* magazine religiously. His parents were completely befuddled by what their son was doing. 'I had this very expensive Ivy League education at Yale and then spent the early part of the seventies being a radio DJ and a meditation teacher,' Kapor says. For a period he worked as a councillor at a local community psychiatric hospital, before deciding that the field would be better off without him and throwing in the towel. Channelling Benjamin Braddock in *The Graduate*, he was a little worried about his future and agonised over whether he should go off to graduate school and find himself a proper job.

There were always computers, of course. Kapor had had his first introduction to computers in 1963, when he had built a simple binary counter for adding and subtracting numbers up to 1,000 for a school science fair. Later on he read Ted Nelson's *Computer Lib / Dream Machines*, which married his passion for electronics with his interest in the counterculture. He was sold. He started looking around at the various personal computers on the market, but found that all of them were way out of his budget. At the time, Apple had just dropped the price of the Apple II by around $300. Kapor still couldn't quite afford it, so he sold his stereo for extra cash and drove to New Hampshire where there was no sales tax. 'I bought one and I didn't know why,' he remembers. He spent the next two nights playing about with the machine, imagining the possibilities that it represented. Not long afterwards the unemployed Kapor was hanging out in a computer store when a smartly dressed man in his forties entered the shop and walked up to the front desk. 'I'm an ophthalmologist,' he began, and explained to the sales clerk that he wanted an Apple II to carry out certain functions of his job. Kapor edged over and started eavesdropping. When there was a break in the conversation he jumped in.

'Sir, I think I can help you with your problem,' he said. 'I'm a computer consultant.'

As soon as he had spoken, he was taken aback by his boldness. 'I don't know what got into me,' he recalls. 'That was not my usual personality; I don't normally have that kind of chutzpah.' But he got talking to the ophthalmologist and the man agreed to hire him for $7 per hour to set up his computer system. Before long, and having broken up with his girlfriend, Kapor moved to California and took a job as product manager for Personal Software, Inc., the company which published VisiCalc. It was in this capacity that he first met Steve Jobs. 'Apple was already a big deal and Steve was in transition from his hippie days,' Kapor recalls. Like almost everyone who encountered Jobs, he found the Apple co-founder to be highly opinionated, but strikingly charismatic. To a middle-class Jewish kid from Long Island, Jobs may as well have been a rock star. 'He was a force of nature,' Kapor says. 'He was at a higher level than the rest of us – that was my initial impression. He was just a demigod.'

If Steve Jobs was en route from bare-foot hippie to high-tech demigod, then the latter years of the 1970s found him at something of a personal crossroads. There would always seem to be some ideological problems with the conflation of counterculture and big-business culture but in later years Jobs would simply ignore them. Here at least he dwelt on the problem. In 1977 Jobs had invited Dan Kottke, his old friend from Reed College, to join him at Apple. Kottke still had no idea what he wanted to do with his life; just that he didn't want to spend it working for some big corporation. He was won over by Jobs' exuberance. *Remember your second year at Reed when we cleaned the cobwebs out of that crawl-space in your girlfriend's house and we put in Indian prints and that carpet on the floor, and that little incense burner and candle, and went and dropped acid in there? It will be just like old times.* Jobs set Kottke up with a minimum-wage job testing logic boards for the Apple II and filing the keys on the keyboard down to size, so that they would fit in the right slots. Together the pair rented a four-bedroom house near De Anza College in Cupertino, along with Jobs' high-school girlfriend Chrisann Brennan. Kottke nicknamed it 'Rancho Suburbia' as if it were their personal Hotel California. For his own amusement he filled one of the rooms with discarded foam packaging, taken from the myriad boxes of components which arrived at Apple each day. 'I thought, why just throw this stuff away?' Kottke says. 'So for a couple of weeks I would put them in the trunk of Steve's car, because I didn't have a car of my own.' Eventually Chrisann's cats got into the room and the stench of urine became so unbearable that he was forced to abandon that plan. Kottke's position in the company was typical. Many of the other roles in the company were held by friends of Jobs and Wozniak – particularly Wozniak. Randy Wiggington, then in his last year of high school, worked for $2.50 an hour, getting up at 3.30 in the morning and working until it was time for him to go to class, then pulling a late shift at the end of the day.

But the more Apple turned into a big business the more Jobs seemed unsure of it. For a while he half-heartedly entertained thoughts of packing it all in, moving to Japan and joining a Zen monastery. He looked at men like Mike Markkula and Scotty and ... *God, could you imagine being like that?* 'I didn't want to be a businessman because all the businessmen I knew I didn't want to be like,' he lamented. Jobs' anti-business rhetoric was hardly uncommon. In 1976 the nearby Lockheed Corporation had almost capsized when news emerged that its board members had paid $22 million in bribes to foreign officials in order to gain contracts. The following year, Jimmy Carter was sworn in as the thirty-ninth President of the United States, with one of his first acts being to pardon the Vietnam War draft dodgers. The times were a-changin', it seemed. Jobs decided to stick around and see how it would all play out. One problem that he clearly saw at Apple however, was the presence of Mike Scott, better known as Scotty. Jobs viewed the gruff, authoritarian Scotty as a square. Scotty saw Jobs as an irritation; someone who would buzz around at a million miles an hour, stirring up chaos. 'He needs to be sat on,' he complained. He particularly liked to tease Jobs about his oddball habit of taking his shoes off and standing in a toilet bowl, then flushing it to relieve tension after a hard day.

As far as Jobs was concerned, he was the one who should be in charge of Apple, a line of thinking that few others agreed with. 'What twenty-one-year-old knows how to run a company or how to build a company?' Randy Wiggington says. 'That's what the two Mikes had in common: managing Steve Jobs.' For his part, Mike Markkula was able to use Scotty as a buffer; someone to play the bad cop when it came to dealing with the temperamental Jobs. Since Jobs' operating power within Apple was limited, most of his arguments with Scotty – which came to be nicknamed 'Scotty Wars' – were over comically trivial issues. On one occasion the two almost came to blows over the issue of who should sign a set of purchase orders. It wasn't so much about signing the purchase orders themselves, which was an immeasurably dull task, but, rather, what doing so represented in terms of authority. Not all of the Scotty Wars were so vicious. Occasionally they could manifest themselves in more humorous forms. On Jobs' birthday one year he arrived at his desk in the morning to find that someone had deposited a funeral wreath of white roses, complete with an unsigned note reading 'R.I.P.

Thinking of you'. Jobs – who even at a relatively tender age already had an unenviable list of those with whom he had fallen out – had no idea who had placed it there. He received a good clue when, shortly after, Scotty took on the white rose as his personal emblem.

In the summer of 1979 Apple sold $7,273,801 worth of stock. Jobs parted with just over a million dollars' worth of his own holdings. At twenty-four, he was a millionaire. With the money he bought a house up in the hills of Los Gatos, one of Santa Clara County's oldest communities, and the first in what would prove to be a long line of Mercedes coupés. To keep the rebel chic going he invested more money buying a BMW motorbike – even if the orange pom-poms on the handlebars didn't exactly smack of the kind of bike Dennis Hopper would have ridden in *Easy Rider*. Although he might have moved out of 'Rancho Suburbia' for good, Jobs stayed close with Dan Kottke. After work the two would drive up through the mountains and winding back roads behind Stanford to drop acid or skinny-dip in one of the reservoirs. With Jobs out, Kottke brought in some other tenants to fill the vacant rooms in the rental house: a couple of strippers and a cowboy with the disconcerting habit of firing twin Colt .45 pistols in the air. Jobs, for his part, was done with that kind of zaniness. He more or less resigned himself to being a straight businessman, someone happy to wear a suit whenever the occasion called for it. He had broken up with Chrisann, and was now dating a strikingly beautiful half-Polynesian, half-Polish girl who worked at Regis McKenna. Finally, in an act of personal symbolism every bit as potent as when he had shaved his head while in India, Jobs cut his hair short, never to regrow it.

□ □ □

Apple was rapidly changing, too. The sales figures were staggering. Who could have predicted that this many people would want personal computers? Revenue had jumped from $770,000 in 1977 to $7.9 million the following year. In 1979 it would leap again to $49 million. Such was the demand for Apple products that new projects – dozens of them – were added on what seemed to be a weekly basis. With new projects came new engineers, poached from the top engineering companies in the area with the promise of lucrative stock options for those who got in on the ground floor. A standing joke among Apple employees stated that, such was the ferocity of the hiring sprees, the company phonebook was all but obsolete within twenty-four hours of it being printed. One of Apple's chief hires was a man named Tom Whitney, a star engineer who had previously been Wozniak's boss at Hewlett-Packard, and whose claim to fame was helping to develop the HP-35, the world's first handheld scientific electronic calculator. Whitney became Apple's new executive vice-president of engineering.

All these new employees, of course, needed somewhere to work, and for months at a time Bandley Drive became a non-stop construction site as one new office after another was erected. These were built using what is referred to as tilt-slab construction, whereby a horizontal plastic barrier is set up, on to which a bed of wet concrete can be poured. Once this had dried it was just a matter of using cranes to lift the slabs vertical and – hey presto – a new wall! The chaotic feeling of Apple at this time was conveyed by the naming of its new buildings, which made no sense to anyone who hadn't been there from the start. Instead of reading Bandley 1, then 2, then 3 … the buildings were christened in the order in which they had been occupied, so that Bandley 2 stood between Bandleys 4 and 5. It all added to the impression that Apple, while a company, was not *really* a company. 'Apple was very much like a club,' Phil Roybal, the head of product marketing, later recalled. 'We would have these management retreats at spectacular resorts, like Pajaro Dunes, south of Monterey, right on the ocean. There would be a couple of days of meetings and at night we would open the bar and dance until we dropped.'

Among those who worked there, there was a strong sentiment that Apple should not lose its Homebrew hacker roots. Other successful companies might fall into the trap of being big, unfeeling corporations, but Apple was a place started up by a couple of freewheeling hippies, man. This was evidenced when a new manufacturing head joined from National Semiconductor. His first suggestions included making Apple more security-conscious; monitoring who entered through the computer-assembly area, and installing security guards and hidden microphones. 'No way,' came the response. The man found himself heading out of the door before he could even decorate his cubicle. To help keep Apple on the straight and narrow, a number of employees, including William 'Trip' Hawkins, who would later go on to found Electronic Arts, worked together on a hippie-dippy-sounding endeavour entitled the Apple Quality of Life Project. This was designed to put into easy observable words the idealistic tenets the company had been founded upon:

One person, one computer.

We are going for it and we will set aggressive goals.

We are all on the adventure together.

We build products we believe in.

We are here to make a positive difference in society, as well as make a profit.

Each person is important; each has the opportunity and the obligation to make a difference.

We are all in it together, win or lose.

We are enthusiastic!

We are creative; we set the pace.

We want everyone to enjoy the adventure we are on together.

We care about what we do.

We want to create an environment in which Apple values flourish.

Once this was completed the company set about answering the bigger question: what are we going to do now? Although sales of the Apple II showed no signs of slowing down, it was inevitable that sooner or later there would need to be a computer to replace it. Several suggestions were made for logical next steps. A few people set about working on esoteric research projects which might yield results three, five, ten years down the line. Wozniak led a team of engineers in creating what would be a logical upgrade of the existing Apple II architecture, that would come with more RAM and an improved version of the Applesoft BASIC programming language packaged as standard. Then there was the Apple III. This was exciting for several reasons. First off, it would be the inaugural computer designed from the ground up by Apple as a company. The Apple II had been put together on a tight budget by one engineer working in a garage. Wait until the world got a load of what Silicon Valley's best and brightest could do in fully equipped labs with no financial constraint! Steve Jobs, eager to be part of any project that had the potential to be the next big thing, wasted no time in climbing on board. He announced his devotion to the cause early on by distributing to employees glossily printed posters reading: THE DECISIONS YOU'RE MAKING NOW HELPED SHIP 50,000 APPLE IIIs IN 1980.

Right from the start the Apple III was envisaged as a machine targeted at businesses. Rumours were already starting to circulate about an IBM personal computer in the pipeline, and the people at Apple knew that it was crucial to beat this to market. While the Apple II had shown that there were perhaps more hobbyist-hackers in the country than anyone had previously imagined, it was evident that getting computers into the Fortune 500 companies was where the real money was going to be found. Was selling computers to big business losing sight of the libertarian ideals that Apple had been founded on? Of course not, Jobs figured. If anything, getting personal computers into a world previously occupied by IBM mainframes was a subversive triumph. Everyone would benefit. With a computer, a lowly secretary could grow into an area associate. So what exactly would the Apple III feature? For one thing it would need a better, high-resolution display. And it should have a more advanced operating system. And it would need more memory to handle the large spreadsheets businesses would want a computer for. And people would want to store those, so there should be a built-in floppy-disk drive. And it should have a full upper- and lower-case register keyboard. And it should be back-compatible with the Apple II, so that all the software already created for that computer would also work on this one. The list went on and on. 'Everybody had certain ideas about what the Apple III should do and unfortunately all of them were included,' complained Randy Wiggington, summing up the attitude of the hacker engineers.

The problem with building a computer by committee was exactly that: it was by committee. With the Apple II, Wozniak had been the sole creator of the machine. Aside from the cost of the components that he could afford, the only real limitation he felt was the ingeniousness of the solutions that he could devise for certain engineering problems. That in itself was more of a challenge than a problem. When it came to a committee, however, so many compromises were needed that what could too often occur was a dumbed-down, middle-of-the-road machine that appeased a lot of people but satisfied no one. But, hey, Apple was a democratic company, wasn't it? Wasn't it good to try and appeal to as wide a base of potential customers as possible?

The man chosen to somehow take all the disparate ideas about what should constitute the Apple III and turn them into a functional computer was a brilliant hardware engineer by the name of Wendell Sander. Sander was another ex-Fairchild man, who had won his way into Apple after impressing Steve Jobs with a souped-up version of the Apple I. When he was asked to come up with a codename for the project (every Apple project before and since has had a name by which it is referred to inside the company, which differs from its release name) he decided to call it Sara, after his daughter. He would have his work cut out for him. As was indicative of the whole project's direction, the most important decision – what microprocessor to use – wasn't even Sander's to make. While the engineers unanimously thought it a terrible idea, the marketing department was keen that the Apple III, while being a demonstrably superior machine in every way, should also have the ability to run programs designed for the Apple II. By this time so many developers had created software for the platform that throwing it away just seemed like a gigantic waste, they pointed out. In fact, what sounded eminently sensible in marketing meetings (where other topics of conversation included whether the computer should be stylised as the Apple III or the Apple ///) proved something of a nightmare on an engineering level. What it effectively boiled down to was that the Apple III would require some kind of 'emulation mode' where it would have the ability to forget momentarily that it was Apple's next-generation business machine and assume the pose of Wozniak's hobbyist Apple II. To make implementation of this idea easier it was decreed that the Apple III would therefore contain the same 6502 microprocessor that had given life to its predecessor. 'Would that be it?' Sander asked. After all, no matter how you dress it up, reusing a relatively ancient microprocessor that had already been doing the rounds for coming up to almost half a decade was hardly the technological leap that had been discussed in early meetings. 'Don't worry,' he was told, 'of course not.' Apple's big plan was to turn the Apple III into a dual-processor machine, essentially giving it a second brain, with the second, primary microprocessor the more powerful one. But then a bit of number-crunching ruled that out, and it turned out that Sander *would* have to make do with the 6502. With a bit of engineering wizardry he came up with a brilliant solution to squeeze a bit of added oomph out of the 6502's meagre 64K of memory. For this he would use a technique known as bank switching, which entailed having several banks of 64K, which the operating system would keep track of, so as to remember which piece of information was stored where. When the operating system required information from another bank, it could simply shimmy over and retrieve the data that it needed. The microprocessor itself would behave as if the Apple III only had 64K, but any software applications were able to act as if the computer were handling 128K or 256K.

Helping Sander was Daniel Kottke, who had graduated from his minimum wage job on the Apple II production line to the role of technician on the Apple III. Early on in the project, Kottke did much of the point-to-point wiring for the wire-wrap prototypes of the circuit boards. Wire wrap is a method used to construct circuit boards without having to create a printed one. While it was understood that the Apple III would use printed circuits later on, in order to get software working on the machine in the meantime it was essential to get some prototypes up and running. Kottke put on the headphones of his brand-new Walkman cassette player and got to work. When he was finished, the software engineers began working on the computer's new operating system. It may just have consisted of green characters glowing against a black backdrop, but it was a step up from what Apple II users had had to work with. Full of pride, they called it the Sophisticated Operating System.

□ □ □

As if it wasn't tough enough balancing the views of the people inside Apple, it now appeared that those outside the company wanted to chime in with their five cents about how the Apple III would turn out. That was the problem that Jerry Manock, now Apple's corporate manager of product design, was inwardly cursing about when he heard that the Federal Communications Commission was about to issue stipulations specifying the amount of electromagnetic emissions that all digital devices would legally be allowed to throw out. 'The FCC had started getting a lot of complaints from people living in apartment blocks, who would all of a sudden find that their TV reception was going south because someone in the next apartment was using this new thing called a home computer,' Manock recalls. A computer for every person in America may have sounded like utopia to the Homebrew hackers, and rang like a mint to the people who stood to make money from selling them, but it seemed like a nightmare to those concerned about electromagnetic pollution. This is where the new guidelines would come in. That was all well and good, Manock thought, but what exactly *were* the limitations? Nobody seemed to know, and the FCC was not forthcoming with the information. Like building an all-purpose bomb shelter, the only possible solution left to Manock was that he would have to build a case that could withstand damn near anything. He certainly didn't have time to wait around for exact numbers to be announced.

Playing it safe, he decided that the Apple III would give off a grand total of zero emissions. To do this he decided that it would be fitted with a die-cast aluminium chassis. He approached a company named Dolar-Jarvis, who normally provided their services to the automotive industry or defense department. To test the results, Apple even went so far as to lease an ocean-facing property in an otherwise deserted valley, north of Santa Cruz. In it they constructed a Faraday cage, a mesh enclosure designed to block out all outside interference, and surrounded it with sophisticated measuring equipment. When Manock received the various potential castings from Dolar-Jarvis, he would put a few test circuit boards inside, package it up and send it out to the coast. Eventually they came up with a design that would satisfy even the most aggressive FCC regulations.

The downside to a piece of solid sheet metal sealing off the electronics, naturally, was that the inside of the Apple III rapidly grew hotter than a furnace. This is where a fan would come in. Of course, the Apple II had had no fan, thanks to a piece of ingenious engineering by engineer Rod Holt, but then it also hadn't had—

'No fan,' Steve Jobs said. He may have stopped going to his weekend retreats at the Los Altos Zen Center, but he still firmly believed that working on a computer should be every bit as soul-soothing as sitting in a Japanese rock garden. What was tranquil about trying to concentrate with a constant whirring noise in the background? No, Manock would have to think of another answer. He went away and came up with a solution that hinged on a complex series of external fins and flutes, which would channel the heat to the outside of the case by way of convection currents. To test this, Manock required a method by which he could discover the points at which the rising hot air inside the computer was meeting with resistance. It was possible to spend a lot of money on this kind of investigation, bringing in special cameras with state-of-the-art thermal-imaging technology. Manock came up with an altogether more elegant solution. Accompanied by several colleagues, he would take the Apple III into a small room at the back of the engineering labs, with a torch and a pack of incense sticks. He would then plant the incense in the computer's power supply, turn the lights off and, with the aid of the torch, watch as the smoke slowly meandered its way out of the machine in tiny jet streams. It worked a treat, although it proved difficult to explain to anyone unfamiliar with what they were doing. 'With three or four engineers, all the lights out and the smell of oriental incense, somebody walking into the room would really wonder what the hell we were smoking in there,' Manock recalls.

Certainly one of the people wondering what Apple was smoking was Steve Wozniak. For a while he worked on the souped-up Apple II – now renamed the Apple II Plus – but found that it didn't keep his interest. With the influx of new employees he felt that the hobbyist-hackers of Apple's early days were being phased out in place of degreed engineers and MBA business-school graduates. To relieve his boredom he played pranks. One day he sneaked into the cubicle of one of his colleagues and deposited a live mouse inside his computer. When the programmer returned he couldn't for the life of him figure out where the incessant squeaking sound was coming from.

Wozniak was certainly far from impressed by the Apple III. It wasn't an Apple II, that much was for sure. It seemed to him like a computer dreamed up by marketers.

Few dichotomies illustrated just how far Apple had come in a few short years than the difference in scale between the launch of the Apple III in comparison to its predecessor. Where the Apple II had been brought, eyes blinking, into the world as one of myriad miscellaneous computers at the West Coast Computer Faire, the Apple III would be proudly treated like the newest-born son of a $120 million-per-year company, which in 1980 Apple was. The grand unveiling took place on 19 May at the National Computer Conference in Anaheim, California, with a shipping date announced as autumn that year. Sander thought that the launch was a few months premature, but was shouted down. As long as there was an Apple III 'in workable shape' they may as well ship. To celebrate, Apple loaded 7,000 NCC conference attendees on to specially chartered London double-decker buses and shuffled them down the road to the nearby Disneyland resort, which had been rented out for five hours at a cost of $42,000. When the wining and dining was over, the initial reports from publications given a work-in-progress sneak preview of the computer were good. *InfoWorld*, in particular, gushed over the considerate nature of Apple to make the machine back-compatible with existing Apple II programs: 'This software orientation is typical of the design philosophy behind the Apple III ...' In all, it appeared that Apple was on to another winner – even if Steve Jobs privately griped about people mispronouncing the name of the Apple III's Sophisticated Operating System. 'It's SOS, as in apple *sauce*,' he would point out, although that did little to stop people pronouncing it as S-O-S, as in the international Morse-code distress signal. He didn't realise at the time how apt that would prove to be.

It didn't take long after the first batch of Apple IIIs had shipped for the bad news to start flooding in. According to customers up and down the country, the new Apple computer was a dog. Simply put, it didn't work. The first problem came from a special, much advertised National Semiconductor chip that was designed to be used as an internal clock. 'Included in the basic unit is a battery-powered clock/calendar, which the company says will run for over three years, and which keeps track of day, date, and time to within one millisecond,' wrote *InfoWorld* in its hands-on report. Now it turned out that it was faulty. How had no one spotted that? Through gritted teeth Apple agreed to drop the price of the machine, and handed out a rebate of $50 for everyone who had purchased an Apple III up to that point. But the disasters kept coming. Dealers began to bombard the company with messages complaining that one in every five Apple IIIs was dead on arrival, and that others mysteriously broke down at regular intervals. Steve Wozniak fielded embarrassed phone calls from his brother, who ran a computer store in Sunnyvale, stating that he had never received a single Apple III in working order. In the end, Apple management ordered the first 14,000 units be recalled and replaced. Customers loyal to the Apple brand said that they were willing to give the company the benefit of the doubt – until it became apparent that the replacements didn't work any better.

Everyone was quick to point the finger at Jerry Manock, accusing the intermittent failures of being the result of a thermal problem. Why on earth had he built a solid metal chassis, they asked. Manock argued he had done his due diligence and that there was enough internal surface area and external flutes for heat to be properly dissipated. Nonetheless, he went into his backroom with more incense sticks and conducted another round of tests. He even got a thermal-imaging camera, but still no one could find out what was wrong. It all added up to one big disaster for Apple. Developers stayed away in their droves, which meant that there was barely any software for even those computers that *were* working. 'I spent $4,000 and all I got was

VisiCalc and a paperweight,' opined one unhappy customer.

□ □ □

There are few more exciting times in the life of a Silicon Valley start-up than its public offering. Shrugging off the failure of the Apple III as best it could (the Apple II continued to sell like silicon hotcakes in the meantime), Apple had its highly anticipated IPO on 12 December 1980. 'Not since Eve has an Apple posed such temptation,' read the *Wall Street Journal*. One Merrill Lynch analyst commented that even her brother 'who invests in the stock market only on Tuesdays in leap years' had called her to ask what was going on with the hot little computer company from Cupertino. The week of the offering turned out to be a strange few days for Steve Jobs. On the Monday, along with the rest of the world, he heard the tragic news that John Lennon – his favourite member of the Beatles – had been shot and killed by a crazed fan outside his New York apartment. On the Friday he became rich beyond his wildest dreams.

Of the 237 companies that made initial offerings in 1980, Apple was by far the biggest – the biggest, in fact, since the Ford Motor Company had made its own public offering in 1956. The atmosphere at Apple was jubilant, as 12 December became an unofficial public holiday. Mike Scott wheeled in several crates of champagne to celebrate, while a few employees tried to garner support for a scheme to rig up a mock thermometer in the road that separated Apple's two main buildings, upon which they would mark notches of 'heat' as the stock rose and rose throughout the day. Originally filed to sell at $14 per share, Apple's stock instead opened at $22 and rose to $29 by the end of the first day of trading. Thanks to the stock options that had been handed out, more instant millionaires were created than by any other company up to that point. President Mike Scott made $95.5 million. Mike Markkula received a return on his investment – and then some! – to the tune of $203 million. So did venture capitalist Arthur Rock, whose $57,600 gamble netted him the not inconsiderable sum of $21.8 million. Rod Holt – the chain-smoking Marxist engineer who had built the Apple II power supply, and who brewed coffee so strong that it kept the team buzzing throughout the day – found himself sitting on a socialism-challenging personal fortune of $67 million. Some freshly minted fat cats noted that with great wealth came greater headaches. Since selling stock was seen as showing a lack of faith in Apple's continued world domination, engineers who had never previously had any interest in something as unpredictable as the stock market suddenly found themselves glued to the financial pages of the newspaper. Bruce Tognazzini complained of erratic mood swings as his personal finances rose and fell with the stock market: 'I went through a year of being totally whacko because my mood was entirely tied to the Dow Jones.' Meanwhile Apple's two co-founders did better than anyone. Wozniak took home $116 million; even taking into account his having generously distributed a percentage of his stock among Apple employees who otherwise wouldn't have qualified for options, in a scheme referred to as the 'Woz Plan'. Steve Jobs – whose 7.5 million shares made him the single largest stockholder – became a 25-year-old with a net worth of more than $217 million. It was, in short, a day to remember.

It proved a memorable day for Jobs' old Reed College friend Dan Kottke, too, although not for such cheerful reasons. A class divide had entered Apple with the introduction of stock options. Salaried employees such as engineers received stock options, while hourly, unqualified employees such as technicians did not. It was into this latter category that Kottke fell. He was left hurt and confused. 'I looked at Woz and Jobs and saw that they were self-taught engineers and neither of them had a degree,' he says. Since first being issued, the Apple stock had split two-for-one five times. What this meant was that all the people who had originally received 1,000 shares originally ended up with 32,000 shares by the time the company went public. 'I didn't have visions about becoming wealthy,' Kottke says. 'It was more about fairness ... There were one hundred engineers at Apple at the time we went public. Every one of them became a millionaire. That was quite a remarkable event and I missed the boat. Steve wouldn't talk to me about it at all, so it was an unhappy time. It was really cruel of him.' Many within Apple questioned why Jobs would deny his friend the chance to get rich along with everyone else. On one occasion Rod Holt confronted the Apple co-founder about it directly.

'Dan's been here a lot longer than most people and I think he deserves to do well when this company goes public,' he said to Jobs. 'I've got stock options myself, so if you'll give Dan some of yours, I'll chip in with the same number.'

'Okay,' Jobs agreed.

'So how much will you give him?' asked Holt.

Jobs thought about it. 'Zero,' he said.

Later on, Jobs shared his thoughts on Kottke with one of his confidants. 'Daniel generally tends to overrate his contributions,' he said coldly. 'He did a lot of work that we could have hired anybody to do, and he learned an awful lot.' As far as Jobs was concerned, Kottke had got more out of Apple than he should have ever hoped for. He had a full-time job, didn't he? Without Jobs, Kottke wouldn't even be *working* in Silicon Valley. There was another reason why Jobs was denying stock options to his old friends: revenge. Steve Jobs had fathered a daughter with Chrisann Brennan, and was now denying paternity. His friends were appalled. Jobs had grown up knowing that his own parents had given him up for adoption, and now he was denying the existence of his own daughter. Many who knew of Jobs' control-freak personality speculated that the man who subjected himself to gruelling diets and demanded the final say on every detail of

Apple's computer designs, down to the shade of beige on the case, simply couldn't accept that the choice for him to bring a daughter into the world had not been his own to make. When Kottke had spoken out, claiming that 'friends don't let friends deny paternity', Jobs cut him out of his life, refusing even to look at him when they would meet in the hallways at work. 'I felt unworthy, that was the phrase I came up with,' Kottke says. 'Steve put a cloud of unworthiness over me, because he wouldn't talk to me. I couldn't make sense of it. It sucked.' Kottke wasn't alone in being denied stock. Randy Wiggington, Chris Espinosa and Bill Fernandez were all denied options on the grounds that they weren't of a sufficiently high rank within the company. When he heard the news Fernandez quit. (He would later rejoin.) 'We missed out on the American dream because we were too nice to grab a part of it,' Espinosa said. 'Kottke was too nice. Fernandez was too Buddhist and I was too young ... We weren't obnoxious enough to make ourselves millionaires.'

On 7 February 1981, Steve Wozniak – the man who had never crashed anything more severe than a computer in his thirty years of life – found himself in a far worse predicament: he crashed a plane. Like many Silicon Valley hobbyists with the means to advance his love of gadgets beyond simple circuit boards and Heathkits, Wozniak had taken up recreational flying once he struck it big. After much thought, he decided to buy a turbocharged single-engine, six-seat Beechcraft Bonanza A36TC – a snip at a mere $250,000. This was a turbulent time in Wozniak's life; one that saw the Apple co-founder drift away from the company he had helped start, as he watched it change into something he no longer recognised. 'Woz was never really interested in Apple as a company,' Jobs noted in a later interview with *Playboy*. 'He was just sort of interested in getting the Apple

II on a printed circuit board so he could have one and be able to carry it to his computer club without having the wires break on the way. He had done that and decided to go on to other things. He had other ideas.'

Wozniak's personal life had received its own recent shake-up. He had just divorced his first wife and several weeks later started dating a woman named Candi Clark, who worked as a secretary at Apple.* On their first date, he took her to see a schlocky science-fiction film at a cinema he had bought with some of his money from the IPO.† ('My lawyer said to diversify so I ... bought a movie theatre.') The next day the pair drove to the Malibu Grand Prix track, near the San Francisco airport, and raced bumper cars around the circuit. 'I beat her really well,' Wozniak proudly recalled in his memoir. The two decided to get married and Wozniak offered to fly Candi, her brother and her brother's girlfriend from Scotts Valley to San Diego, where Candi had an uncle who had offered to design her a customised wedding ring. However, on takeoff, something went horribly wrong. Wozniak climbed the plane too abruptly, stalled and careened through two fences into the car park of a skating rink. The cause of the crash remains unknown. The National Transportation Safety Board investigation put it down to premature takeoff, compounded by Wozniak's lack of experience as a pilot; he had flown just fifty

---

* Wozniak's first wife, Alice Robertson, did fairly well out of the marriage. When she got hitched to Wozniak in 1976 he had just quit a good job with a steady paycheque to start up a crazy new company selling computers that he built in a garage. Five years later her share of the Apple money netted her $42.4 million. She showed off her newfound wealth by buying a gold-coloured Mercedes. The licence plate read '24 CARAT'.

† The cinema itself was in a low-rent area with plenty of petty vandalism. When Wozniak took it over, the bathroom was so badly covered in graffiti that, rather than scrub it off, he ordered that the walls be painted completely black. This proved to be a short-term solution when the graffiti artists switched to using white spray paint instead. All in all, the place proved a bad investment. Just about the only film he ever had any success running was the 1979 gang movie *The Warriors*. 'That made sense, considering what part of town we were in,' Wozniak recalled.

hours in total. Wozniak thought it possible that Candi, who was sitting next to him at the time, had accidentally leant on the controls. Regardless, Apple's co-founder wound up in hospital. He spent his recovery time playing video games and convincing his old Homebrew Computer Club friend Dan Sokol to smuggle in pizzas and milkshakes for him. But he didn't go back to Apple.

□ □ □

If people at Apple were shaken by the news that the more beloved of its two co-founders had narrowly escaped death and was going to be out of action for the foreseeable future, they didn't have long to recover before the next disaster struck. Less than three weeks later, Mike Scott made the decision to fire a large number of the company's engineers. With almost 2,000 employees, he felt that the company had simply grown too big, too fast. A-players might hire other A-players, but B-players hired C-players. All the non-stop hiring had resulted in what some referred to as a 'bozo explosion'. Scott had tried dealing with the problem the previous summer when, for the first time in Apple's existence, it had been organised into separate divisions. There was the personal computer systems division, headed by Tom Whitney, which comprised the Apple II and Apple III product lines. Then there was the professional office systems division, headed by a man named John Couch, which would work on Apple's next-generation business machine. And then there was the accessories division, which was responsible for printers, disk drives and other peripherals. That had done a bit of good in Scotty's opinion, but not nearly enough. Heads would have to roll. He started by asking each departmental manager for a list of names they thought could stand to be cut loose. He then compiled these names on to one memo and circulated the list, asking that forty names be nominated. Using this, on the morning of 25 February, he began calling the relevant people into his office, one at a time, and telling them that they no longer had a job at Apple. The mass layoffs quickly became known by employees as 'Black Wednesday'. Most confusing was the timing of it all. 'Usually shake-ups within companies happen when things are going badly,' recalls Andy Hertzfeld, then a systems programmer. 'Black Wednesday was one of a number of shake-ups which took place at Apple when things were going great. Sales were doubling almost every month, so that was a little unusual, I would say.'

Hertzfeld had been at Apple for eighteen months. A young computer-science graduate from Berkeley, he had ditched a Ph.D. program to join the company. He tended to work later and sleep in in the morning, and so had missed Scotty's crack-of-dawn first round of dismissals. He was left trying to piece it all together. 'As soon as I came in you could tell that something big had happened,' he remembers. The atmosphere was funereal, with people standing around huddled in small groups. No one was doing any work. Why would they? For all anyone knew, they could be next on Scotty's chopping block. As Hertzfeld was wondering who to talk to, one of the engineers, Don Denman, who worked in a cubicle not far away from his own, walked past.

'What's happened, Don?' Hertzfeld asked.

'Andy, didn't you hear? Scotty fired almost half of the Apple II engineering team this morning,' Denman replied. 'I think more than thirty people have gone so far. No one knows why, or who's going to be next. Apparently there's going to be a meeting around noon when he's supposed to tell us what's going on.'

Not everyone was caught quite as off guard as Hertzfeld. Some, like Jerry Manock, had sensed that something was brewing several days earlier after espying Mike Scott meeting with some of the engineering managers in a conference room. Because of the room's glass walls, he could clearly see a list, detailing every project Apple was working on, written on a whiteboard. About three-quarters of them had been crossed out. 'If you had five projects that you were working on as an engineer and all five were cancelled you were out of a job,' Manock says. In many ways Black Wednesday marked the end of the utopianism that had characterised Apple's earliest days, when people worked hard, played hard and no one seemed at risk of losing his or her job. As Bruce Tognazzini explains, 'It was the end of Camelot. It was the end of loyalty. If they could treat us like that, why should we care about Apple?' Although Mike Scott was the man handling the dismissals, Steve Jobs spent the day striding around, making little effort to appear anything less than joyful to the swathes of nervous employees who had either lost their jobs, or were uncertain as to whether they might. Having seen enough, Chris Espinosa furiously squared up to the Apple co-founder.

'This is no way to run a company,' he said.

Jobs looked at him blankly.

'How do you run a company?' he replied, genuinely interested. At midday, a ceasefire was announced and the remaining employees were invited to a 'beer bust' in the grand Apple tradition. Normally this would have been held outside, but sporadic rainfall throughout the day meant that instead it took place in the garage of a new building along De Anza Boulevard which Apple had recently leased. The garage itself was built on a continuous slant, so that employees were shunted deeper and deeper underground as more and more survivors arrived. Realising the pressure that everyone was under, Mike Scott started by announcing that everyone in the garage was safe, the firings were over and that things were now going to return to normal. To lighten the mood he even cracked a joke, which thirty years on still registers as distasteful.

'I used to say that when being CEO at Apple wasn't fun any more, I'd quit,' he said. 'But now I've changed my mind – when being CEO isn't fun any more, I'll just fire people until it *is* fun again.'

The gag was in even worst taste than it initially appeared. 'Meanwhile there are managers circulating through the crowd, tapping people on the shoulder, because as it turns out they *hadn't* finished firing people yet,' Tognazzini recalls. 'So people are getting pulled out of the garage one by one and told they no longer have a job. It was one of the ugliest things I've ever seen. If you were at the bottom of the pyramid you typically got fired because your boss did. People were made to feel that they were useless, when they weren't. It was a just a terrible thing.' Several days later an anonymous flyer was posted on various bulletin boards around the company:

APPLE EMPLOYEES!!

Are you tired of being pushed around by an arbitrary management?

Bothered by the lack of morale?

Disgusted at the recent capricious mass firings?

Angry at the way it was done?

Annoyed at being treated like children?

Then DO SOMETHING about it! We are forming the COMPUTER PROFESSIONALS UNION (CPU) so that we can keep Apple's management IN LINE. The thing they fear most is concerted employee action, the tactics they use are divide and conquer, and threats of economic reprisal.

THEY CAN'T GET AWAY WITH IT IF WE UNITE!

Apple was once a good place to work, management preaches to us about the 'Apple Spirit'. Let's show them what a real bit of spirit is like and ram it down their throats!

We have real issues: job security (don't you think some of the recent firings were spiteful rather than motivated by concern for productivity?), working conditions (a majority of workers and the committee of employees DID NOT WANT open offices and NOW they tell us we can't talk unless we're sitting down – what is this, a grade school?) and paranoid management (how do you like being yelled at in public?).

Soon a meeting place and time will be set up. <u>YOU CAN'T BE FIRED FOR GOING TO A MEETING. TELL YOUR FRIENDS</u>.

That was the last anyone heard of the Computer Professionals Union. For whatever reason the meeting never took place, although many of the Apple employees held on to the notice for years. It was immediately clear to all involved that any effort to unionise was going to meet with failure. This was where the other part of the stock options came in; the element not designed to make people feel like they were part of the family. Because of the transient, job-hopping culture of Silicon Valley, stock options could be used to keep employees in place, by way of the so-called 'golden handcuffs'. Stock options carried three dates on them: the date when the options were granted, the 'vesting' date when the employees could get their hands on the options and the date of conversion, when the options could be converted into actual stock. Since there was generally one year separating each of these phases, and options were paid out incrementally, it meant that Apple produced a lot more hypothetical millionaires upon going public than it did literal ones. If an employee was sacked before he or she could vest their options, they would be entitled to nothing. In other words, people fired on Black Wednesday were losing more than their jobs – they were potentially having millions of dollars wrenched from them. At Apple, the impermanence of the particle-board partitions which made up its cubicle 'offices' took on both metaphorical and all too literal meaning. One day a person could have a seemingly secure job with their own office. The next day the walls might, very genuinely, come tumbling down and they could be out on the street with nothing. Rather than being concerned about employee rights, most people would much rather keep their heads down and hold on to their stock options for dear life, than risk running foul of management.

As it turned out, even Mike Scott himself wasn't safe from the swinging executioner's blade of gainful employment. Such was the extent of the bad feeling directed his way over the layoffs that Mike Markkula was compelled to demote his friend to the role of vice-chairman and to step into the breach himself in order to keep the peace. Scotty stayed at Apple for a couple more miserable months, before finally storming out for good, dashing off a grimly worded memo to the executive staff before doing so:

> *I believe that a company is a business, not a*
> *social institution. It must demand competence and*
> *contribution from its employees in exchange for fair*
> *compensation. A company is an interactive system*
> *and must have proper controls and teamwork.*
> *Hypocrisy, yes-men, foolhardy plans, a 'cover your*
> *ass' attitude and empire builders can kill it. I don't*
> *believe in an 'executive social club' ... I am no longer*
> *proud or having fun. So I am having a new learning*

experience, something I've never done before. I quit, not resign to join a new company or retire for personal reasons (including as an elected member of the board of directors). This is not done for those who fear my opinions and style, but for the loyal ones who may be given false hope.

He signed his name 'Michael, Private Citizen'. Then he left Apple for good. During Scotty's time as president he had seen the company rise to spectacular heights – become *the* place to work in Silicon Valley – before crumbling away into what he saw as tragic ruin. He needed something to take his mind off the epic melodrama of it all. He decided to buy tickets to the opera.

□ □ □

One day Dan Kottke was working on his Apple III. Although the product line was well on its way to being regarded as a dismal market failure, the company was insisting on installing one of the machines on every desk, much to employees' chagrin. Suddenly Kottke's screen started showing garbled data. The computer had died. After the year he had had, this was all he needed. Out of frustration he lifted the Apple III a few inches off the desk and slammed it back down. Immediately the monitor flickered and returned to normal. *How the hell did that happen?* Kottke wondered. Normally when an Apple III behaved as this one had, it meant that it was done for. He thought about the problem for a minute and realised that the chips must have become dislodged from the logic board. The sudden impact of dropping the computer had reseated them. For a few seconds he thought about telling someone, but decided against it. Why would they listen to him, he figured. He was only a lowly technician. So he stayed quiet and waited for one of the real engineers – the millionaire engineers – to figure it out. This was Daniel Kottke's revenge.

6

GREAT ARTISTS STEAL

'... In the electric age there is no longer any sense in talking about the artist's being ahead of his time. Our technology is, also, ahead of its time, if we reckon by the ability to recognize it for what it is. To prevent undue wreckage in society, the artist tends now to move from the ivory tower to the control tower of society.'

Marshall McLuhan,
Understanding Media (1969)

A couple of years earlier, in December 1979, Steve Jobs had paid his second visit in as many months to number 333 Coyote Hill Road, Palo Alto. The address in question was home to the Palo Alto Research Center, better known to those in the valley as PARC. PARC was the research and development wing of the Xerox Corporation, the East Coast giant that had made its mark on the world with its hugely profitable line of industrial photocopiers. Xerox's Palo Alto research lab had been started up in 1970, with the aim of preparing the corporation for a dystopian future in which people in offices no longer wanted or needed the ability to make paper copies of their documents. Xerox had come knocking at the perfect time for the geeks of Silicon Valley.

Computer science had long been sponsored by the defense department, but growing antiwar sentiment, combined with the militarisation of all DARPA (Defense Advanced Research Projects Agency) research, meant that many of those working in the field were on the lookout for greener pastures. Working for an East Coast corporation – the soulless home of the grey-flannel-suit-wearing man – wasn't ideal, of course, but it was certainly an order of magnitude better than helping the government bomb helpless Vietnamese men, women and children. Located a country-sized buffer away from Xerox's bosses in Rochester, New York, PARC employees immediately set about dreaming up the office of the future, one of the most liberating mission statements ever handed out by a large corporation. After all, with a clever bit of lateral thinking practically anything could be folded into the theme. *Nobody knew what offices in the future might look like, or what work would be done in them.* 'It was completely up to us to come up with it at every level,' recalls one former PARC scientist.

Over the next decade PARC would develop many of the innovations that would go on to revolutionise computing during the next forty years. It was to high-tech what Neils Bohr's institute in Copenhagen was to quantum physics in the 1920s, or what Greenwich Village was in the post-war years for artists inspired by Abstract Expressionism. Most indicative of the kind of person PARC attracted was one of its first hires, Alan Kay, a bushy-haired computer scientist with a Tom Selleck moustache and a mischievous grin. Half of Kay's ideas were off the wall, utterly unworkable. The other half were nothing short of brilliant. 'Computing is just a fabulous place ... because it's a place where you don't have to be a Ph.D. or anything else,' he once said. 'It's a place where you can still be an artisan. People are willing to pay you if you're any good at all, and you have plenty of time for screwing around.' A child prodigy, Kay's had been anti-authoritarian – a rebel without a cause – as far back as he could remember. 'I had the fortune or misfortune to learn how to read fluently starting at the age of three,' he recalled. 'So I had read maybe 150 books by the time I hit 1st grade, and I already knew that the teachers were lying to me.' He hired others just like him. According to him the average scientist at PARC was 'about as straight as you'd expect hotrodders to look ... They're kids who tended to be brilliant but not very interested in conventional goals.' Kay liked to claim that fifty-eight (or was it, as he sometimes stated, seventy-six?) of the world's top one hundred computer researchers were employed at PARC. Most were hippies to a fault; no-good long-hairs with genius IQs, who slept in late, rode their bicycles indoors, and wore Birkenstocks and 'Question Authority' t-shirts to work. They set about creating a world in their image. 'The best way to predict the future is to invent it,' Kay famously noted in an article written for *Scientific American* in 1977, entitled 'Microelectronics and the Personal Computer'.* The first book PARC owned was *The Whole Earth Catalog*, which Kay saw as a summation of what it was that they were trying to achieve. Even the PARC building itself appeared to be in keeping with the topsy-turvy image of a world turned upside down. Since it was built midway up a hill, half embedded into the ground, the facility's main entrance was actually on its top floor, which was described as ground level. To get to the second floor one went *down*.

Such was PARC's geek rock-star image that, in 1972, *Rolling Stone* magazine published a story about the place, written by *The Whole Earth Catalog*'s Stewart Brand. An accompanying image, snapped by then 23-year-old photographer Annie Leibovitz, showed PARC's prized scientists lounging on the floor like

---

* This, in turn, is a rephrasing of the Italian poet Cesare Pavese's notion that 'to know the world one must construct it'.

college students in the dorm-like Bean Bag Room. 'That is the general bent of research at Xerox,' wrote Brand. 'Soft, away from hugeness and centrality, toward the small and the personal, toward putting maximum computer power in the hands of every individual who wants it.' The way he phrased it, PARC was more or less a salaried version of the Homebrew Computer Club, complete with less than subtle parallels between the technological world of high-technology and the 'inner exploration' of psychedelics. Bosses back in New York were understandably perturbed. *Didn't PARC scientists realise that they had a corporate outfit to run; one which was almost exclusively aimed at conservative big business?* Xerox's Chief Scientist Jack Goldman took the first flight out to California and scolded the group like naughty schoolchildren, warning them that if anything like this happened again Xerox would permanently shut down the laboratory.*

For all the ground-breaking work that was being done at PARC, however, Xerox management remained unsure what exactly to do with it. A good indication of the corporation's lack of comprehension was given in 1973 when CEO Peter McCullough flew in from New York City for a demo of the newly built Alto computer. The Alto was years ahead of anything being developed anywhere else in the world. Instead of its screen being filled with numbers, letters and other command lines as was the norm, it featured squares of varying sizes which the team called 'windows'. These were dotted with dozens of tiny pictures, dubbed 'icons', and, rather than interacting with them using a keyboard, Xerox scientists utilised a small box – referred to as a 'mouse' – which connected to the Alto by way of a

---

* When Stewart Brand later approached Xerox PARC's co-manager of the Computer Systems Laboratory about turning his *Rolling Stone* article into a book, the compromise reached was that he would not use the centre's real name, but rather refer to it in print as the 'Shy Research Corporation'.

thin wire. Next, McCullough was given a look at two pieces of software, called Bravo and Gypsy. Bravo was the world's first WYSIWYG (pronounced 'wiz-ee-wig', which was Alan Kay-ese for 'what you see is what you get') document editor, that would later be revised, sold to Microsoft and repackaged as Microsoft Word. Gypsy was a word processor, which included a host of innovations that went on to become ubiquitous in such programs. Xerox's CEO watched as both programs were shown to him, the demonstrator's hands flying over the keyboard, highlighting blocks of text and moving them about as if by magic.

'What did you think?' asked PARC's resident graphics expert, Robert Flegal, afterwards. McCullough considered the question.

'I think I've never seen a man type so fast,' McCullough said.

In other words, he drastically missed the point. To those in the know in Silicon Valley, it was as if Xerox's bosses were sitting on a goldmine and quibbling over the price of drill-bits. Technological development at PARC might have been zooming ahead, but Xerox was going nowhere fast.

Which partially explains why Steve Jobs – flanked by seven Apple employees, including three hardware engineers, one software engineer, the vice-president of software, the executive vice-president of marketing and company president Mike Scott – was driving to Coyote Hill Road on that particular December morning, and why a small group of three PARC researchers was anxiously waiting to meet them. The three employees were Adele Goldberg, Dan Ingalls and Larry Tesler. Although the trio were there to present a unified front on behalf of the Xerox Corporation, they were far from united in their individual thought processes. Each had disparate motivations and rationalisations that shaped their view towards Apple. Each was asking the same question, albeit for very different reasons: *What on earth was Xerox doing?* For Goldberg the answer was a move tantamount to suicide. How could her bosses, having given PARC free rein for ten years based on long-haul progress, now be willing to sell it all for short-term gain? It was mindboggling, and she had categorically refused to be involved unless her employers put their foot down and insisted. They did, but Goldberg felt that she had made her point. For Ingalls the feeling was one of excitement. He had no real interest in the business side of proceedings and was just happy about having the opportunity to show PARC's ground-breaking work to another interested party of visitors. Larry Tesler came down somewhere in the middle. He shared Goldberg's appreciation of how this might impact on Xerox's bottom line, but felt that the decision was a positive, not a negative.[*]

In his article 'Creation Myth' for the *New Yorker*, Malcolm Gladwell characterised Apple's raid on PARC as essentially a predatory one. 'Jobs was the fox,' Gladwell writes, '... and PARC was the henhouse.' A henhouse with extraordinarily developed, golden egg-laying hens, that is. Of course, this is a retrospective piece of analysis. At the time – and with the notable exception of Adele Goldberg and a few like-minded colleagues – Xerox thought it was brokering itself quite a deal, as indeed Apple *knew* that it was. Both parties, at least at the top levels of management, felt that they would benefit from the confluence. Apple, for its part, would get a peek inside the Xerox wonderland, while Xerox saw Apple as an opportunity to test-drive some of its products in front of a wider audience. Through Jobs' hobbyist line of computers Xerox would be able to dip its technological toe into

---

[*] As one of the few Xerox employees to recognise immediately the importance of the personal computer, Apple meshed well with Tesler's countercultural sensibilities. Far better than another visitor from the same time period, the CIA, who had shown an interest in several PARC innovations. On that occasion Tesler had been far less personable; showing up for work kitted out in a trench coat, fedora and pair of dark glasses, behind which he spent the day glowering at the new arrivals. Adele Goldberg, conversely, was very enthusiastic about those particular visitors, whom she viewed as more like Xerox's traditional paying customers than Jobs' motley crew.

the personal-computer market to find out exactly how warm the waters were. Then they could make their move. Because of the sensitivity of the operation there were very specific ground rules regarding exactly what Apple would and would not be allowed to see during the PARC visit. Beyond that there were agreed terms of business. These broke down as such: one year before Apple's IPO, Xerox would be allowed to purchase 100,000 shares of Apple stock at $10 a share, for a total cost of $1 million. (When Apple went public the following year, those shares were instantly worth $17.6 million.) In addition Xerox would use its international business connections to help Apple crack Latin America; a potentially lucrative market in which Xerox would represent Apple's business interests by acting as its distributor and middleman. In hindsight, Goldberg was absolutely right – and Jobs, while he would never have outright admitted it at the time, entirely agreed with her. 'Basically, [Xerox] were copier heads that just had no clue about a computer or what it could do,' he said years later in a televised interview with Robert Cringely for the documentary *Triumph of the Nerds*, adding that this was probably one of many decisions taken by the organisation which ensured that Xerox 'grabbed defeat from the [jaws of the] greatest victory in the computer industry'.

The Apple party arrived and was led into a square conference room near the main entrance of PARC, where an Alto computer had been set up for the day's demonstration. Although the room was a reasonable size, there was just one chair in it, to seat only whichever Xerox employee happened to be conducting the demo at the time. Everyone else would have to stand, crowded around the monitor to see what was going on. Not that Jobs felt like sitting down. Caffeinated to the gills, he stalked back and forth like a caged animal. When he finally did stand still for any length of time he rocked his head slowly on its axis, as if trying to enhance the flow of blood to his cerebral cortex to better handle the bombardment of new information. 'You could see the ideas running through Steve's brain really fast,' Tesler remembers. 'He could make connections at an incredibly high rate.' Tesler felt that Jobs had a different kind of intelligence from most of the people he encountered. 'We were technologists with very logical minds,' he says. 'But Steve also knew about marketing, distribution, finance – every aspect of the business you could think of.' Jobs' team had done their research. When one of Goldberg's engineers tried to distract the Apple lot with the fanciness of Xerox's Gypsy word processor, like diverting a child's attention with a flashy toy, Jobs decreed that they were wasting his time.

'Cut the bullshit,' he snapped sharply.

Ted Kaehler, employed as a researcher at PARC, was impressed by how much the Apple group appeared to be aware of already. 'They instantly knew what they were looking at,' he says. 'Someone said afterwards that this was the most dangerous group of people that you could possibly bring in, because when they saw something on the screen they knew exactly what it was.' The Xerox lot had done their homework on Apple, too, but all their reconnaissance had suggested that it was still a company staffed by Homebrew Club hackers. What they didn't realise was that in the interim period Apple had hired some incredibly talented computer scientists. There was no generality at all in their questions. They cut right to the technical core of what it was that they were seeing. At several key moments the tour ground to a halt as the PARC employees giving the demos grew nervous about how much they were being asked to disclose.

'You're still not showing us what we want to see,' Jobs complained bitterly. Although he hadn't personally been up to PARC previously, he had been briefed in advance about what it was that they were working on. Phone calls were made to Xerox's venture-capital division. 'Tell them what they want to know,' came the directive from back East. Adele Goldberg stormed off in a huff. In the end, the Apple team wound up being shown three innovations. The first was the Alto's graphical user interface. The second was Smalltalk, Alan Kay's object-orientated programming language. The third was PARC's astonishing Ethernet network, which saw more than one hundred Alto computers networked together with the ability to email one another. Jobs, however, was so spellbound by this first one that the other two didn't even sink in. As a non-engineer he always responded best to visual stimuli; never really getting the excitement of a well-executed line of code or experiencing the buzz associated with the feel of a soldering iron in one's hand. Despite its flaws, he thought that the Alto's graphical user interface was, quite simply, the greatest thing he had ever seen in his life. 'Within ten minutes it was so obvious that every computer would work this way someday,' he recalled in a 1994 interview with *Rolling Stone*. 'You knew it with every bone in your body. Now, you could argue about the number of years it would take, you could argue about who the winners and losers in terms of companies in the industry might be, but I don't think rational people could argue that every computer would work this way someday.'

If there was something that *could* be questioned it was how exactly Apple planned to use this information. As far as the Xerox employees were aware, Apple was in solely the Apple II business (the Apple III would not, in 1979, see the light of day for another six months). What use could the company possibly have for technology that, even if they understood it, would never in a million years fit into their current line of machines? The tentative answer to that question had been hinted at in an earlier meeting between Apple and Xerox. Jobs had been asked the question and had looked agitated, as if he was itching to say something. Eventually he did.

'Come on,' he pleaded with the other Apple reps in the room. 'We have questions that we can't even start to ask without telling them.'

The Apple lawyers huddled around and whispered something to Mike Scott. Mike grumbled for a moment and then acquiesced.

'Okay, Steve.'

Without wasting a second, Jobs excitedly turned his attentions back to Larry Tesler and the other Xerox representatives.

'Guys,' he said, 'I need to tell you about the Lisa.'

□ □ □

The Lisa was, in fact, cause for excitement. Apple had started working on it the year before, in 1978, when Steve Jobs and Trip Hawkins, the manager of marketing planning, began discussing the creation of yet another new machine that would pick up where the Apple II had left off. Jobs wanted a 16-bit computer and, for a while, passed the project over to Steve Wozniak, who began sketching out on a paper napkin possible ways to adapt a technique he had seen used in minicomputers. Nothing really got moving, however, until July the following year when a former Hewlett-Packard employee named Ken Rothmuller was hired as the project's manager. Jobs was upset by the appointment. After all, the purpose of the Lisa was to transform Apple into a $500 million-per-year company. Everything up until this point was going to look like a dry run, an afterthought. As Jobs enthusiastically told visiting *Time* magazine journalist Michael Moritz (now a successful venture capitalist), the Lisa 'will sell 12,000 units in the first six months and 50,000 in the first year'. Now it was being taken away from him. Having been put in charge of sniffing out the direction in which popular computing was heading, Jobs felt that this was proof positive that management didn't have enough faith in him to see his mission through to its conclusion. *Not trusted to run the company, and now not trusted to do this either. Screw that!* 'After setting up the framework for the concepts and finding the key people and sort of setting the technical directions, Scotty decided I didn't have the experience to run the thing,' Jobs later recalled. 'It hurt a lot. There's no getting around it.' When Rothmuller came aboard, the Lisa existed as nothing more than a series of specifications. It was to be a computer aimed at businesses. It was going to cost $2,000. It would ship in March 1981, giving them exactly twenty months lead time. It might as well include the Motorola 68000 processor Wozniak had tried using. And, at least for the time being, it would have a built-in phosphor display, keyboard and traditional user interface. The reason why the Lisa would have to be more powerful than the previous generation of Apple machines was down to an engineer named Bill Atkinson. Atkinson had lobbied Jobs to build the Lisa around what was referred to as a 'paper' paradigm: meaning that the computer would feature black characters against a white screen instead of the classic green characters glowing against a black backdrop. Although this didn't sound like a major change, it meant that in order to avoid a continuous epilepsy-inducing screen flicker the monitor would have to refresh at a much faster rate than most machines. Hence the power. Hence the cost. As he waxed eloquent about the machine to the Xerox lot, Jobs was now busy mentally scratching out much of this. What he was saying, what was important to convey, was that the Lisa would be physically capable of carrying out what it was that the people at Xerox PARC were working on – or at least what Jobs thought they were working on.

A few months after the PARC visits, Hawkins was asked to draw up a list of revised specifications for the Lisa. Now the Lisa would have a graphical user interface, a mouse, a local area network, file servers, and a host of new and innovative software applications. It would still cost $2,000. And it would still ship in

March 1981, now only twelve months away. Rothmuller balked.

'There's just no way we can incorporate all of that and still stick to the original pricing and due date,' he complained.

Apple fired him for his pessimistic attitude, replacing him with another former Hewlett-Packard employee named John Couch, who had been working with Jobs as Apple's director of new products. If anyone could captain the Lisa to sure-fire success it would be him.

□ □ □

What should Apple call its new computer? That was the question being asked on Bandley Drive as work on the project continued. Lisa was an acceptable codename for internal usage, but naturally it would have to be changed for the official launch. For the first time in Apple's history it was decided that it might be of benefit to call in the experts. $75,000 was duly siphoned off and given to an advertising agency, along with a deadline of eight weeks, in order to invent the kind of name that only paid creatives could come up with. Two months later the day arrived for the grand unveiling. A group of Apple employees gathered in the conference room of the Lisa building on Bandley Drive. At the front of the room the ad agency had rigged up a red velvet curtain, which covered the much anticipated new name that had been selected. Just before the meeting was scheduled to begin, Jobs entered and sat down in the front row. The president of the agency started with a brief introduction, in which he explained that the chosen name had been trialled in brainstorming sessions, surveys and focus groups, and had scored excellently every time. It would, he said, be as close to a guaranteed winner in the business-computer market as was humanly possible. Without further ado he pulled a cord at the side of the stage. The curtains opened to reveal –

The Apple Applause.

Before anyone could react, Jobs spoke frankly.

'We're not going to call it the Apple Applause,' he said. 'If there are any problems with the computer, everyone's going to call it "The Clap".'

With that, he got up and stalked out of the room. Just twenty seconds had elapsed since the unveiling. The computer continued to be referred to as the Lisa.

□ □ □

Shortly after the Apple demonstration, Larry Tesler quit his job at Xerox. If one believed in karma this was meant to be. One year earlier Tesler had been travelling around Europe. In a rural French village he had his fortune laid out for him by a tarot reader. When she concluded her reading by telling Tesler that he would leave his current employment within twelve months Tesler looked at her doubtfully and said, 'But I have the best job ...' Now he was leaving. For one thing the pay wasn't great (it was actually less than he had been making several years earlier as a software consultant). More importantly, however, Xerox was failing to turn any of PARC's bright ideas into products. It was one thing to have freedom to operate, with deadlines stretching as far as the eye could see, but it was quite another never to see a finished product hit the shelves. To no avail, for several months he tried to get bosses to turn the NoteTaker portable computer the team had been working on into a commercial release. When they ignored him, Tesler left.* He began putting

---

* The Xerox NoteTaker, which resembled a portable sewing machine and would have carried a hefty price tag of $50,000, never did enter production. Ten prototypes were made and the team moved on to something else, although some of the design components later popped up in the Osborne 1 and Compaq Portable computers. For all Xerox's talk of an 'office of the future' in which paper was all but obsolete, the creation that eventually wound up saving them was the invention of the laser printer, dreamed up by PARC researcher Gary Starkweather. Thanks to him, offices today produce more excess paper than they ever did before.

out feelers to colleagues in the industry to find a new position. Most of the roles they talked about sounded monumentally dull. Then he thought of Apple. 'It's funny because Apple was really the trigger for me wanting to leave Xerox, but I'd never seriously considered it as a career option,' he says. 'Even though I had been pretty impressed by the people who attended the PARC demo, I still thought of them as primarily being a hobbyist computer company. It wasn't like I was against going to work for them, but I also didn't come out of the demo thinking, "I've got to go and work for Apple".' A meeting with an old friend, Phyllis Cole, who was Apple employee number 32, persuaded him to give the matter some thought. Cole put Tesler in touch with John Couch, who in turn introduced him to several other members of Apple's research and development team. What Tesler didn't know at the time was that Apple was desperately trying to recruit Tesler's PARC colleague Dan Ingalls. It was Ingalls who had given the Smalltalk demo to Jobs. When Apple approached him about joining them, however, he turned them down flat. Just then Tesler had come along and Apple bosses figured that, hey, if Ingalls wasn't coming, Tesler wasn't a bad fallback option. They hired him.

Larry Tesler is a fascinating figure in the history of personal computing: one who perfectly encapsulates the convergence of hippie and high-tech culture that made the whole scene possible, and gave it its personality. A slim man with a beard and a shock of red hair, in later years Tesler would rise to the level of vice-president at Apple. Before that, prior to his job at Xerox PARC, he had participated in antiwar demonstrations and taught classes at San Francisco's Free University. One of these, commencing in the autumn of 1968, was entitled 'How to End the IBM Monopoly' – and, oddly, turned out mostly to be attended by employees of the monolithic computer firm itself. Another course, with a distinctly less political bent, was called

'Taurus Party' and was marked in the school's catalogue as being strictly for those 'people born with Sun in Taurus only. We'll overeat, overdrink, overdance, oversex, oversleep, and hangover in true Taurian *[sic]* fashion. Please bring food and drink, but no non-bulls.'

Somewhere among all this oversexing and oversleeping, Tesler found time to teach himself programming and invent the concept of computerised cut, copy and paste. He was passionate about modeless computing: something of a forgotten war today, but a source of great annoyance to early users who would constantly run into errors by having to switch between different input states. So passionate, in fact, that his Dodge Valiant bore a customised number plate reading 'NO MODES' and – in smaller writing along the top and bottom – 'How Do I Get Out Of This Mode?' A t-shirt that he wore with some regularity advised colleagues 'Don't Mode Me In'. In 1970, at the age of twenty-five, Tesler had announced to the world that he was dropping out and tuning in. 'I'm going to move to the land with my daughter and we're going to grow vegetables,' he said. Along with a group of friends – one of whom was a former employee of *The Whole Earth Catalog* – he chipped in money to buy a piece of cheap land in rural Oregon, for the founding of a hippie commune where he might live out his remaining forty-plus years. A month later Alan Kay came along with a job offer he couldn't refuse.

Tesler started at Apple on Tuesday 15 July 1980. Although it was already a large – and still growing – company, he noticed the difference immediately when compared to working for the sprawling giant that was Xerox. 'We weren't spread out around the country and around the world,' he recalls. 'Apple was just in a couple of buildings. At Xerox I would make an appointment with a VP, and that appointment would be three months or six months from now. We'd be trying for years to get a meeting. And then when you saw that person they'd advise you to have a meeting with a different VP, which would take another six months to be set up. Nothing could get decided. At Apple, I'd walk a few doors down from my office and talk to Steve Jobs. If Steve wasn't in, I'd talk to his secretary and have a meeting with him four hours later or else I'd catch him in the hall. It was just a totally different situation in terms of the ability to reach people and to get decisions made.' There were other, less welcome differences. Apple employees were constantly working to impossible deadlines, which would invariably fail to be met. Early on, several of the former Xerox employees banded together to form a research group similar to the one which had worked so well at PARC. They dissolved it a week later when they realised that Apple was nowhere near ready to concentrate on the future, while it was all hands on deck to finish *this* generation's product line. The excessive hours – an aspect of Apple life which subsists today – came as a shock to the Xerox lot. 'People in Silicon Valley work long hours anyway, but PARC was relatively light hours most of the time,' Tesler says. 'There was one project where for several months we were working twelve-hour shifts, but that was the exception to the rule. At Apple it was the rule.'

As with other Xerox employees who jumped ship to Apple around this time, Tesler was sent to work on the Lisa project. An early ethical and legal question that occurred to him was how exactly he could lend his considerable experience to Lisa without disclosing trade secrets developed at Xerox. He made a beeline for a man he recognised, Tom Malloy, who had previously worked with him at PARC. Malloy, it transpired, had come up with a system to get around just this problem. He would pose questions to the Lisa team which had occurred to him at Xerox (*How do we create this? How do we get around this without changing the dimensions of that?*) but would avoid answering how Xerox had, in fact, solved those problems. If a member of the Apple team just happened to come up with an idea that followed along lines Malloy knew were promising, he would strongly encourage them without revealing exactly why he was doing so. It was Silicon Valley charades.

□ □ □

The person in charge of the Lisa's physical exterior design was a man named Bill Dresselhaus. After Jerry Manock, Dresselhaus had been employed as the second full-time member of Apple's industrial-design group. He had trained as a chemical engineer and was making in excess of $40,000 per year working in Los Angeles, but decided that he wanted to change direction with his life. Through an acquaintance, Dresselhaus was recommended to the staff at Stanford's product-design school and went in to interview for a graduate post.

'I'm a former chemical engineer,' he told his interviewer, 'but I don't like chemical engineering. I want to be an artist and I want to design things.'

The meeting went well. Dresselhaus was offered a teaching-assistant role and advised to start as soon as possible. His interviewer promised that working at Stanford was going to spin his head around. 'I had no idea what that meant,' Dresselhaus recalls. 'I was just this greenhorn chemical engineering kid from Nebraska. But I went to Stanford and, sure enough, they twisted my head on backwards.'

Stanford was one of the most progressive product-design courses in the country. Instead of old-fashioned engineering, departmental head Robert McKim emphasised such concepts as 'ambidextrous thinking' and 'seeing by visualising'. 'Stanford was really a school of innovation,' Dresselhaus says. 'If you've studied innovation, you've really *got* to think differently. You can't be too rigid or too straight arrow to come up with new ideas.' In his teaching-assistant job, one of Dresselhaus's responsibilities was to lead the engineering students in 'directed fantasies', in which they were encouraged to imagine themselves as various inanimate articles in order to help them see design from an object's perspective. According to legend, this radical concept had been dreamed up McKim in consultation with LSD exponent Dr Timothy Leary.* McKim insisted that it was the best way to encourage students to flex their 'imaginative muscles'. To this end, he ordered the building of a miniaturised cardboard geodesic dome in the Stanford classrooms. Dresselhaus would enter this with between five and ten engineering students at a time. While the students lay on their backs in a star formation on the wooden floor, he led the fantasy from the position of narrator. To begin with he would project mood-establishing slides or a sequence from a film on to the ceiling of the dome. Then he started speaking.

'Imagine that you are an apple hanging from a tree,' he was told to say, enunciating softly as if to induce a hypnotic trance. Then he opened the lid of the jar of fruit essence to release a pungent waft of Gravenstein apples. 'You are one of thousands of apples in an orchard. It's a warm day in late summer and there's a gentle breeze rustling through the trees. You've been ripened by the sun and – in fact you're so ripe that you've fallen to the ground. You bounce, once, twice, and lay still. Now here comes a worm ...'

There was nothing quite so far out in life at Apple, but the experience prepared Dresselhaus for some of the more unusual incidents – and characters – that would punctuate his tenure with the company. When it came to the Lisa's design, Steve Jobs was incredibly hands-on. Although his preferences as an

---

* Professor Robert McKim was every bit a product of the sixties counterculture. On one occasion he fed half of his class mescaline to see whether they would prove more creative than the ones not on hallucinogens. He rented one of the rooms in his home out to a spiritual guru. 'Picture this,' says another of his former students, 'I went to my professor's house one day and sitting on one of the beds in the lotus position was this guru. I just said, "who the hell is that?"'

aesthete were not yet as developed as they would become, he was very specific about what he was after. It proved to be a long and exhaustive list. 'Steve had many, many ideas about what he wanted,' Dresselhaus recalls. The form factor of the machine (tech-speak for a device's physical design) was markedly different from the direction that Jerry Manock had gone in with the Apple II and III. Here, the circuitry, display and two disk drives were to be contained in one single, microwave-like unit, with the keyboard connected via a cord at the front. Because Dresselhaus was told that the chief competition in the marketplace was not going to come from other high-priced computers, but, rather, the humbly pervasive IBM Selectric typewriter, he borrowed the typewriter's design for the look of the Lisa keyboard.

Some things stayed the same, of course – not least of all Jobs' insistence that the machine have no inbuilt cooling fan. When the Lisa was ready to be prototyped, Dresselhaus called for a design review meeting, involving the heads of sales, marketing, manufacturing, purchasing, engineering and electronics, along with several Apple managers. Dresselhaus's chief engineer, Ken Campbell, talked the group through the blueprints, using a concept drawing he had hung on the wall. Suddenly Jobs sprang out of his chair and pointed animatedly at a small black box on the drawing.

'What is that?' he demanded.

There was an awkward silence.

'Well, uh ... that's a fan, if, IF, we need one,' Campbell said. 'We're trying very hard not to have a fan. But that's where it would be if we *have* to have one.'

Jobs stormed out. Dresselhaus followed him and found Jobs leaning against a wall, his head down. He was crying. Dresselhaus was terrified. He had become 'the man who had made Steve Jobs cry'. But Jobs' outburst had the desired effect. As the Apple III had done, the Lisa wound up relying on convection currents for cooling. There would be no fan. 'So he got his wish,' Dresselhaus says.

Jobs' demands for perfection didn't just mean making the Lisa the easiest machine to use; it also meant making it the easiest to repair. He wanted it completely serviceable without the need for tools. 'The power supply, the two disk drives, the two floppy drives, the keyboard, the PC boards in the back, the mother board, the baseboard, the card cage – everything you could take apart with these simple thumb-screws or snap fits to swap out all the components,' Dresselhaus recalls. 'In many ways the Lisa was one of the first green products, even though we didn't design it to be green.' Even with all this neat functionality, however, he readily admits that it was not the outside industrial design which made the computer revolutionary. One small exterior clue, though, did hint at the PARC-inspired wonders that lay within. The Lisa came with a mouse.

Today, when pointing and clicking on an icon comes as naturally as underlining a sentence with a pen on a sheet of paper, it is difficult to remember just what a seismic shift it was in the first half of the 1980s to ask people used to working on a keyboard to suddenly switch to using a rolling box the size of a cigarette packet in order to control what was happening on screen. Like the best Apple solutions the mouse was simple, elegant and, most importantly, *intuitive*. After all, as defenders of this new technology were quick to highlight, pointing at an object to indicate it as an area of interest is a universally recognised shorthand. 'If I want to tell you there is a spot on your shirt, I'm not going to do it linguistically: "There's a spot on your shirt fourteen centimetres down from the collar and three centimetres to the left of your button",' Steve Jobs told *Playboy* when later quizzed about his decision. 'We've done a lot of studies and tests on that, and it's much faster to do all kinds of functions, such as cutting and pasting, with a mouse, so it's not only easier to use but more efficient.' According to technology writer Phil Patton the mouse was not so much a neat addition to Apple's personal-computer rebel yell as a complete articulation of it:

> The mouse, a device that symbolized the small
> computer's aim to give power to the weak, its cutesiness,
> its connection with moving objects, became almost
> universal ... The mouse had become more than a simple
> functional device; like the car or the computer itself, it
> was a symbol of individual power and potential, whose
> shape and contours should suggest speed, strength and
> some abstract intangible future.

The company chosen to bring this glimpse of an intangible future to artificial life was Hovey-Kelley, a local upstart design firm, founded by two twenty-something graduates, also from Stanford's product-design course. Hovey-Kelley operated out of a tiny office – no larger than a bedroom – in downtown Palo Alto. Their first machine shop was confined to a small rooftop area, not altogether in accordance with local planning permission, which they could reach only by clambering out through the office window. Cheques payable had to be written out to 'Hovey Design' – a hangover from Dean Hovey's side business making bike frames in high school. Since neither could afford to open a new bank account they just figured Kelley's name could be added on at a later date. Only in Silicon Valley could starting a business be read as an act of rebellion. 'Growing up in Ohio, to me a company was something like General Motors,' says David Kelley, one of the co-founders. 'Who wakes up and says, "I'm going to start General Motors"? But in Silicon Valley it didn't seem like a risk. I didn't want to work in corporate America. It just felt like I was pursuing my passion.' Hovey-Kelley's consulting fee came to $35 an hour, and by both men's admission they were scared to death of not being able to scrape together the three hours of paid work each month to pay the $90 property rental fees. Their first client had been a medical company, which shelled out for an expanded version of Hovey's master's thesis: an electronic white-cell differential counter for determining whether an infection was present in blood. Next came a project for Telesensory Systems, who wanted a book reader for blind people. Apple became their third customer.

Hovey-Kelley Design had just celebrated its six-month anniversary when Hovey drove to Bandley Drive in Cupertino for a meeting with Steve Jobs. It was a Friday afternoon and he had a list of ideas prepared for potential new computer accessories to sell. As soon as Hovey started his pitch, however, Jobs held up his hand imperiously and said, 'Stop! I know exactly what you need to do.'

Jobs told him all about the world-changing experience at Xerox PARC, about the eye-popping genius of the graphical user interface, about the incredible feats of engineering being created in a lab but squandered by management. Most importantly, he told him about the mouse.

'What's a mouse?' Hovey asked.

'It's a device that you could put in your hand and roll around on the desk,' Jobs explained. 'It has a ball inside it and the mouse picks up information off the ball and turns that into an electrical signal, which controls a cursor on screen. That's how you interact with software on the computer.'

There was a pause.

'Do you think you can build one?' Jobs said.

Hovey admitted that he wasn't sure, but asked for a weekend to try and work it out. Before then, Jobs had several additional spanners to throw in the works.

'It needs to have less than a $15 dollar manufacturing cost,' he said. 'It needs to last a couple of years. It needs to be able to work on a regular Formica tabletop, and it needs to be able to work on my blue Levis.'

One of the main areas stressed by the Stanford product design programme was the importance of making a physical prototype early on. This would invariably offer up several clues about what to do and what not to do in the final design, that would otherwise have remained invisible on paper. The constraints Jobs had placed on the mouse weren't a problem for Hovey. He was used to working within limitations, and from an ambitious designer's perspective a total lack of constraints was as concerning as the blank page is to a writer. One thing which *was* bothering him as he left Jobs' office and headed to the parking lot, however, was a disconcerting question that he was not altogether sure how to answer: *where on earth am I going to get my hands on mouse parts?*

Hovey got in his car and drove to Walgreens pharmacy at the corner of Grant and El Camino in Mountain View. 'I went in and headed straight to the underarm deodorant section, because I knew there would be these round balls in the deodorants that I could use as the ball that would roll around on the tabletop,' he says. 'I also went into the house-wares area and bought a few cups and plates and butter dishes, thinking that they would be quick things that could simulate the size of something that would fit in the hand.' Next it was a trip to TAP Plastics on Castro Street to buy epoxy casting resin and Teflon. Then he went home. By Monday morning when he drove into work he had assembled four different mouse prototypes.

'What's that?' David Kelley asked when Hovey placed the proto-mice down on his desk.

'Forget the other ideas,' Hovey answered. 'Steve wants us to do this. It's called a mouse. I've done some experimenting and this is what it might look like. I think we ought to do it.'

They phoned Apple that morning to accept the contract.

The mouse as a concept traces its origins back to 1964 and a device designed by Doug Engelbart at the Stanford Research Institution. His mouse (so-called because of the long rodentine tail which came out of one end) was nothing more than a wooden shell with two small metal wheels to control it. PARC had refined the concept somewhat, replacing the metal wheels with a roller ball supported by ball bearings. It worked better than the Engelbart mouse, but was expensive to manufacture – around $300 – and broke after a fortnight of use. The Apple mouse, representing the next step in the device's evolution, needed to live a long and harder-wearing life, and it needed to do it for one-twentieth of the cost. 'The biggest difference was the way we got information off the ball,' Hovey says of his own variation. 'In the Xerox mouse the roller ball acted as somewhat of a support, so the ball bearing against the surface of the table it was rolling on was what you pressed on. What we discovered was that we wanted to let the ball roll as lightly and freely as possible, and for it to grab information by a gentle touch of the ball as opposed to using it as a support. That was the breakthrough. We called it a free-rolling ball.' To test this floating-point concept Hovey built a robotic exerciser in which the mouse was attached to a motorised bar, which in turn rolled it around non-stop on a desktop for three weeks at a time. Eventually he was satisfied that the mouse not only worked, but could stand whatever perils Formica or a pair of Levis might be able to throw at it. At this point he showed it to Larry Tesler.

Having the most experience with a mouse of anyone at Apple, Tesler had been placed in charge of refining the concept. One of the early decisions made was for the mouse to have just one button, which would carry out every task. While the original Engelbart mouse had also been one-button, the researchers at PARC had adopted a three-button mouse for their work, with the rationale that a user would utilise two fingers – one on each side – to steady the device, while leaving the middle three free for clicking. Try as he might, Tesler could never convince the PARC engineers to build a one-button prototype. 'They said, if you want to experiment with using one button just ignore the other ones', he remembers. Employing a one-button mouse at Apple became a symbol of the company's dedication to creating a user-friendly product. When Dean Hovey handed him the prototype mouse, Tesler scrutinised it, slowly turning it over in his hands as he did so. He set it down on a table and began clicking the button, trying to put himself into the position of the user. *Was the clicking sound too loud, so as to be distracting? Was it too quiet, so that there would be no auditory indication you had clicked at all? How did the button feel? Was there too much resistance? Was there not enough resistance? How many times could you click before your finger started to get tired?* This led him to his next question.

'How many clicks before the button starts wearing out?' Tesler asked.

Hovey looked pleased. 'We estimate up to fifty thousand times.'

Tesler was aghast. 'I said, "fifty thousand? That's not nearly enough",' he recalls. 'I explained that, let's say you're typing on the keyboard and you want to click a menu, you pick up the mouse and you click. Then you go back and type on the keyboard some more. They saw it as a device you would use once in a while when you wanted to navigate to another part of a document. Maybe once every five minutes. I said, "no, there are times when you don't even touch the keyboard". I had to sit there and act out what it would be like for me to use a mouse. I was clicking away, and he stood there counting the clicks. And it worked out at around three million clicks per year.' Hovey went away and continued his search. The problem was finding a microswitch with the necessary duty cycles to sustain the equivalent of seven clicks per minute, eight hours a day, over a two-year period. Eventually he found one that fitted the criteria. Its inclusion added 20¢ to the price of production.

The next problem was the cord that connected the mouse to the computer. The length and thickness were fairly important, but what was essential to get right was the cord's tension. If it was limp, like an old-fashioned lamp flex, it would move freely without interfering with the onscreen cursor's motion, but the mouse would run over the cord more often than was acceptable. The alternative was making the cord stiff. In this instance the mouse would never run over it because, as the user moved the device, the cord would retain its rigidity and move away from the mouse. The problem came when the user stopped moving and wanted to click on an object. At this point the tension would cause the mouse to move on its own, often away from the area of the screen the user had been pointing towards. In some instances this would trigger the software to perform an unwanted action. Clearly this was an equally unacceptable result. Hovey kept bringing one cord after another, until Tesler finally asked him to bring, perhaps, a dozen at a time. Ultimately one Apple employee recalls 150 potential mice being tested, in long 'wine-tasting' sessions that, like all good wine-tasting sessions, ended when nobody was able to think straight. It took a lot of work – and earned Hovey-Kelley Design around $100,000 – but Apple got its mouse.

Of course the mouse worked hand-in-hand with another development that the Lisa brought to the masses for the first time: the graphical user interface. Not every user understood the revolutionary nature of the GUI (pronounced the 'gooey') as thoroughly or as technically as Jobs and his team did, but almost any user could recognise that it represented a considerable step forward. With the graphical user interface, concepts like the moving of a file or the opening of a document suddenly became recognisable in real-world terms. For the first time the computer appeared as an environment through which the user could travel. The majority of people might not have known their BASIC from their BIOS, but they did understand the metaphor of working on a desktop, or of tearing off a sheet of paper to write on, or taking a page out of one folder and placing it into another. Although the reality is that it does nothing of the kind, the graphical user interface appeared to empower the user; giving sovereignty to the person carrying out the tasks, rather than subjugating them to the way in which the computer itself chose to operate. It was about human beings using machines as peripherals, and not the other way around. Some years later the Italian writer and philosopher Umberto Eco wrote an article in which he compared the coldness of the PC's MS-DOS system to the warmth of Apple's graphical user interface through ecclesiastical parallel. Apple's system, he argued, was Catholic: aiding its faithful congregation in their quest to reach the kingdom of heaven (or at least the opening or printing of the correct document) with a hand-holding, step-by-step approach. 'It is catechistic,' Umberto writes, 'the essence of revelation is dealt with via simple formulae and sumptuous icons. Everyone has a right to salvation.' PCs, on the other hand, with their restrained, spartan white-on-black C:\> prompt, were Protestant, or possibly Calvinistic. DOS allows for free interpretation of the scripture, demands handwringing personal decisions (where *did* you store that file?), imposes a subtle hermeneutics on the user, and takes as gospel the notion that not all can achieve salvation.

If the GUI was akin to a religious experience, then the man most synonymous with the role of all-seeing Creator was a shaggy-haired, mustachioed computer scientist by the name of Bill Atkinson. Atkinson was responsible for LisaGraf, the graphics routines on the Lisa display which was at the heart of the machine's graphical interface. He had joined Apple in 1978, coming from the University of Washington, where he had been a graduate student. When Atkinson was invited down to Cupertino to see the work the company was doing, he was initially reluctant to go. However, curiosity, along with the offer of his airfare being paid both ways, got the better of him and he flew to California to meet Steve Jobs. Jobs wasn't particularly interested in Atkinson's scholastic record (as a college dropout he was scornful of academics his entire career) but he did want to convince Atkinson to come and work for Apple.

'You know,' Jobs said, turning on the charm, 'you're up there in Seattle and you're reading about brand-new products coming out, and you think that's hot new news, but really they were made two or three years ago. You're just at the end of a delivery chain. If you want to shape the future, you have to be ahead of that lag time. You have to be where the things are being made. If you come to Apple, you'll be able to help invent the future.'

Atkinson was taken aback by Jobs' confidence. At first he thought that the company's co-founder was laying it on a little thick, but quickly came round to the idea that there might be some truth in what was being said. 'The thing that drew me to Apple was this notion that you can do something with your life,' Atkinson recalled years later in an interview with the Computer History Museum. 'You can have an impact for the positive if you are where things are being created.' That weekend he phoned his wife back in Seattle and asked, 'What would you think about living in the Bay Area?' Since he had followed her to Seattle, she agreed to return the favour, and made arrangements to finish up law school at the University of Santa Clara. Atkinson quickly proved his worth to Apple. He was fantastically bright and an exceptionally fast study. One of the things that had first struck him when he joined the company was its lack of experienced programmers. 'Most of the people learned backyard hacking kind of stuff, and nobody even thought about ideas like structured programming or modular programming,' he recalled. By the time that Steve Jobs visited Xerox PARC in late 1979, Atkinson was among his favourite employees; firmly entrenched in the camp Jobs regarded as 'heroes' rather than their opposite number: 'shitheads'. When Jobs asked him, on the drive back from Xerox PARC, how long it would take to build a graphical interface like the one they had just seen demonstrated, Atkinson weighed up the question for a few seconds, did some mental calculations and answered, 'Probably around six months.'

This had been an optimistic estimate, but Atkinson was an optimistic and ambitious person. It would have been easy enough for Apple simply to copy everything about the Alto – which is to say to *Xerox* Xerox – but the Lisa team didn't do that. While they borrowed the windows, pop-up menus and neat scroll bars of Smalltalk, they also radically improved the concept by adding a menu bar, pull-down windows, a trash can (from which deleted files could be resurrected) and other innovative features. George Pake, the physicist and research scientist who had founded Xerox PARC, likened it to the Soviet Union creating their own atomic bomb within several years of America doing the same. 'They developed it very quickly once they knew it was doable,' he said. That isn't to suggest that there were no teething problems. While computer scientists like Bill Atkinson were undoubtedly brilliant, they could also, on occasion, overlook some of the more basic aspects of the work they were doing. In other words, while they had no problem envisioning the whole forest, sometimes they lost track of individual trees.

One glaring example of this came when it proved time to test the Lisa's graphical user interface on the proverbial 'person on the street'. Various computer-virgin test subjects were brought in to find out whether a high-tech novice could decipher the means by which to carry out a simple task without having access to an instruction manual. The task selected was the creation and saving of a document. A room had been set up for the experiment, equipped with a chair, desk and Lisa terminal. Two video cameras had also been arranged, with one pointing down at the keyboard and mouse, and the other towards the computer monitor so that the user's face could be seen reflected in it. Then there was a microphone, positioned in such a way as to pick up any words uttered by the test subject. Atkinson and his team stood quietly in an adjoining darkened room, obscured from view by a two-way mirror. They watched, fascinated, as their hard work was put through its paces. The experiment appeared to be going well until the point at which the user was required to save the document. When the command box appeared on screen – its two options reading 'do it' and 'cancel' – a look of extreme annoyance flashed across the user's face and he clicked 'cancel'. Atkinson couldn't figure out what had happened. Another person was asked to come in and carry out the same test. Proceedings again appeared straightforward until it was time to save the file. This time when the command box came up, a barely contained look of rage passed over the test subject's face and she recoiled from the screen. Atkinson turned up the microphone's volume so that he could better hear what was being said.

'What's this *dolt?*' the user said, staring at the button labelled 'do it'. 'I'm no dolt. Why are they calling me a dolt?'

She clicked 'cancel'. It turned out to be a simple matter of comprehension. The spacing was ever so slightly off and this had caused users to read the phrase as one word and to mistake the 'i' of 'it' for an 'l'. Since the users were novices, they had simply assumed that they had made a mistake and that the computer was mocking them for it. It taught a valuable lesson that no element of a computer's design – even something as seemingly trivial as the typeface used – was below taking a vested interest in. It was also a warning about the perils of being too clever. Atkinson and his team hadn't wanted to use the term 'OK' since it was both a colloquialism and an abbreviation – and not even a correct abbreviation at that. They had chosen 'do it' instead, but it had just wound up confusing people. Once the necessary changes had been made, users sailed through the remainder of the beta tests. All was good again in the world. Or so it seemed.

□ □ □

If there was one member of the Lisa team who was proving to be disruptive it was Steve Jobs. 'It was perfectly obvious that Steve was brilliant,' says Bruce Tognazzini, who had peripheral involvement with the project. 'It was also perfectly obvious that he was obnoxious, and also that he was just plain young. He would poke his nose into things, and he poked it into the Lisa project.' Despite lacking the power and unquestioned support of top management, Jobs was still a major shareholder in the company and not to be taken lightly. He was, in many ways, the equivalent of the boss's son: wanting in any real authority but with an abundance of it implied. 'He had a passion for perfection, and many of us didn't understand that at the time,' says Bill Dresselhaus. 'Fortunately, or unfortunately, one of his ways of pushing people and inspiring them was to tell you that something looked like hell or looked like shit.' Dresselhaus most notably experienced this when Jobs was presented with the final prototype for the Lisa outer case. Although he wasn't there at the time, he later heard from Tom Whitney, Apple's executive vice-president of engineering, that Jobs' reaction had been to cast a disdainful eye over the mock-up, utter some choice words and storm out of the room. 'The point is, if he really thought that, he would have fired me the next day, which didn't happen,' Dresselhaus says. Indeed, the designer eventually heard through the grapevine that Jobs loved the way the Lisa looked, although he never conveyed this information directly himself. 'There were some days when Apple couldn't possibly live one more day with Steve because he was so disruptive,' says Paul Dali, general manager of the company's personal-computer division. 'On other days I would go to work and say to myself, there's no way this company could do without Steve, he's so dramatically instrumental.' On occasions he could still overstep the mark. One night Larry Tesler was sound asleep in his bed at home when the telephone rang. He rolled over and looked at the clock. It was 2 a.m. His first thought was that there must be some kind of emergency. It turned out to be Jobs, calling to ask his opinion about some technical aspect of the Lisa – so minor that Tesler no longer recalls what it was. 'I can't remember whether he even apologised for the late hour,' he says. 'I only minded a little bit. I was kind of flattered that he would call me at that time, but I thought that it was a little off-base and I jokingly mentioned it to my boss the next day. It turned out that that wasn't the right thing for Steve to do at all. A lot of people had mentioned similar experiences and they were saving these up to build a case against him.'

Before long the *Steve Jobs versus Reasonable Working Hours* suit wound up in front of senior management, who decided that they had had enough of the 27-year-old co-founder's meddling. The verdict reached was that he was to be removed from the Lisa team, effective immediately. A man scorned, Jobs suddenly found himself at a loose end.* He had already severed ties with

---

* There is an additional reason why Jobs may have found his exclusion from the Lisa team to be personal. In Apple's early days many of the projects were given female nicknames, often after the wives, girlfriends or daughters of members of those involved. In the Lisa's case, it was rumoured that the machine was named after Jobs' illegitimate daughter, Lisa Nicole. The evidence against this interpretation of the project name is that at the time that development on the Lisa was underway, Jobs was vehemently denying his daughter's paternity, even going so far as to take a blood test in an attempt to prove that he was not the father. In Walter Isaacson's authorised biography, however, Jobs answered the question of whether the Lisa was named after Lisa Nicole by saying, 'Obviously it was named for my daughter.' Others dispute whether this was the case. 'He ... didn't have the power to name a project,' said one member of the Lisa team. 'He was just kind of a hanger-on.' In any event, the ensuing scuffle to deny that the project had anything to do with Jobs' offspring saw Apple reverse-engineer the acronym 'Local Integrated Software Architecture'. Others joked that this simply stood for 'Let's Invent Some Acronym'.

the Apple II and III teams, referring to their respective products as 'a horse and buggy versus a jet plane' next to the technical superiority of the Lisa. Excluded from three of Apple's four main projects, he took the only path left to him. He headed across the way to a small office where an engineer named Jef Raskin, along with a few associates, was working on a paid research project, which barely registered as a ripple on Apple's radar. It wouldn't stay that way for long.

Steve Jobs was about to join the Macintosh team.

7

YOUNG MANiACS

'Very few of us were even thirty years old. We
all felt as though we had missed the civil
rights movement. We had missed Vietnam. What
we had was the Macintosh.'

Macintosh team member,
as quoted in 'The Second Coming of
Steven Jobs' *Esquire* (1986)

Growing up on the East Coast, Jef Raskin was the archetypal computer geek. Unusually for someone who went on to work with user interfaces, his early interests lay in hardware, not software. 'When I was a kid, I was the only kid in high school with a Tektronics oscilloscope in his shop, when none of the other kids even knew what an oscilloscope was,' he proudly recalled in a 2000 interview for the Stanford archives. As an undergraduate at the State of New York University he got into a heated debate with his philosophy adviser over whether computers could put human usability over simple number-crunching. Raskin said they could. His adviser said they couldn't. Raskin created a program which seemed to suggest that he was right, and went back to prove his point. His adviser refused even to look at the computer output. 'I felt like Galileo with the church fathers refusing to peer through the telescope,' Raskin said. He duly switched from philosophy to computer science. For his college thesis he wrote a paper arguing that computers should be completely graphical, rather than text-based, in nature. Few seemed to agree, though, and he dropped out of the computing field for several years, finding vast impersonal mainframes 'really dull'. Raskin moved to California at the height of the 1960s and distracted himself with other activities. For a while he directed guerrilla theatre, conducted the San Francisco Chamber Opera Company and worked as a painter, even having some of his work displayed in the Museum of Modern Art. He took up a teaching post at UC San Diego, which ended when he climbed into a hot-air balloon and floated past the residence of the chancellor, shouting down his resignation while blowing melodies on a recorder.

However, he was ultimately directionless, jumping from one endeavour to the next in search of his true calling. Then, in 1975, the Altair 8800 came along and his life irrevocably changed. 'Aha! This is what I've been waiting for,' Raskin said. He threw himself headlong into the emergent hobbyist scene and began writing for magazines like *Dr. Dobb's Journal* and the *Silicon Gulch Gazette*. In 1976 he was approached by the Two Steves about doing some work for them. At the time Raskin was running a small consulting business called Bannister and Crun – named after two characters from *The Goon Show* – and Jobs and Wozniak thought he would be perfect to write the user manual for the Apple II. Having heard about the two college dropouts' anti-academic bias, Raskin chose to keep that aspect of his past to himself. 'As far as Jobs was concerned ... I was a musician, a bedraggled street musician from San Francisco, and a music teacher, who wrote articles,' he said. 'I felt knowing about my formal background would be a hindrance.' What seemed more of an initial impediment was the amount of money the overqualified Raskin was being offered to write the manual.

'$50 per page?' he said doubtfully when his two would-be employees announced the figure.

'$50 for the whole thing,' Jobs corrected him.

Nevertheless, Raskin was intrigued enough to join Apple as its thirty-first employee, making a brief addendum in his contract that allowed him to schedule work obligations around opera rehearsals. Having spent some time at PARC during a sabbatical year at Stanford, it had actually been Raskin's idea that Jobs consider visiting Xerox's research laboratory in late 1978 to investigate the ground-breaking work that was being done there. But Jobs' worldview was all too often a binary one, and he could only conceive of people as heroes or bozos. At that moment his perception of Raskin had been locked firmly in the 0 (off) position for whatever reason, and he dismissed him as 'a shithead who could do no good'. Later on, of course, Jobs turned around and decided to go to PARC after all.

In 1979 Raskin went in for a meeting with Mike Markkula to discuss his future. Markkula wanted him to develop a $400 games device of the kind that Atari was very successfully manufacturing. This wasn't an area that particularly interested Raskin, and he countered instead by laying out an idea for a new category of machine, which would provide the power of a personal computer, but be considerably cheaper and infinitely easier to use. 'You can tell how open Apple was because I could go to the chairman of the board, and say, "your product strategy for the company is wrong, and I propose this instead", and he listened,' Raskin recalled years later. Markkula was intrigued by Raskin's vision and gave him permission to go away and look into it as a research project.

On a Monday in late May – the same month that American Airlines Flight 191 crashed during takeoff at Chicago's O'Hare International Airport in the deadliest aviation accident in US history, graduate student John Harris was wounded by a bomb sent from 'Unabomber' Ted Kaczynski, and Margaret Thatcher became Britain's first female prime minister – Raskin sat down at his desk and began to write out a paper entitled 'Design Considerations for an Anthropophilic Computer'. In it he discussed creating a computer for the so-called 'person in the street' (which he rather unflatteringly abbreviated to the PITS). It should, he suggested, 'be truly pleasant to use [and] ... require the user to do nothing that will threaten his or her perverse delight in being able to say: "I don't know the first thing about computers".' The way Raskin saw it, in order for computers to reach the same critical mass of public acceptance as a device like the television or telephone, it was of paramount importance that it wasn't assumed that every potential customer was a nascent engineer. *People used a toaster without knowing how to open one up, didn't they?*

By this time Raskin had rechristened the project. Originally it had been codenamed Annie, after the pneumatic comic-strip heroine Little Annie Fanny, who appeared in the pages of *Playboy* each month. Raskin thought the name was sexist and decided to come up with something more appropriate. He settled on a phonetic spelling of McIntosh, after his favourite cultivar of apple.* His vision for the Macintosh was markedly different from the machine that would eventually ship. As he saw it, the interface would be graphical, but, rather than being made up of individual files and folders, the user's productivity would be accumulated in one large multi-purpose structure, not dissimilar

---

* Because there was already a McIntosh Laboratory in existence, making high-end audio equipment, Raskin changed the spelling to Macintosh. 'If it conflicted with the overcoat, who cares?' he said. Later on in 1982, Apple ran into problems when trying to trademark the name. At that point Steve Jobs penned a letter to the president of McIntosh Labs, Gordon Gow, claiming, 'We have become very attached to the name Macintosh. Much like one's own child, our product has developed a very definite personality.' For a while Apple considered shortening the name to MAC, standing for 'Mouse-Activated Computer' (or 'Meaningless Acronym Computer') until a cash settlement with McIntosh cleared the way for them to use the name.

to a single personal web page. Turn the computer on and start typing a letter, and you were word-processing. Write an equation and it would become a calculator. Raskin didn't much like the mouse, which required that the operator continually move his or her hands away from the keyboard. He jokingly referred to users of the technology as leading a 'hand to mouse existence'.

Raskin's view of computers may have been utopian, but the real revolution as he saw it was going to relate to price. Simply put, computers weren't going to be for everyone until everyone could afford them. 'The most important goal ... in my opinion is for this computer to have a selling price of $500 or less,' he wrote in a collection of papers he cumulatively titled the Book of Macintosh, after the Book of Mormon. Unlike the Lisa computer, which Raskin saw as destined to become too large and too expensive, every feature of the Macintosh would be a carefully thought-through trade-off between price and performance. A colour monitor might be ideal, but a monochrome one would do nicely. The 68000 microprocessor of the Lisa was his first choice, but it was nowhere near as affordable as the less powerful 6809E. A floppy-disk drive like the Disk II was very popular, but a digital cassette drive was an acceptable substitute to save money. Starting the other way around – beginning with the ideal features of a computer and scaling it backwards – was no way to work, he reasoned. After all, if he had his way and there was no limit on time or budget, his dream computer might as well include speech recognition and the ability to synthesise music, right down to simulating Caruso singing alongside the Mormon Tabernacle Choir. No, Raskin concluded in a memo, 'we must start both with a price goal, and a set of abilities, and keep an eye on today's and the immediate future's technology'.

To aid with the creation of the Macintosh, Raskin brought in two of his former colleagues to work with him. The first was Brian Howard, a friend who had graduated from Stanford

University with a degree in electrical engineering. Like Raskin, Howard fused a love of science with a fondness for music. His father had been a physics professor at the University of Oklahoma, while his mother was a classical pianist. For a while, Raskin and Howard had been roommates, sharing a tiny apartment in Palo Alto. To make ends meet they played gigs at weddings and birthday parties, and cooked up spare lab rabbits from the Stanford biology department in the evenings. The other new hire was Marc LeBrun, a LISP programmer who had, within a single decade, transitioned from high-school dropout to a valued member of the team at Stanford's acclaimed Artificial Intelligence Laboratory. The trio was joined by Burrell Smith, a cherubic 23-year-old who had been working in Apple's repair shop, fixing the broken logic boards on faulty Apple IIs returned by customers. Smith was a diminutive computer geek, with the lopped-off blond curls of a fourteenth-century monk. He had been a regular attendee at the Homebrew Club and, while lacking in training (with the exception of a handful of classes at the nearby Foothill community college), showed an almost preternatural ability to solve complex engineering problems. As part of his process Smith would take home the parts of whatever project he was working on, lay them out across his bed and stare at them intently for several hours. Sooner or later patterns would begin to emerge. *Wasn't it obvious – this would go with this, and that could connect over there.* 'You thrash around the design space long enough and you learn the idiosyncrasies,' Smith said. By the time he went in to work the following day he would invariably have his answer.

Smith's big breakthrough had come when he was able to impress Bill Atkinson by suggesting a way in which more code could be fitted into the 64K of memory that the Lisa software was being written on. Since there was no room for Smith on the overstaffed Lisa team, Atkinson mentioned Smith to Jef Raskin, commenting that he would be the perfect hardware designer for the Macintosh project.

'We'll see about that,' Raskin responded.

Smith was elated. Nowhere else in Apple would a person with such a bare-bones résumé get the chance to design the main logic board for a computer. 'It was the one chance of a lifetime to go through the cracks of the corporate culture,' he would later say. '[I] raced through the elevator door just before it shut.' Over the Christmas break in 1979, when everyone else had gone home, Smith lived almost permanently inside the Apple building, scavenging parts for the Macintosh from the various offices to carry out his task. He finished during the first week of 1980, although remained unsure whether any graphics routines would run on his creation. Before he left for home one evening, he threw down an informal challenge to a friend in the engineering department. When Andy Hertzfeld returned from dinner that night, he took up the task and, along with some help from an older technician named Cliff Huston, managed to decipher Smith's scribbled instructions and get an image of Scrooge McDuck to appear on the screen. Because the embryonic Macintosh had 256 lines of resolution compared to the Apple II's 192 there was still a bit of space remaining at the bottom of the picture. Hertzfeld thought about it for a few seconds and then added the message 'HI BURRELL!' in evenly spaced, 24-point font.

□ □ □

None of this was of particular interest to Steve Jobs. Deeply immersed in the Lisa project, he had read the memos Raskin circulated and thought that, by and large, they were nonsense. *A low-end computer that would be seen as a sub-Apple II? What kind of a bozo thought up an idea like that?* For the most part the Macintosh team rarely came into contact with Jobs. Raskin was already having to stave off various threats of cancellation and the best solution seemed to be to stay clear of those who could bring down the axe at a moment's notice. Indeed, the team saw so little of Jobs that Marc LeBrun didn't even recognise him when the two first came into contact. Jobs was in a conference room, with his feet up on a table, berating a group of cowed, although clearly competent, engineers who were mumbling their apologies as he ranted and raved. LeBrun thought Jobs was immensely arrogant, labelling him a 'wet-behind-the-ears marketing puke, dressed in a ridiculous chalk-pinstripe, complete with banker's vest, shoes off, stinky feet up'. He certainly wasn't aware of how much power this man – 'with that uniquely dripping disdain the ninth floor at Tech Square can sometimes inculcate' – had over the work he was doing at Apple, and just considered him an abrasive punk in need of a slap. At that moment Jef Raskin happened to pass by and, noticing the look on LeBrun's face, yanked sharply on his arm.

'Don't you know who that is?' he hissed.

'Who is it?' LeBrun said blankly.

Meanwhile, the Macintosh group was ticking along well enough to gain a fifth member. Guy 'Bud' Tribble was a self-taught software whiz and a typically offbeat personality. He had first met Raskin a decade earlier when the latter was staging a piece of performance art, entitled 'Happenings and Events', in San Diego. They had stayed in touch ever since. Tribble was busy working on a joint MD/Ph.D. about neural disorders in cats at the time, but decided that here was a project too interesting to miss out on. He certainly wasn't wrong.

In the autumn of 1980, Jobs was kicked out of the Lisa team. Full of vitriol, he was determined to find some way, any way, of wiping the Lisa off the face of the earth. It didn't matter that he stood to make more money than anyone if the computer was a success. As far as Jobs was concerned, he had been slighted and that was indefensible. The people running the project didn't understand Apple's culture, he told himself. They thought of the Apple II – *the computer he and Woz had slaved over* – as nothing more than a silly toy. Jobs had seen himself as part of this new breed of slick engineer, but his expulsion from the team had brought him sharply back down to earth.

By this time, the Macintosh team had moved out of the main Apple building and into a small, second-floor office on Stevens Creek Boulevard, behind a hippie health-food restaurant called The Good Earth. Inside, Jobs found something resembling an Island of Misfit Toys: a cramped time capsule that represented everything that Apple had been just a few years earlier. The group worked hard and then relaxed by playing games of 'it' with a Nerf ball that was tossed around the office. Every afternoon they would head across the road to a pizza joint called Cicero's, where they took turns eating slices of pizza and playing 25¢ games on the restaurant's Defender arcade machine.

These, Jobs realised, were his people. Just like him they were outcast artists, who had taken a circuitous route before arriving at personal computing. Joanna Hoffman, the team's only marketing employee, for example, was the daughter of Eastern European refugees. After earning a physics degree at MIT, she had changed tack and attended graduate school at the University of Chicago, where she studied archaeology. The overthrow of the Shah of Iran in 1978 had resulted in her dig being shut down and prompted yet another directional rethink. *Back to California!* 'I decided I'd been living in the past so long that I felt I now wanted to be in the future,' she claimed. She moved in with a young engineer from Xerox PARC and then joined Apple as the sixth member of the Macintosh project. Jobs' own journey to personal computing (grew up in Silicon Valley, dropped out of college, befriended Steve Wozniak, started a company) contained nowhere near as many twists and turns, but that didn't matter to him. More than anything it was the idea which appealed to him.

Re-energised, Jobs took an increased interest in Raskin's Book of Macintosh. Containing enough 'change the world' proclamations and bite-sized Marshall McLuhan-esque sentiments to pique his interest, the Book of Macintosh caused Jobs to consider that maybe there was more to the concept of a stripped-down, mass-market computer than he had originally thought. Although Raskin and Jobs would later clash vehemently over the project's execution, what is remarkable is just how similar the ideas which shaped both men's vision in fact were. In his notes Raskin preached how 'if the computer must be opened for any reason other than repair (for which our prospective user must be assumed incompetent) even at the dealer's, then it does not meet our requirements'. Seeing the technical innards of the machine, he explained, 'is taboo. Things in sockets is taboo ... Billions of keys on the keyboard is taboo. Computerese is taboo. Large manuals, or many of them (large manuals are a sure sign of bad design) is taboo ... Ten points if you can eliminate the power cord ... It is better to offer a variety of case colours than to have variable amounts of memory.' All these were ideas that Jobs himself would later champion. Despite its quasi-academic research-project status, Raskin's vision was also profoundly commercial. 'If it is anticipated that fewer than 100,000 of these anthropophilic computers will be sold in a 2½-year period, the project should not be undertaken,' he noted.

If Jobs agreed with Raskin, however, he did a good job of hiding it. As soon as he started dropping by the Macintosh office with increased regularity, he set about trying to put his own stamp on the project. First of all came the casual challenge of asking Burrell Smith whether he could find a way to incorporate the more powerful, 16-bit Motorola 68000 microprocessor in place of Raskin's slower, cheaper 6809E. Almost exactly one year on from Smith's design of the original Macintosh logic board, he completed the task with flying colours. 'Burrell came up with this genius way of hooking up a 68000 chip to only eight memory chips, to do it in a much, much cheaper way than anyone thought possible,' recalls Andy Hertzfeld. 'Once he did that we had a computer that cost just a quarter or less of what the Lisa did, but was actually twice as fast. It was like capturing lightning in a bottle.' If not lightning, it was certainly enough to capture Jobs' attention. Having previously mocked the Macintosh project, he now did an about-face. On 20 January 1981 – the day Ronald Reagan was inaugurated as the new President of the United States – Jobs announced that he was officially on board HMS Macintosh. Raskin's emotions were mixed. Certainly, he was pleased that the company's co-founder was so enamoured by his vision for an anthropophilic computer, but he was also concerned about what it meant for his own role in the project. After all, a ship could only have one captain.

Raskin's fears proved well founded. Jobs was never one for sharing credit. If he was truly going to do something that would make its mark on the world, he would first have to find a way to separate Raskin from his baby. Part of it was certainly that Jobs saw an opportunity to finally make his mark on Apple, having been denied a role of any real importance up to this point. He may have regarded himself as a Silicon Valley poet, his sensibilities forged in the fiery youth protests of the counterculture, but most people simply regarded him as a slick, yuppie marketer. Jobs was not an experienced corporate infighter, but he was better equipped for it than than the artistically temperamental Raskin. 'Jeff had strong ideas but he certainly was a lover not a fighter,' says Marc LeBrun. Even without Jobs, Raskin often found himself rattled by some of the more boisterous employees at the company. He had a stammer and Burrell Smith had wasted no time in honing a pitch-perfect impersonation of it. Steepling his fingers and leaning back in his chair, the young hardware engineer would affect his best Raskin voice for one of his favourite jokes.

'Wh-wh-why, I invented the Macintosh,' Smith mimicked.

Then sitting up straight and adopting the voice of a television interviewer, he said, 'But I thought Burrell Smith invented the Macintosh?'

Finally, switching back to Raskin's voice for the punchline, he said, 'Wh-wh-why, I invented Burrell.'

Almost as soon as Raskin had resignedly recalculated the Macintosh production costs to include the new addition of the 68000 microprocessor, Jobs announced that he had another change he wanted to make: he now wanted a mouse. Since the 68000 would allow the Macintosh to run many of Bill Atkinson's LisaGraf routines and now he was adopting the same WIMP interface (that is, windows, icons, menus and a user-controlled mouse pointer), it seemed clear to Raskin that Jobs was simply trying to make a cheaper version of the computer project he had been booted out of. Jobs, however, refused to listen to Raskin's complaints. He could be obnoxiously boorish when he wanted to be, speaking over people whenever the mood took him (which was often). On one occasion a visiting speaker had been giving a presentation to the Apple troops.

'The linearity is .25%,' he was saying, 'the overall accuracy – which is not important in these operations – is about 2%.'

Jobs loudly chipped in, saying, 'What we at Apple are interested in is not accuracy! What is the *linearity* of the product? That's what's important.'

The speaker looked baffled, then repeated his earlier statement. It left many Apple employees speculating over the cause of the company co-founder's sociopathic tendencies. At the same time, as had been the case on the Lisa, Jobs would often surprise people with the originality of his ideas. One thing was for sure, though, and that was that working for Jobs was never relaxing, as evidenced by this 12 December 1980 memo from Brian Howard to Jef Raskin:

> *Your ability to mix play and work further helped reduce*
> *the tension and increase the fun in the job. I came to*
> *work on the Macintosh largely because I wanted to be*
> *in that work environment again. However, since the*
> *Macintosh has lost its research status, and especially*
> *since management has begun being split between*
> *you and Steve Jobs, I have felt this relaxed, free-*
> *to-concentrate-and-work-hard atmosphere to be in*
> *jeopardy. Steve Jobs seems to introduce tension, politics*
> *and hassles rather than enjoying a buffer from these*
> *distractions. I thoroughly enjoy talking with him, and*
> *admire his ideas, practical perspective and energy. But*
> *I just don't feel that he provides the trusting, supportive,*
> *relaxed work environment I need.*

Raskin was the wrong person to complain to. It wasn't long before he was run off the Macintosh team altogether. (Bud Tribble, while one of Raskin's chosen few, decided to support whoever looked likely to see the project through to its conclusion. He switched allegiances to Jobs, carried on working, and tried to ignore 'these bad, political things happening ... above my head'.) In his attempts to get Raskin to step down, Jobs demonstrated a coldly calculating ability to embarrass or humiliate. Apple regularly gave informal lunchtime presentations on a variety of topics to whichever employees wanted to attend. On 17 February 1981, Raskin was scheduled to give a talk on his ideas for the Macintosh project. On that morning Jobs told him that the meeting had been cancelled and that he was not to talk to any Apple employees about the project. Upset, Raskin wandered over to the auditorium where the meeting had been due to take place, in case there were any stragglers who hadn't heard about the cancellation. Instead he found a room full of a hundred interested, but slightly impatient, Apple employees who had turned up to hear Raskin speak. The meeting hadn't been cancelled at all: Jobs had just been trying to make him look bad. Raskin announced the cancellation himself, then gave an off-the-cuff presentation about the work that interested him at Apple at that moment. Two days later he sat down and penned a furious letter to Apple's president. It ran to four pages and catalogued a laundry list of complaints about Jobs, including the fact that he routinely missed appointments; that he acted without thinking and with bad judgement; that he was emotionally combative and often resorted to personal insults; that he failed to give credit where it was due; that he interrupted people and was generally inconsiderate; that he made wastefully absurd decisions by trying to be paternal; that he didn't keep promises; that he was unrealistic about how long certain tasks would take; and that he was a bad manager of software projects. All were, to some degree or other, true. But it also didn't matter. Management had sided against Jobs with the Lisa. Unless they wanted to risk further upsetting the company's largest shareholder they would have to give an inch at some point. The Macintosh looked a very low-risk project; more like a sandpit for Jobs to get his power trip from commanding. Fine, it was agreed, he could have it. Jef Raskin would have to go.

To cement his hostile takeover of the Macintosh project, Jobs began to bring in all his old Apple II associates. In came Jerry Manock, Randy Wiggington, Rod Holt and others. Holt now owned a nearby pub called Eli McFly, named after a fictitious Victorian inventor and bedecked with geeky adornments such as a light which threw off electrical sparks, and posters for time-travel vacations ('Professor McFly's Temporal Displacement Excursions') to destinations like the island of Lesbos in 645 BC or the sci-fi staple Alpha Centauri in the year 2090. On the wall was a cartoon portrait of maverick scientist Nikola Tesla, a pioneer in the field of electromagnetism whose contributions had been downplayed next to those of his arch-rival, Thomas Edison. There was something fittingly anti-authoritarian about the scene. In the steampunk, alternate history milieu of Eli McFly's, Tesla's unorthodox theories about electricity had won out against Thomas Edison's. And back in the second-floor office on Stevens Creek Boulevard, Steve Jobs had now invented the Macintosh. In order to erase Raskin from memory completely, Jobs even tried to rename the project Bicycle, a moniker so horrible that it was roundly ignored until he forgot about it.

Another new person brought into the project was Andy Hertzfeld. Hertzfeld, who lived in a cluttered bungalow in Cupertino, represented the first generation of Apple employees that had discovered the products as a consumer before joining the company. In other words, he was among the first true Apple fanboys. Hertzfeld had grown up in Philadelphia on the East Coast, and then moved to California to study computer science at UC Berkeley. Never reacting well to authority figures, he more than anything feared the kind of job that would have him churning out code for the man in the proverbial grey flannel suit. Then he had attended the West Coast Computer Faire and seen the Apple II. 'As I started playing with it, the design style was so unusual and amazing that it took over my life,' he says. 'I idolised the creator even though I didn't know who it was. And ultimately that style of creativity and imagination and sense of humour took me to Apple. The product was so fantastic I had to follow it back to its source.'

'Are you any good?' Jobs had asked Hertzfeld in an impromptu job interview. 'We only want really good people working on the Mac, and I'm not sure you're good enough.'

Hertzfeld responded that, yes, he considered himself to be fairly good at what he did. He started reeling off the work he had done previously, before Jobs interrupted him.

'I hear that you're creative. Are you *really* creative?'

Hertzfeld said that he probably wasn't the best person to give an objective opinion on that, but that he loved what he had seen of the Macintosh project and that he'd give it his absolute best shot. A few hours later, when he was doing some low-level coding for the Apple II's DOS 4.0, Jobs appeared again, looking over the wall of Hertzfeld's cubicle.

'You're working on the Mac team now,' he said.

When Hertzfeld thanked him, and said that he needed a couple of days to wrap up the work he was doing, Jobs simply unplugged the Apple II he was coding on – causing him to lose his work for that afternoon – picked up the computer and its monitor and walked off. 'Come with me and I'll take you over to your new desk,' he said. He plonked the machine down on a desk in the Macintosh office. 'Hey, whose is this stuff?' Hertzfeld asked, noticing that the workplace still contained the belongings of its previous occupant. It turned out to be Jef Raskin's former desk. He had been shunted out the door so quickly that he hadn't even packed.

As its main software engineer, Hertzfeld became one of the driving forces in turning the Macintosh into the machine that it ultimately became. In a sense, his appearance marked the final step in the transition from the Macintosh being a Jef Raskin project to it becoming a Steve Jobs production. Raskin had never thought very much of Hertzfeld, dubbing him a 'hacker in the worst way, a patcher of programs'. Even though Hertzfeld had been the one responsible for getting the first graphics routine running on Burrell Smith's original logic board, Raskin had remained resolutely unimpressed. Certainly Hertzfeld, the archetypal rough-and-tumble, new-generation Silicon Valley engineer, who kept up to date with the latest pop music, was a world away from the sensitive artist that was Jef Raskin. If Raskin was first-generation hippie, then Andy Hertzfeld was punk rock. Since the Macintosh was now close to resembling a pared-down Lisa in terms of its graphical user interface, one of Hertzfeld's jobs was to decide just which elements to keep and which to lose. He saw his work as streamlining the Lisa user interface, which was ground-breaking but also sometimes needlessly complex. Some of Hertzfeld's simplifications came down to providing a more straightforward user experience. For example, he eliminated the concept of triple-clicking with the mouse, which on the Lisa had had various applications dependent on which program you were using at the time. Other modifications, such as the ability to only run one application at a time, mainly centred on saving on hardware costs like extra RAM and a special memory-mapping chip.

As a personality Hertzfeld meshed perfectly with Burrell Smith and the two quickly became close. They wore t-shirts with improbable nudge-nudge-wink-wink slogans like 'I'm User Friendly' and chatted back and forth in a sort of techie Nadsat; referring to situations being 'meta-stable' or calling an attractive woman a 'good prototype'. When Rod Holt's secretary, Peggy Alexio, was given the task of ordering new business cards for the team Hertzfeld insisted that his labelled him as a Software Wizard. When Smith heard about it he asked that his own make mention of him as a Hardware Wizard; something that annoyed Hertzfeld who felt it diminished the originality of his own title. For the most part, however, they were an inseparable double act: the hellraisers of Bandley Drive. At restaurants Smith would often toy with waitresses, talking them into letting him divide his pizza order into thirds, or even fifths, with a different topping on each section. After a bit of back-and-forth negotiation, the waitress would sigh and say that she'd see what the chef could do.

'Is that all?' she would ask before leaving.

'For the pizza? Absolutely.'

'Can I get you something to drink with that?'

'I'll have three-quarters Coke, and one quarter Sprite,' came Smith's response.

With the likes of Hertzfeld and Smith representing the average team member, the mean age of person working on the Macintosh now skewed years younger than the more mature academics that had started the project. 'We were all pretty much within plus or minus a few years of each other,' says Bruce Horn, another new member of the team. 'I was probably one of the youngest guys. When I joined Apple in 1981, I was twenty-two and just out of college. Andy Hertzfeld was a few years older. Steve was four years older. Larry Kenyon and Steve Capps (both programmers) were both in their early to mid-twenties.' Like Mafia wiseguys, members of the Mac team would hang out only with other members of the Mac team. 'Steve Capps would talk about getting us one of those gigantic cabs that would hold twelve people, so that we could all get in one car to go to lunch and dinner,' says Horn. The vast majority of people on the team had no family commitments, no girlfriends, no hobbies that didn't involve sitting in front of a computer screen for hours at a time. With Jobs' constant reminders that they were making history (every history-making milestone would be celebrated with the popping of champagne corks) the team seemed all too happy to put their lives on hold until the Macintosh finally shipped. 'I was only in my thirties, but I was still the second-oldest guy on the project,' says George Crow, an analogue engineer who defected from Hewlett-Packard to join the Macintosh team. 'It was such a youth-oriented culture.' Crow had been at UC Berkeley during the Free Speech Movement and even covered the protests as a photographer for the local newspaper. In other words, he had lived through the 1960s. For younger members of the team, men like Andy Hertzfeld and Steve Capps, the Macintosh was their Vietnam and Woodstock all rolled into one.

□ □ □

The Macintosh team was growing fast and before long it was relocated again. The new offices were on one floor of a nearby building that Apple was leasing. It was hardly the grandest of headquarters, but the team gave it a grand name nonetheless. Since it was near to a petrol station they dubbed it Texaco Towers.

This period of the Mac's development was marked by an influx of new talent. Along with Hertzfeld, Susan Kare was another new arrival. A graphic designer from Philadelphia, Kare had known Hertzfeld since high school. The Mac group had been searching for someone to design icons and fonts for the computer and, as multi-talented as many of them were, none had the kind of graphical chops to pull off what it was that Jobs was asking for. When Hertzfeld approached Kare, she was living in Palo Alto, having recently resigned from a curatorial position at San Francisco's Fine Arts Museum. At the time, she was in the middle of working on a slightly unusual sculpture project – welding a life-size razorback hog for a museum in Hot Springs, Arkansas. Kare had no experience of working with computers, but when Hertzfeld explained just what it was they were looking for, she said that she would be willing to give it a go. To start with, Kare designed the Macintosh icons on sheets of graph paper, plotting them out by hand like 32 by 32 mosaics (giving just over 1,000 possible combinations). Before long Hertzfeld created a program for editing icons, which meant that she was able to begin working directly on to a prototype Macintosh. She loaded up her cubicle with books like Henry Dreyfus's *Symbol Sourcebook*, which came complete with a foreword by Buckminster Fuller. Her favourite part of the book dealt with hobo signals: simple chalk-on-stone marks that would be left by transients as they moved from place to place to indicate what they thought of the people they had encountered in a certain area. A top hat and triangle, for example, indicated 'these people are rich'. Nowhere in the book was there a symbol for 'these people want to change the world'.

Kare's icons lent the Macintosh a friendly, whimsical personality. The first image a user would see when turning on the machine was a tiny portrait of the Macintosh computer with a smiling face to indicate that everything was working correctly. To access MacPaint, the ground-breaking paint program developed by Bill Aktinson, the user simply double-clicked on a picture of a hand painting on a sheet of paper. If there was a fault and the computer crashed, a dialogue box would appear showing the image of a bomb. This was not the Apple IV (or V, depending on whether the Macintosh would ship ahead of the Lisa) but, rather, a sequel to the Apple II, continuing that same thread of playfulness that Wozniak had imbued in his first mass-market computer. Kare was also chief font wrangler for the Macintosh, designing each typeface after various genre archetypes, so that London had the kind of old-fashioned serif 'blackletter' feel one might have imagined Charles Dickens favouring, while Geneva possessed a cleaner Swiss sans-serif look. For reasons never fully explained, the font named San Francisco mimicked the look of a ransom note, each letter looking as if it had been snipped individually from various books or magazine articles. It was only later that more recognisable names, such as Helvetica and Times New Roman, were added to the mix.

As someone who lived outside the engineering bubble, Kare proved useful in other ways. At one point Bill Atkinson was creating a menu feature containing all of the additional tools that would fit nowhere else. He told Kare that he planned to call it the 'Aids Menu'.

'You know, there's this thing starting to happen – there's this disease, and I think you better stay away from that word,' she said.

Atkinson reluctantly gave it up. He renamed it the 'Goodies Menu'.

□ □ □

Another unofficial task allocated to Kare was painting a large pirate flag (with an Apple logo for one of its eyes, naturally) which proudly flew above Bandley 3. In the same way that the brash, borderline arrogant personality of the Mac team would later become the dominant Apple culture, so the Macintosh project represented the first time – outside of the garage in which the Apple II had been built – that Apple would put together the kind of small, dedicated team that would produce some of the company's greatest products in later years. Jobs referred to this company-within-a-company approach as returning to 'the metaphorical garage'. If Apple was a big corporation, able to move only marginally faster than its larger rivals, then the Macintosh team still had all the piss and vinegar of a start-up. 'Innovation has nothing to do with how many R&D dollars you have,' Jobs told *Fortune* magazine some years later. 'When Apple came up with the Mac, IBM was spending at least one hundred times more on R&D. It's not about money. It's about the people you have, how you're led, and how much you get it.'

One of the major advantages of the small team that was working on the Macintosh was the cross-pollination that resulted, with members of different departments who would otherwise be unlikely to communicate regularly opening up channels of conversation. On one occasion, for example, Jerry Manock was speaking with Andy Hertzfeld. Hertzfeld was sitting at his desk, working on the graphic design for the Mac desktop. Manock noticed that there were numerous sharp square edges on the screen. In his mind, sharp edges said 'don't touch me' so he had added chamfers and bevelled edges to the Macintosh case to give it a softer look. He explained all this to Hertzfeld.

'If I gave you that radius would you be able to carry it through to the software, so that everything looks like it fits together?' he asked.

'Oh yeah, I could do that,' Hertzfeld responded.

It took him two minutes to implement, while Manock stood there. 'Those little details tied in what he was doing with what I was doing, which made for a unified expression of function,' Manock says. He, too, showed a remarkable lack of preciousness when it came to the industrial design of the Macintosh. Hit with a burst of inspiration, one lunchtime Manock popped out of the office, headed down to a nearby office-supplies store, and bought one of the poster-display racks which allow customers to flick through the different poster designs on sale. Taking it back to work, Manock stuck on the Macintosh design blueprints – one per page – and then attached a red pencil on the end of a string. 'I told people that if anyone had an idea as to how we could make this a better design to write it on these drawings along with their name and telephone extension,' he says. 'It didn't matter to me whether it was the janitor or the chairman of the board – if I saw a new entry there I would call the person up and discuss it with them.'

Of course, all of these ideas eventually wound up being filtered through Steve Jobs, whose 'maniacal focus' (in the words of marketing manager Mike Murray) meant that he absolutely had to have the final say. Fitting the idea of the Macintosh being an 'information appliance' (a term coined by Jef Raskin) Jobs insisted that it look less like a computer than the kind of domestic appliance one might normally find in a kitchen. One weekend he visited the department store Macy's in Palo Alto, and when he returned to work on Monday morning he insisted to Manock and fellow industrial designer Terry Oyama that the Macintosh should visually resemble a Cuisinart food processor. Raskin and Oyama dutifully went to the shop and looked at one. Jef Raskin had assumed that the Macintosh would prove so transformative in people's lives that they would feel compelled to take it with them everywhere they went. Portability had therefore been part of his remit: designing the computer so that it could fit under an aeroplane seat and incorporating a built-in battery. Jobs didn't yet care about portability; he just wanted the computer to look good. 'When you're a carpenter making a beautiful chest of drawers, you're not going to use a piece of plywood on the back, even though it faces the wall and nobody will ever see it,' he said. 'You'll know it's there, so you're going to use a beautiful piece of wood on the back. For you to sleep well at night, the aesthetic, the quality, has to be carried all the way through.' Manock may have been a perfectionist in his own right, but he recalls an altogether more practical reason for ensuring that the machine looked aesthetically pleasing from whichever angle it was viewed. 'The idea was that the Macintosh would sit on the CEO's desk,' he says. 'Therefore anyone coming into his office would be sat looking at the back, which should therefore be as attractive as the front.'

As always, the more Jobs cared about a certain aspect of a project, the brusquer he would be in his manner. And he cared about the look of the Macintosh very, very much.

'What's that crap you're working on?' he snarled at Manock on one occasion, announcing himself in the most abrasive manner possible.

Flummoxed, Manock began stumbling over his words.

'I don't like it,' Jobs interrupted, peering over Manock's shoulder at his designs. 'Has to be smaller.'

With that he walked away, disgusted. Feeling he hadn't conveyed himself properly, Manock took several hours to write down the reasoning behind what he was doing, and then went to see Jobs in his office. 'Steve, when you came by earlier I wasn't able to answer you because I wasn't prepared,' he said. He began explaining the rationale behind his decisions, talking Jobs through the myriad trade-offs between size, cost, manufacturing processes and human factors that every designer is forced to make. Once Jobs had heard the full presentation, and realised that Manock's compromises were down to design rather than laziness, he acquiesced. According to Manock, figuring out this psychological mechanism for dealing with Jobs was key for any Apple employee hoping to enjoy a long and happy career with the company. Recalls another former research scientist, 'If he came into your office and you told him what you were doing, and he said, "this is shit", what he was really saying is, "I can't deal with it today – let's talk about it tomorrow". You had to understand that or you couldn't work with him.' Many employees, however, never learned this. Especially the younger members of the team, or simply those too intimidated or overawed by Jobs to stand up to him, would just scrap what they had been working on and go with whatever spur-of-the-moment suggestion Jobs had made to them. Of course, when other problems invariably surfaced, Jobs would no longer be in any rush to claim responsibility. The poor saps who had gone along with the ideas would inevitably wind up leaving the company – either fleeing it of their own accord or being shown the door – or else staying around and being looked on unfavourably. Manock feels that Jobs was simply looking for interaction, and would provoke it in whatever way he could: 'He didn't know how to say, "Let's sit down and discuss what you're doing. I really want to know why you're doing things the way that you're doing them. I want to learn." He didn't know how to say that. Instead he would make a rash statement and then see how you reacted to it.'

Those who knew Jobs simply shook their heads and wondered how a student of Zen Buddhism could be so impatient in his dealings with people. They referred to his habit of unexpectedly dropping in on people as 'swooping', and his management style as 'Zen Crazy'. In fact, Jobs was just passionate about the work he was doing. He was also competitive. Despite the headstart and far greater resources given to the Lisa team, Jobs bet John Couch, the Lisa's project manager, $5,000 that the Macintosh would be the first of the two machines to ship. He lost.

□ □ □

Jobs may not have won his wager with John Couch, but at least he didn't lose $12 million doing so. As such he was still one step ahead of Steve Wozniak. Following his plane crash, Wozniak had opted to take some time off. At first he couldn't remember anything about the accident and was convinced that he had dreamed it. Eventually he realised that, while he could remember every detail of the flight until takeoff, he couldn't actually remember landing. At this point everything else – the cards from well-wishers, the photographs of him playing computer games in hospital, all of it – started to fall into place. When he went to break the good news of his recovery to his therapist, his therapist suggested that he was suffering from manic-depressive disorder.

'But I'm thirty,' Wozniak protested.

'It starts at thirty,' his therapist replied.

Rather than return to Apple straightaway, Wozniak had decided that he would go back to college and finish the degree he had dropped out of a decade earlier. He enrolled at UC Berkeley, taking his old engineering classes, in addition to psychology for majors and two classes about human memory.

In an effort not to draw any unnecessary attention to himself, he signed up under the name Rocky Raccoon Clarke: a combination of the first name of his dog and the surname of his fiancée. One day when he had been driving to class, Wozniak discovered a local radio statio, KFAT, playing what was referred to as progressive country music. It suited his mood perfectly. This was a transitional period in his life. He had just turned thirty, got married for the second time and was fast falling out of love with the company he had helped start. '[Apple] had become big business, and I missed tinkering,' he told *People* magazine. 'I just wanted to be an engineer.' At the same time, he had more wealth than he had ever dreamed possible. He decided to do what any aspirant hippie with a spare $116 million to his name would attempt in his position. He would stage his own version of Woodstock.

Wozniak quickly found himself out of his depth planning the so-called 'Woodstock West'. In Apple's earliest days, he had had complete autonomy. The only constraints were the engineering solutions that he himself could devise and the cost of the central microprocessor he was able to afford. This time Wozniak could afford almost anything, but found himself in uncharted waters, dealing with music-industry heavy-hitters completely unlike the Silicon Valley geeks he was used to working alongside. He decided that he needed a real qualified promoter to bring his dream to life. The man he settled on was Bill Graham, a legendary rock promoter famed both for his silver-tongued ability to convince bands to play his venues and his willingness to throw his weight around when necessary. Graham filled the card with rock luminaries, ranging from the Grateful Dead and the Ramones, to the Kinks and Fleetwood Mac. Because of the scale of the proposed festival, many artists were able to ask for, and receive, obscene amounts of money to perform a single set. 'I would see the money going out – my God, so much more money than these bands have ever been paid!' recalled festival controller Carlos Harvey years later. Wozniak himself was allowed to choose one progressive country act to appear. He opted for Jerry Jeff Walker, the singer who had had a 1960s hit with the song 'Mr. Bojangles'.

The location Graham chose as a venue was a 500-acre stretch of undeveloped land in Devore, California. Wozniak was insistent that the festival should not take place in a stadium or arena, but, rather, that it echo the outdoor atmosphere of Woodstock, where people could turn up and camp for the weekend, mingling freely while having picnics and barbecues on the grass. Diggers and bulldozers moved in and began landscaping the area. Eighteen hundred Portaloos were installed, while two helipads were created for those whose mode of transport since the summer of '69 had taken the leap from Love Bug to luxury chopper. The constructed staging area was particularly impressive. At 560 feet wide it was the largest ever built for a concert. The amphitheatre alone set Wozniak back $10 million. 'We have some very professional people,' Wozniak said proudly. 'We're not wild hippies with wild ideas who are just doing it for the sake of doing it. We want to make sure that it comes off well, that we look good, and that the community benefits from it. I hope it makes San Bernardino county look really good.'

The first 'Us Festival' sprang to life on a September weekend in 1982. The name had been chosen to celebrate the passing of the 'me decade' that had been the seventies. In a later retrospective, *Spin* magazine would describe it as a conflation of 'sixties idealism, seventies decadence and eighties cash into one big, glorious mess'. Bill Graham rode around the concert area on a giant Harley-Davidson motorbike, resembling a regal Hell's Angel surveying his kingdom, barking instructions at roadies and any revellers unfortunate enough to get in his way. As a last-minute spanner in the works, Wozniak's wife had gone into labour the night before the festival was due to open. She gave birth to a healthy baby boy, whom Wozniak initially wanted to name Jesse James, after the Wild West outlaw who had been reimagined as a countercultural rebel icon. (Eventually he had a change of heart and settled instead on Jesse John.) Wozniak being who he was, technology featured heavily in the festival. If Woodstock had been content to be 'three days of peace and music', then this was a three-day 'celebration of contemporary music and technology'. Robert Moog was in attendance to give a demonstration of his famous synthesiser, while jazz legend Herbie Hancock led a discussion about the Apple II/Alpha Syntauri music system and wowed the crowd with a few impromptu riffs. Elsewhere a large hot-air balloon, emblazoned with the Apple logo, floated by the main staging area. 'It was trippy, it was fun, there were people running around naked and there were people smoking dope,' recalls Phil Roybal, Apple's then manager of communication programs, who was the company's representative for the event. Apple took the opportunity to hand out thousands of free stickers to interested parties.

Not all of the technologies on exhibition proved successful. A presentation about UFO research had to be shut down when dust blowing from the campsite caused it to glitch. Meanwhile, a tent described as the 'Sensadome', promising a 360-degree multimedia light show in a Buckminster Fuller-style geodesic dome, started late and was generally agreed to be less impressive than a commercial planetarium display. Festival-goers were similarly bemused by an L. Ron Hubbard *Battlefield Earth*-inspired exhibit which featured several men dressed in Scottish garb guarding what they claimed was a fifteen-foot-tall alien creature that had, thankfully, been sedated.

Several members of the Macintosh team were in attendance for the weekend, although Jobs himself was not among them. All had been issued with gold passes which were stamped with the magic words WOZ GUEST, enabling them to head backstage. When they did so, Wozniak came bounding over, looking, as *Time* reporter Jay Cocks described his appearance at the event, like a Steiff teddy bear on a maintenance diet of marshmallows.

'Do you guys want to introduce a band?' he said. 'Which one? I've got it all worked out with Bill Graham so my friends can introduce their favourite band if they want to. We've still got plenty of slots left.'

Andy Hertzfeld considered asking whether he could introduce the Kinks, but realised that he couldn't imagine a scarier scenario than getting up on stage in front of so many people.

'Could I introduce Santana?' Burrell Smith said. 'That would be the coolest thing ever.'

'No problem,' Wozniak replied, with devil-may-care generosity.

That evening when he returned to the campervan the Macintosh team was staying in, Smith wrote out a short, humorous introduction for Santana and started memorising it. The next day he used his gold pass to get backstage and waited patiently for his moment to shine. Unfortunately, that moment was interrupted by Bill Graham.

'Who the fuck are you?' the promoter growled. 'What are you doing on my goddamn stage?'

Smith pulled out his pass and the handwritten notes, and waved them around.

'I'm introducing Santana,' he said. 'I'm a friend of Woz.'

Graham looked around to see if Wozniak was there to verify this unlikely story, failed to spot him, and turned his glare back upon the diminutive programmer.

'Sure you are,' he snorted sarcastically. '*I'm* going to introduce Santana. And you're going to get the fuck off my stage, this instant!'

The next thing the rest of the Macintosh team knew, one of Graham's burly, heavily tattooed bodyguards had lifted Burrell off the ground by his shirt collar and carried him in the direction of the festival gates. They didn't see him again for six hours.

On the last day of the Us Festival, Wozniak got up on stage and sang a duet with Jerry Jeff Walker. The song he chose was entitled 'Up Against the Wall, Redneck Mother'. The event ended with a piped-in recording of Frank Sinatra's 'My Way' – readable either as a celebratory hurrah from Wozniak or a derisive snort from the exasperated Bill Graham. Afterwards Wozniak was on cloud nine. He told his friends that it had been the greatest three days of his life. It was only when the ticket receipts were totted up that he realised how much money he had lost. Although the audience for all three days had been impressive, it transpired that a large percentage of those that had turned up had done so without paying, either by using counterfeit tickets or simply by climbing over the perimeter fences. The following year Wozniak tried again. This time his losses came to $13 million. Even worse, a man was beaten to death with a tyre iron, while a twelve-year-old girl was run over while she slept in her sleeping bag. Two and a half thousand people required medical attention, many for drug overdoses. Wozniak decided to call it quits on his concert-promoting days.

'If I do this for another fifty-five years, I'm in trouble,' he said.

When sheer brute force wasn't enough to get what he wanted, Steve Jobs found other ways to motivate his team. 'In the Mac group the perks got pretty wild,' says Bruce Tognazzini, who, as an outsider working in another part of Apple, felt envious when he saw how some of his colleagues were being treated. 'In the Mac building Steve Jobs had bought a piano, just in case anyone felt like playing. It was a Bösendorfer, which is one of the finest pianos in the world. They cost $75,000 to $80,000 and it was just sitting out there in the lobby. That was only playable by the Mac team, and only during development time.' Andy Hertzfeld objects to this suggestion. 'I don't even know if you'd call some of those perks,' he says. 'They were things that I would refer to as *symbols*. Steve was always thinking about how best to inspire the team and how to reflect our values, which were really *his* values. So he started getting these beautiful objects for the Mac building ... You could call those perks. I mean, they were expensive, but really – at least in Steve's mind – they were symbols of excellence.' Other symbols of excellence included a BMW motorcycle belonging to Jobs and a series of black and white prints by Ansel Adams, the iconic photographer whose portraits of the American West implicitly cast the Bay Area as Pioneer Country.

Motorbikes and fine art may have been one thing, but what really turned the Mac team on was the addition of a Defender arcade machine – which they would now no longer have to trek across the road to Cicero's pizza bar to play. This was proof positive that they were on to something good. Games on Defender quickly became competitive, with the various team members competing to see who could rack up the higher score. As the project wound on, and a ninety-hour working week became more the norm than the exception, it also proved a good way of working out who really needed to go home and get some sleep. If cocaine is God's way of telling you that you have too much money, then an atypically low score on Defender was His way of saying that your brain, your body, or both, were in the process of meltdown.

Even then many of the team kept working. As far as Jobs was concerned, people would have the chance to go on vacation, have children, or raise families later on in life. But the Macintosh would only ship once. At any time of the day it was almost guaranteed that there would be at least one person in the office. Bill Atkinson and Bud Tribble worked alternate nights and days, beavering away on the same tasks but at wildly different hours.

They communicated with one another through Post-it notes and Polaroids.

□ □ □

As 1982 came to an end Apple geared up to launch the Lisa. Jobs begrudgingly paid John Couch the $5,000 he had bet him regarding whose machine would make it to market first. To show off the Lisa in all its graphical interfaced glory an annual sales meeting was arranged, to take place south of the border in Acapulco. Because there was instability in the Mexican government at the time, contingency plans were made whereby, in the event of a coup, Apple would rent a boat to transport the pre-release Lisas out to sea, so that they could be dumped rather than be seized by the military.

In January the following year the Lisa finally went on sale. It was, according to most who came across it, a revolutionary computer. 'It blew me away,' says Al DiBlase, a vendor who first encountered the Lisa at a New York Business Expo. 'It was like an alien technology. I had never seen anything like it before in my life.' In a breathless article, *Popular Science* urged its readers to 'forget all [they] know about running a computer'. The magazine found the user interface a revelation. 'The arrow darted across the video screen, keeping perfect rhythm with the palm-sized device I pushed and pulled over the desk top,' wrote journalist Jim Schefter of his efforts to get to grips with the mouse. 'In moments, the idea of controlling an incredibly complex computer with nothing more than hand motion and a single button seemed perfectly natural. The computer was Lisa ... and with it, computers suddenly stopped being mysterious, forbidding machines that boggle the mind. Lisa is easy; you'd have to work at it to make a mistake. It is a radical new development in computer technology.'

Even with these gushing reviews, however, the Lisa floundered. For a machine aimed at businesses it was considered woefully incompatible with the already ubiquitous IBM PC, equipped with MS-DOS. Furthermore, its speed was glacial. 'When you moved a window on screen you didn't move the actual window, but an outline of the window, two horizontal lines and two vertical lines, and it *still* couldn't keep up,' says Bruce Tognazzini. 'It was about twenty pixels behind. It just didn't have the horsepower. It didn't have the software. It didn't have the distribution channels. It was too expensive. There were all kinds of things wrong with it.'

Back at Apple, Jobs poked fun at the Lisa, pointing out how the prominent brow above its monitor gave the machine a Neanderthal Cro-Magnon appearance. Amidst all the laughter at no point did anyone point out that this design consideration had come at Jobs' own request.*

On 8 April 1983, John Sculley was announced as Apple's new president. Sculley was a distinctly left-field choice for the role. Mike Markkula had been doing such a good job running the company since Scotty's departure that most people saw no reason why he would want to step down. Markkula, however, had only taken the post out of necessity and was eager to bow out of the limelight as soon as possible and go back to serving on the board. Many expected that Earl Floyd Kvamme, Apple's executive vice-president of sales and marketing, might be Markkula's natural successor. Another friend from Markkula's National Semiconductor days, Kvamme was a Silicon Valley veteran and a traditional marketing guy with a calm, authoritarian manner. A staunch Republican, later in life he would be appointed co-chairman of the Presidential Council

---

* The Lisa would struggle on for the next several years, and one more incarnation. The last computer in production rolled off the assembly line on 15 May 1985. Several years later, and under the watchful gaze of armed security guards, 2,700 Lisas were buried in a landfill in Logan, Utah, so that Apple could receive a tax write-off.

of Advisors on Science and Technology by the George W. Bush administration. John Sculley was altogether different. He was not Steve Jobs' first choice for the post (that would have been Jobs himself, although the Apple board deemed him too young. As a trade-off, he was allowed the option of vetoing any suggestion the board made that he did not agree with.) After failing to find anyone suitable on the West Coast, David Rockefeller – an early Apple investor – suggested that they take the search nation-wide. This is when they came across Sculley, who was then president of PepsiCo.

The downside with Sculley was that he had no computer background whatsoever, although he had recently – and entirely coincidentally – purchased 525 new Apple II computers, complete with Disk II floppy drives, for Pepsi's US bottlers. 'They could have it for free by earning off the cost if every week they would send in to us their sales report on a floppy disk through the post,' Sculley recalls. 'That was pretty revolutionary, because it used to take us almost a month to get the sales reports from the bottlers, and we would only get it on a monthly basis, so there was a huge delay – a couple of months in some cases – before we were able to put together how the sales were doing. This was a successful program with the bottlers, and that was the first indication to me that personal computers did something useful.'

Despite having no background in high-tech, Sculley fitted the profile remarkably well. With Pepsi he had proven that he could take an underdog brand and transform it into a marketplace contender.* One of the best ideas he had initiated during the

---

* Sculley told me of one particular piece of market research Pepsi had conducted, which stated that if a customer was 'going to serve a Coca-Cola to someone they would go out to the kitchen, grab the Coke bottle from the fridge, and bring the glass with them to pour it out in front of the guest in the living room. If it was a Pepsi, they would go out to the kitchen, take the Pepsi bottle and pour it into the glass in the kitchen, leave the bottle in the kitchen, bring the glass of cola out, and just hope the person didn't ask what brand it was.'

so-called 'Cola Wars' with rival Coca-Cola was the Pepsi Challenge: a blind taste test designed to see whether customers preferred Pepsi or its perceived higher-quality older brother. (Most chose Pepsi. Sculley, so legend has it, chose Coke.) Jobs liked Sculley, and Sculley was beyond impressed by Jobs, who lured him in with the now-famous pitch, 'Do you want to sell sugar water for the rest of your life, or do you want to come with me and change the world?'

The two men were opposites in almost every regard, but their differences complimented each other, as had been the case with Jobs and Wozniak. While Jobs at the time favoured Porsches, which he could race at high speed through the winding mountain roads of Silicon Valley, Sculley was driven around New York in a limousine, complete with bar, television and a chauffeur named Fred who was on call twenty-four hours a day. He was fourteen years older than Jobs, and while he had been in his mid-twenties during the 1960s, and was thus the right age to tune in, turn on and drop out, he had felt no great compunction to do so. 'I didn't identify at all with the hippies,' he says. 'I found Haight-Ashbury a strange place, and I couldn't understand how these people my age were wasting their lives smoking pot. I actually never tried pot. My generation liked alcohol, not drugs.' Where the long-hairs attended courses in Maoism at Free Universities, Sculley had been to Wharton Business School and graduated with an MBA. Where the hippies ventured into the wilderness to build communes so that they could be at one with nature, Sculley constructed traffic-flow patterns to determine where best to locate a future cinema, shopping centre or restaurant. In 1969 he didn't join the throngs of flower children who gathered to hear Jimi Hendrix mangle the national anthem in a feedback-drenched guitar solo at Woodstock. Instead, his defining moment that year came from a lecture he attended at New York's Museum of Modern Art, presented by the anthropologist Dr Margaret

Meade. 'Being a young marketer at the time, one of the things she said that pricked my interest was, "The most important fact for consumer marketers is going to be the emergence of an affluent middle class which will be characterised by today's teenagers",' Sculley recalls. 'She named them the baby boomers: the people who were then in their late-teens and early-twenties. She said that this is the first group, demographically, to have so much discretionary income to spend. They were the kids of returning World War II veterans who had quickly got married and had children back in the early 1950s.' He realised that this moneyed consumer audience was going to be key to the new business markets which would increasingly dominate the latter half of the twentieth century.

The 1960s were a good time to be an acid-dropping hippie, but they were perhaps an even better time to be a young business executive, especially one involved with marketing. For the first time, youth was considered a valuable commodity, as companies struggled to retune themselves to the wavelength of a generation who considered themselves too smart to be sold the kind of mass-produced stuff the older generation had got off on. A new market had emerged for packaged sixties idealism, that saw customers treated to cars like the Dodge Rebellion (which would lead 'the charge on Dullsville'), while Booth's Gin '[Protested] Against the Rising Tide of Conformity' and Columbia Records pondered why 'If you won't listen to your parents, The Man, or the Establishment ... should you listen to us?' Traditional advertising, by comparison, was almost comically unhip. Coca-Cola was the patriotic soft drink that had helped American GIs quench their thirst en route to victory during the Second World War. It was the safe distillation of traditional American values; a representative of a world of white picket fences, fathers in grey flannel suits and Norman Rockwell scenes of familial bliss. In this position it was absolutely unassailable. So unassailable, in fact, that Coca-Cola never changed. America did. Suddenly the post-war Age of Affluence – complete with its gleaming Fordist factories churning out one identical product after another – was not a comforting reminder of America's pole position as a universal leader, so much as it was a symbol of the kind of stiflingly banal conformity that people were so keen to escape from. Coca-Cola's marketing people had already dismissed rock and roll as 'dirty and low-class' and deemed that it would never be the soundtrack to fizzy-drink consumption. This was where Pepsi could capitalise. Under Sculley's watch PepsiCo relaunched a 1960s advertising campaign called the Pepsi Generation: a series of aspirational advertisements that cast Pepsi consumption as an embodiment of the kind of hip lifestyle that young people around the country dreamed of. 'They were about everything from skateboarding to skin-diving to surfing to dirt-bike racing – all sorts of outdoor high-energy experience-type sports – but we never mentioned anything about the product itself in the commercials,' he says. 'We felt that if we could find a way to appeal to the baby boomers in a unique way, maybe we could get a leg up on Coca-Cola.'

At the same time these were not the acid-dropping hippies that Steve Jobs had so resembled. 'I can remember back in the sixties standing in Piccadilly Circus in London and watching the people with rainbow-coloured hair and peace signs marching against the war,' Sculley says. 'These weren't the people we were advertising Pepsi to. The people we were advertising Pepsi to with the Pepsi Generation were the same age, but the girls were in bikinis, and everything was made up of pleasant beach scenes, with lots of Beach Boys-type music in the background. There was nothing angry about it, no rebellion at all. The people in our commercials were always happy and wholesome; never protesters or hippies. The Pepsi Generation was a much more sanitised version of the baby boomers than anything Steve Jobs was experiencing.' Still, it was hoped that Sculley could weave that same magic with Apple, which was being outperformed two-to-one in the marketplace by Atari and Commodore, and rapidly being caught up by the IBM PC. *If the seventies was the Pepsi Generation, why could the eighties not be the Apple Generation?* 'We liked the fact that he might be able to bring that marketing genius to Apple,' says Paul Dali, who had been part of a small group of Apple executives who met Sculley early on at New York's Four Seasons hotel and, over a lavish dinner, sold him on the idea of taking the position at Apple. 'We knew how to make products, but we didn't necessarily know how to market them to the consumer. We thought John could be the perfect blend to help us in those years at Apple when we were growing so rapidly and beginning to attract more and more individual customers.'

'At this time there was no marketing going on in Silicon Valley,' Sculley says. 'It wasn't that people did bad marketing – there was none at all. It was dormant, like a moonscape, and Apple not only had been controversial in hiring a non-computer CEO to bring in consumer marketing, but most people in Silicon Valley didn't have a clue why anybody would *want* to sell technology the same way you sell automobiles and soft drinks.'

One thing was clear to Sculley and the board: if Apple was going to make computers the likes of which the world had never seen, they sure as hell needed someone who knew how to sell them.

8

THE VOLKSMACiNTOSH

'... From my own point of view, the counter culture, far more than merely 'meriting' attention, desperately requires it, since I am at a loss to know where, besides among these dissenting young people and their heirs of the next few generations, the radical discontent and innovation can be found that might transform this disoriented civilization of ours into something a human being can identify as home.'

Theodore Roszak,
The Making of a Counter Culture (1969)

The French essayist and philosopher Paul Valéry once remarked that a poem is never finished, only abandoned. That goes twice for a computer with the modest aim of putting a dent in the universe. Like the keepers of an animal that has grown to adulthood in captivity, the Macintosh team fretted for months about letting their surrogate child out into the wider world. 'None of us wanted to release it at the end,' Steve Jobs told *Playboy*. 'It was as though we knew that once it was out of our hands, it wouldn't be ours any more.' By the time it neared completion, the project had been rocked by delays. The idea that this was a machine once considered able to beat the Lisa to market seemed like something from a previous lifetime. Bud Tribble had left the project in November 1981, being told by his university advisers that if he did not return to medical school immediately he would lose his place on the program. Andy Hertzfeld had thought then that this marked the beginning of the end, but the team had valiantly struggled on without Tribble, bringing in a new engineer in the form of ex-PARC employee Bob Belleville.

Bill Atkinson referred to the end of the Macintosh project as being like running a race, only to have the finish line moved as soon as it came into view. Nonetheless, the conclusion finally registered on the horizon. The final few weeks were a frantic rush to get everything wrapped up. In a side room littered with half-eaten takeaways and junk food, Steve Capps and Bruce Horn slaved away at a file manager called the Macintosh Finder. As the deadline loomed imminently the pair worked 58-hour stretches, staying cognisant by gobbling vitamin C tablets and playing Dead Kennedys punk records as loud as they could. On the Tuesday night that was agreed as the completion date, the team almost melted down at the last hurdle. 'At two a.m. it seemed like everything was broken,' recalls Randy Wiggington. 'It was one time in my life that I went totally hysterical, I just started laughing. I had to go outside and walk round the block for fifteen, twenty minutes to get a hold of myself because I was under such unbelievable pressure. When we managed to recover from all the bugs by six a.m. and finally sent everything off, it was one of the most incredible moments of my life.'

Despite Jobs' best efforts, word had starting to get out about the industry-shaking computer that was being developed behind closed doors in Cupertino. John Markoff, today a highly respected tech writer for the *New York Times*, was then a rookie staff writer for *InfoWorld*. He had only been in the job for a few weeks when he walked into another reporter's office and saw him holding the phone away from their ear, grimacing. Far though Markoff was from the receiver, he could hear a tinny voice screaming down the phone line. It was Jobs. As it transpired, the older reporter had discovered the existence of the Macintosh and was planning to reveal its name in the next issue. Enraged, Jobs was frothing at the mouth about it, convinced that even mentioning the Macintosh in print was going to give Japanese competitors some kind of advantage.

While he kept schtum about the Macintosh to *InfoWorld*, on other occasions Jobs' emotions could get the better of him.

'What's this "Macintosh" I keep hearing about?' Osborne Computer's CEO Adam Osborne once needled the Apple co-founder. 'Is it even real?'

Osborne, a former Homebrew Club attendee, was well known in the industry for bragging about the upcoming computers his company was developing. Finally Jobs had had enough.

'Adam,' he said, 'it's *so* good that even after it puts your company out of business, you'll still want to go out and buy it for your kids.' He probably wasn't wrong. Among those who had seen even the briefest glimpses of the machine, Macintosh hysteria was running wild. 'I'd sell my mother to get a Mac,' said Mitch Kapor.

□ □ □

Just a few years earlier, another Dick Cavett radio campaign would have been enough to launch Apple's latest computer in style. Not any longer. The man assigned the role of introducing the Macintosh to the world (and vice versa) was a slick marketing director named Mike Murray, with a penchant for mirrored sunglasses. Given the near-religious zeal required by those working on the Mac project, it is perhaps unsurprising that he later went on to become mission president for the Church of Jesus Christ of Latter-day Saints. Murray had joined Apple in 1982, after completing an MBA at the Stanford Graduate School of Business. During that time he had been won over by a visiting Steve Jobs. As part of the MBA course, executives from various companies would come in to give presentations, in the hope that they might convince graduates to come and work for them. Unlike the strait-laced departmental heads from Hewlett-Packard and Shell Oil, Jobs had hopped up on to the desk at the front of the Stanford auditorium and sat in the lotus position while he delivered a typically messianic address to the students, starting with his trademark, 'Hi, I'm Steve Jobs and I started Apple.' Sure enough, Murray had gone along for an interview with the company, which wound up being with engineer Rod Holt. It turned out to be a rather strange meeting. Holt plopped a book of black and white professionally taken photographs down on the table, pointed to one and asked, 'What do you think of that?'

Not wanting to blow his chance at impressing, Murray said, 'Well, that's just great.'

'No, it's not,' Holt said. 'And I'll tell you why ...'

Next he flipped to another page and pointed at a second picture. Murray thought he had started to cotton on to Holt's game.

'It stinks,' he said.

'No, it doesn't,' Holt responded. 'It's great.'

Murray didn't know how to respond by this time, but the important thing was that Jobs liked him. When it came time for him to make an offer, however, he low-balled Murray on potential salary and the young MBA instead took a job at Hewlett-Packard, only to return to Cupertino a year later when Jobs acknowledged that he'd made a mistake. The money may have been a bit more the second time around, but Murray was hardly given the corner-office treatment. Space in Texaco Towers was so limited that his first assigned 'desk' was the shallow tray, designed for holding pens, at the base of a whiteboard. All the same, Murray had quickly proved his worth. When it came time to create a buzz around the Macintosh launch, he was given a $15 million budget to work his magic with. 'No company in the world would spend that much money trying to market a computer product,' he says. 'Steve was willing to bet the ranch on the Mac because he knew that the Apple II would run out of gas sooner rather than later, and he could see from market research that the IBM PC was going to kill us, so it was up to the Macintosh to step up and save the company.'

In marketing meetings for the Mac, the reference point most often brought up was George Lucas's 1977 film *Star Wars*. The art of the wide-release blockbuster was still less than a decade old, but Lucas's space epic held a very special appeal.* During the time that it had been in production, hand-wringing movie executives were so sure that they were on to a flop that they happily signed over all merchandising rights to George Lucas, positive that there was absolutely no value in the concept of Luke Skywalker action figures or Darth Vader Halloween costumes. Conventional studio wisdom in that instance proved to be disastrously wrong. One member of the Mac team had read a short, two-paragraph synopsis of this story and from it determined that this same formula could be followed with the Mac. Says Murray, 'We had to create that same impression at the time of the product launch; that this was already viewed as a fabulous product and was already accepted and used, and if you were a person that knew what was going on in the world, you needed a Mac and were already late to the game.'

Murray approached the publishers of the popular *PC World* magazine and asked if they would be interested in starting up

---

* Up until Steven Spielberg's smash hit *Jaws* in 1975, conventional wisdom stated that films should be opened in one market at a time, so that a picture could drum up interest in a particular city before fanning out to meet the demand, which had hopefully been spreading through positive word of mouth. From *Jaws* onwards, major blockbusters would 'open wide' to best capitalise on the tidal wave of hype.

a second publication, entitled *Macworld*. The response was not overwhelmingly positive. 'If your computer is a success, maybe we could think about doing something a year down the line,' he was told.

'You don't understand,' Murray said. 'We need *Macworld* to be on the newsstand the very same day the Macintosh is launched.'

'That's not going to happen,' came the response.

'Well, what if we can sell off a set number of the pages for software developers to advertise on?' he countered. In the end, Murray struck a deal. The publishers would cover its costs with print ads, and if the magazine failed to sell then Apple would buy up every unsold copy. *Macworld* was born.

□ □ □

If there was a challenge in advertising the Macintosh it was that, after half a decade of production time, explaining it to people was easier said than done. In essence the company became hamstrung; unwilling (or unable) to say just what the Macintosh *was*, but, rather, defining it by what it most certainly *wasn't*. According to a weighty document entitled the Macintosh Product Introduction Plan, it was not merely a home computer, nor an education computer, nor even a fully networked office automation machine. In retrospect, of course, it was all three, although geared to no one market in particular. The Mac was the perfect home computer, although it did not as yet have the wealth of applications to put it on a par with the Apple II. It would be a great tool for education, also, but most universities still featured time-sharing mainframes. Finally, it was brilliant for business, but the Lisa had struggled to get a foothold in that sector, and the Macintosh's networking capabilities, taking the form of the so-called AppleTalk, were considerably slower than the Ethernet connections used by some of the Mac's rivals.

Rather than settle for being second best in any of these areas, the Macintosh was presented as a new class of device, exactly as Jef Raskin had hoped it would be. In introductory materials it was described as 'an advanced personal productivity tool for knowledge workers'. Potential users might include the thirty-five-year-old suburban office worker who wanted to be technologically hip but was intimidated by computers; the university provost looking to jump aboard the information-age bandwagon; the forty-something Rotarian with the vague sense that a computer might help run her small business; and the University of Michigan student trying to decide between conforming (her parents wanted her to become a lawyer) and tuning in, turning on and dropping out (she had a somewhat niche interest in the music of the Carolingian Empire that she wished to pursue).

This term 'knowledge worker' became a buzzword for the Macintosh marketing team. It was a phrase borrowed from management consultant and 'social ecologist' Peter Drucker. In his 1966 book *The Effective Executive*, Drucker had outlined a new breed of hip worker who was busy raising hell in the traditional status quo of the Fordist business world. According to him, 'the centre of gravity has shifted to [this] knowledge worker, the man who puts to work what he has between his ears rather than the brawn of his muscles or the skill of his hands'.* These were the

---

* In his 2001 book *Free Agent Nation*, Daniel Pink elaborated on the rise of the so-called 'knowledge worker' who, for better or worse, was doing for traditional modes of business what the hippies did for society: 'Legions of Americans, and increasingly citizens of other countries as well, are abandoning one of the Industrial Revolution's most enduring legacies – the "job" – and forging new ways to work. They're becoming self-employed knowledge workers, proprietors of home-based businesses, temps and permatemps, freelancers and e-lancers, independent contractors and independent professionals, micropreneurs and infopreneurs, part-time consultants, interim executives, on-call troubleshooters, and full-time soloists. And many others who hold what are still nominally "jobs" are doing so under terms closer in spirit to free agency than traditional employment. They're telecommuting. They're hopping from company to company. They're forming ventures which are legally their employers, but whose prospects depend largely on their own individual efforts.' Naturally, of course, Silicon Valley led the way in this new world order.

people Apple wanted to target – and making a big splash with the Macintosh was simply about finding them and letting them know that this new revolution was taking place. It was decided that the best way really to grab the world's attention was to buy a chunk of advertising time during the half-time show of the 1984 Super Bowl, taking place on 22 January. At first Jobs protested against the decision. He may have been a technological visionary, but he was also a person who spent his day dealing with ungainly computer geeks, many of whom still woke up in a cold sweat at the memory of being picked last for gym class. 'I don't know a single person who watches the Super Bowl,' he said, deadly serious, apparently unaware that 81.8 million people had tuned in to the previous year's game. Eventually he was convinced and Apple shelled out $1,600,000 for a two-minute slot, with the intention of filming one sixty-second commercial and airing it twice, one time after the other.

Finally, to paraphrase Jobs, the company at last had its 'insanely great' product. It had a magazine to confirm to the world how insanely great it, indeed, was. It had an insanely great new class of 'knowledge workers' to sell to. It had an insanely great place to debut its grand unveiling. Now all it needed was an insanely great advertisement to do it.

In many ways Ridley Scott was a perfect fit as the director of the Macintosh commercial. The 45-year-old Englishman came to the project with not just an outstanding track record in the commercials world, but also two classic science-fiction films under his belt. His generation of British advertising director (other examples included the likes of Alan Parker, Charles Saatchi, David Puttnam, Peter Mayle, Hugh Hudson and Adrian Lyne) were a collective of self-styled mavericks who had wrested the industry away from the largely unimaginative good old boys' club which had dominated it for years. Like Steve Jobs, Scott could be a demanding perfectionist – even if it meant abandoning certain social niceties to achieve it. 'He is one of those directors who will come up to you after you've done a scene and say, "Well, I don't fucking believe that",' Sigourney Weaver, who had worked with Scott previously, has said. 'At first I'd be a little taken aback and wonder, "Where's the stroking, where's the diplomacy?" and there just wasn't any.' On the set of his previous film, *Blade Runner*, the director's relationship with lead actor Harrison Ford had broken down so entirely that neither man would speak to the other during filming.

What had ultimately emerged, however, was a messy, convoluted, but ultimately endlessly fascinating piece of cinema, in which the hyper-detailed science-fiction world it portrayed was every bit as intriguing as the somewhat confusing plot. ('What's outside the window?' Scott had repeatedly prompted the film's original screenwriter.) *Blade Runner* was an image of Scott's upbringing in rainy, industrial Northumberland, relocated to Los Angeles circa 2019, whose cityscape owed more than a passing resemblance to the world of soldering irons and breadboards all too familiar to the hobbyist-hackers. 'From the pyramidical building housing the Tyrrel Corporation ... which dominates the landscape, its sides incised with patterns so as to resemble the surface of a microchip, to the street-level prosthetics laboratories, the film presents a visual allegory of a society dominated by techno-science and information technology,' Charlie Gere writes of the film in his book *Digital Culture*.

Up until the Macintosh commercial, advertisements for computers had been so terrified of scaring off potential customers that they had avoided mentioning anything that was too science fiction, too threatening, too ... anything, in fact, that didn't scream nostalgic, happy and approachable. IBM, for example, had utilised the image of Charlie Chaplin's lovable tramp for a series of commercials with the theme 'Keeping Up with Modern Times' which ran throughout the decade.* Apple's own adverts had been consciously unthreatening in their own way. The first wave had included images of the computer in the kitchen and a self-deprecating Dick Cavett. The next wave had portrayed attractive yuppies using Apple computers to make their lives easier and more productive. But nothing had yet attempted to convey the broader Californian vision of the liberating possibilities offered by personal computing. The Macintosh commercial would do just that. Building on the same dystopian 'used future' imagery that Scott had explored in *Blade Runner*, the advertisement channelled a look that would become known in the public consciousness as cyberpunk.

Drawing all of this into a sixty-second running time was the job of writer Steve Hayden, the man credited by many as establishing the 'voice' of Apple that persists to this day. Hayden had first met Steve Jobs back in the late seventies, when he had been the scriptwriter for the early Dick Cavett commercials. Although Jobs liked the tone that Hayden brought to the adverts, their first face-to-face encounter, taking place at a computer show in New York, hadn't seemed to bode well for a long-term relationship.

'Jay tells me you're smart,' Jobs had said, referring to Jay Chiat, the head of ad agency Chiat/Day, which was then handling Apple's account. Hayden was flattered.

'If you're so smart, how come you smoke?' Jobs continued.

How does he know that about me? the copywriter wondered. He quickly realised that he had a pack of Marlboros poking out of his pocket, which Jobs had evidently noticed. There was an awkward pause. Then one of the nearby computer exhibits

---

* Either oblivious themselves, or simply hoping that audiences were unaware, that Chaplin's 1936 film *Modern Times* was created as an indictment of industrialisation rather than a celebration of it.

shorted out, the booth's curtains went up in flames and the New York Fire Department had to be called.

□ □ □

The Macintosh commercial had been filmed over two days in September 1983 at Shepperton Studios in Surrey, England. The plot called for two hundred drone-like workers whose life of greyscale drudgery was supposed to sum up the experience of working on an IBM PC. For his players, Scott recruited a cast of local skinheads. When there weren't enough of these available at short notice, he paid extras £60 for the inconvenience of being shaved bald. IBM was portrayed as an Orwellian Big Brother figure, appearing on a giant telescreen and barking propaganda, as written by Steve Hayden.

Into this scene of buttoned-down conformity and centralised corporate-governmental oppression, runs an attractive blonde woman, dressed in red and pursued by riot police. She sprints towards the screen, screams, and hurls a sledgehammer (originally written as a baseball bat) in the direction of Big Brother, whose likeness explodes immediately in a blinding flash of white light. As the commercial comes to a close, audiences are assured that thanks to Apple's latest computer, 1984 would not be like *Nineteen Eighty-Four*.

It was, in short, a brilliant piece of imagery. In particular the decision to make a woman the emblematic figure of the Macintosh was a bold one. Despite the fact that Ada Lovelace, the person often described as the first computer programmer, was a woman, modern computing was by and large seen as a field that, if not exactly macho, was certainly predominantly male.* The image of

---

* Born in 1815, Lovelace was an English writer best known for her work on Charles Babbage's early mechanical general-purpose computer, the Analytical Engine. Her notes on the engine include what is recognised as the first algorithm intended to be processed by a machine; the result of which being that she is sometimes considered the world's first computer programmer.

the lone female freedom fighter not only set the Macintosh apart from its peers, but brilliantly subverted the IBM advertisement of the early 1970s, which commonly depicted a female or young male subordinate seeking advice from an older male superior. Here the patriarchal figure didn't just get a verbal dose of woman's lib for its insolence; it had a sledgehammer lobbed at it.

The woman in the advertisement was played by a model and former discus thrower named Anya Major. She had earned the role by virtue of being the only actress not to get dizzy when spinning around to throw the six-pound sledgehammer during the try-out. Others hadn't been so lucky. During another audition an errant sledgehammer toss almost killed an elderly lady out walking in London's Hyde Park, where the casting call was being held.

Meanwhile, the Macintosh itself was cast not simply as a more technologically advanced computer than those already available in the marketplace, but as a weapon of mass construction; a digital Molotov cocktail by which users would be able to express their hitherto untapped individuality and throw off the oppressive shackles of the IBM Corporation. Of course, not everyone watching the commercial was going to share, or even be aware of, the anti-IBM sentiment that was felt in some corners of the computer industry. (And nor would some smaller computer companies, such as Commodore, have necessarily agreed with Apple casting itself as the spunky start-up in the face of the despotic institution.) But like the counterculture itself – which was a banner broad enough somehow to incorporate such disparate themes and figures as the civil-rights struggle of Martin Luther King and the eponymous bank robbers in *Bonnie and Clyde,* all under the auspices of getting back at The Man – Apple's rebellion was appropriately vague. It tapped into the fears of a generation by attacking the elephantine military-industrial complex, the dystopia of a corporate world run by men in grey flannel suits, and the still-lingering threat of the Soviet Bloc. *An IBM PC could help you organise your wife's recipe cards. A Macintosh was enough to repel both totalitarian state communism and its capitalist counterpart.* The point was that you, the user, were invited to be a part of it, all for the low, low price of just $2,495.

□ □ □

The commercial was a stunningly effective articulation of the kind of rebel yell that the Macintosh was supposed to represent. Jobs loved it when he saw it for the first time post-production. So did Wozniak. John Sculley wasn't so sure, although it certainly struck him as bold. Then the Apple board of directors watched it. 'Can I get a motion to fire the ad agency?' said Mike Markkula as soon as the advert had finished playing. Another board member, Philip S. Schlein, the CEO of Macy's California, said nothing but simply sat with his head down on the table. Plans were immediately scrapped to let anyone outside Apple see this abomination, and the company's media group started looking around for anyone who would buy up the Super Bowl ad time at the last minute. On the Friday before the Super Bowl was set to air, there were a couple of bites. Hertz was willing to buy up one thirty seconds, while Heinz snapped up another. That left Apple with one minute of paid air time, costing the not inconsiderable sum of $800,000. With no purchasers, it was decided that they would go for broke. Apple would air the commercial.

As it happened, Steve Jobs wasn't simply being contrary when he claimed not to know anyone who watched the Super Bowl. Most of the Macintosh team wouldn't have known the biggest football game of the year if it had hit them in the face. Bill Atkinson was no sports fan and skipped the show altogether, waiting until Monday to find out how it all went. Bruce Horn had been at his mother's house and turned on only for the half-time show. Andy Hertzfeld and a group of friends went to a burger joint named Oasis, which was located near Stanford, ostensibly to watch the game, 'but really to watch people watching the ad'. Even Steve Hayden, the writer of the commercial, missed its airing. With no interest in American football, he was at home alone, washing dishes when the phone rang, following the screening. It was Jay Chiat. 'How does it feel to be a fucking star?' he screamed down the line. 'Great,' said Hayden, still baffled at the recent turn of events that had seen chicken salad turned into chicken shit, and then back again. 'Just don't ask me to do this next year.'

Among people with their fingers on the pulse, the Macintosh arrived at the perfect time to tap into the zeitgeist. Suddenly a subject like computing, previously seen as the remit of greasy, bespectacled troglodytes, became the province of trendy French cultural theorists. Steve Jobs put it poetically. 'Computers and society are out on a first date in this decade, and for some crazy reason, we're in the right place at the right time to make that romance blossom,' he said. He wasn't wrong. The mid-1980s hailed the arrival of techno-music, the dance sound of the post-industrial urban landscape, along with the beginnings of deconstructionist graphic design. (The Macintosh, incidentally, would become a firm favourite with practitioners in both fields.) 1984 also saw the launch of the hipster *High Frontiers* magazine which, in addition to 'gadget pornography', concerned itself with a 'heady blend of ... guerrilla humour, human potential pep talk, New Age transcendentalism and libertarian anarcho-capitalism'. More significant than that was the publication of 36-year-old science-fiction writer William Gibson's debut novel, *Neuromancer*, which quickly came to be regarded as a seminal work in the emergent cyberpunk genre. Gibson himself was an early Macintosh adopter, littering his later books with references to Apple products.

All of the hippie idealism which suggested that psychedelic experimentation could reroute history, lead to the next step of human evolution and create a world free of the kind of repressed behaviour that led to war suddenly came into circulation again; now reconfigured to refer not to dropping acid but to personal computing. Right on cue, counterculture guru Dr Timothy Leary – a former psychology professor at Harvard, who had became one of the strongest proponents of the LSD experience in the 1960s – turned around and announced himself as a Macintosh convert. '[The Macintosh] may be an advance as important as the opposable thumb, face-to-face lovemaking, the Model-T Ford, the printing press,' he claimed in his book *Chaos and Cyberculture*. 'Owning it defines you as member of a new breed – post-industrial, post-biological, post-human – because your humble VM (Volks-Mac) permits you to think and act in terms of clusters of electrons. It allows you to cruise around in the chaotic post-Newtonian information ocean, to think and communicate in the lingua franca of the universe, the binary dialect of galaxies and atoms. Light.'

Leary wasn't the only cult sixties celebrity to embrace the idea of personal computing as a tool of creativity. Andy Warhol announced that he had become a fan of computers, supposedly after using the Macintosh's QuickDraw graphics program at a Manhattan party thrown by Yoko Ono for her son Sean Lennon's ninth birthday. (The following year he would treacherously switch his allegiances to the Commodore Amiga, which he would use to paint a portrait of Debbie Harry.)

Perhaps the most striking evidence that personal computing was catching on came not from a bandwagon jumper like Leary, or someone interested in pushing artistic mediums to their limits like Warhol, but, rather, from the surprising amount of machine terminology that was beginning to find its way into everyday conversation. Where not long before the graphical user interface had borrowed terms from real life to help build its computational metaphor, now the process started to work the other way around. One day Sherry Turkle, the MIT academic whose work revolves around people's relationships with technology, was sitting in a restaurant when she overheard two women talking at the next table. One of them had just come out of a relationship, and had asked her partner to move out of the house they shared. 'The hardest part is reprogramming yourself to live alone,' the woman told her friend. Had he been listening, Andy Hertzfeld would have swelled with pride.

Nobody ever got fired for choosing IBM – or so the saying went. Mike Murray was about to find out how ingrained this sentiment truly was. While appealing to young hipsters was one thing, the Macintosh struggled to make any sort of a dent in the business market. This was painfully highlighted during a focus group which took place in New York City shortly after the Macintosh's release. Standing behind a two-way mirror, a group of Mac representatives watched as a moderator led a discussion with a group of computer users, all of whom worked with IBM PCs in their businesses. After being asked to do tasks such as moving one word from the top to the bottom of a paragraph using the arcane commands of the IBM PC, the moderator demonstrated how that same task could be carried out in a fraction of the time using the Macintosh's 'cut and paste' function. 'The Mac was set up right against the two-way mirror and they had no idea we were just inches away from them,' Murray recalls. 'As these professional businesspeople used a Mac for the first time, they became almost childlike in their excitement. They would laugh and giggle; they were having so much fun. At the end of the focus group the moderator said, 'Which of these two computers was easier to use?' The Macintosh won easily. 'In an office setting, which computer would allow you to be more productive and get more work done?' the moderator asked. The Mac again. The moderator kept asking questions, and each time the answer was emphatically "The Mac, the Mac, the Mac".'

The Macintosh team was becoming increasingly excited, figuring that it would be only a matter of months, if not weeks, before the Macintosh totally eclipsed the IBM PC. The final question the moderator put to the focus group was this: 'If you were put in charge of purchasing computers for your company, which of these two machines would you buy?'

'The IBM, of course,' everyone said, without pause.

'We were absolutely stunned,' says Murray. 'These people had become Mac fanatics in a half-hour in that room, but when the question of what they would purchase came up, their demeanour just changed and they became serious. It had nothing to do with ease of use or practicality, it was that in their minds IBM belonged in business, and here was this fun Mac product and they couldn't even perceive it being there.'

'That's not hard to understand,' John Sculley told me when I asked why Apple consistently failed to crack the business market. 'Businesses weren't looking for a way of empowering individuals. They were looking for cheaper, faster ways of doing the things they were already doing. IBM came in with huge credibility and to them it was just an extension to a very broad product line of things they were selling to business. The idea that you would buy a PC like the Macintosh or the Apple II, which really had no business software, except VisiCalc in the case of the Apple II, and a couple of database programs – it just didn't make any sense.' The lack of immediately available software, while certainly a contributing factor, was not the only reason why businesses failed to embrace the Macintosh. At a time when IBM compatibility was the name of the game, the Macintosh completely shunned it. The Mac wasn't even compatible with the Lisa or the Apple II, so what were the chances that it would work with Jobs' much loathed IBM machines? Certainly the contrast between Apple's offering and IBM's couldn't have been more different. 'IBM already had this humongous, heavy system with slots and separately attachable monitors and Apple comes out with this little, cute, all-in-one system,' says Alan Oppenheimer, who worked on the Macintosh team as a networking specialist. 'A lot of people just brushed it off. Apple's never recovered in the business market because of some of the decisions made on the Macintosh.'

In a strange way, there was a no-nonsense quality about the Macintosh. As with the Ridley Scott '1984' commercial there was a sense of underlying arrogance, suggesting that so long as you agreed with Apple's vision you were one of the good guys. If not, well, with the Macintosh you were going into the future *whether you liked it or not.* One example of this was Jobs' insistence that the machine have no cursor keys on its keyboard, so that users would be forced to get to grips with the mouse. There was no demonstrable reason why users would want cursor keys, save for familiarity, but to deny them altogether caused a stir. More importantly was the fact that, underneath all the sheen of its graphical user interface, the Macintosh itself was woefully underpowered. Despite the fact that one of the first changes Jobs had made upon seizing control of the project from Jef Raskin was to ramp up its memory, it was now clear that he had fallen into the same trap he had accused his 'shithead' predecessor of succumbing to. Jobs felt so sure that he was right that he had shouted down anyone who suggested adding the capability to expand the memory beyond 128K, to a more reasonable 512K. (In one of his only acts of defiance against Jobs, Burrell Smith added this feature anyway, and only let Jobs know when the memory limitations became recognised as a problem.)*

Jobs had also barred the suggestion that the machine come with a hard disk drive, knowing that this would require the addition of one of his hated fans. The combination of these two drawbacks – anaemic memory and no hard drive – was summarised by Alan Kay, who had been lured to Apple from Xerox PARC, in a critique of the Macintosh which he wrote for John Sculley, entitled, 'Would You Buy a Honda With a One-Gallon Gas Tank?' Indeed, to appreciate the Mac in its earliest days was not completely dissimilar to liking the Altair 8800 upon its release. While the Macintosh could undoubtedly do more straight 'out of the box', users were being asked to buy into the same dream of what the computer *could be*, not necessarily what it was at that moment in time. Science-fiction writer Douglas Adams neatly encapsulated this notion when he wrote that:

> *What I (and I think everybody else who bought* [the Macintosh] *in the early days) fell in love with was not the machine itself, which was ridiculously slow and underpowered, but a romantic idea of the machine. And that romantic idea had to sustain me through the realities of actually working on the 128K Mac ...*

The users who formed the Macintosh's cult fanbase in its early years were more forgiving of this kind of thing. They had the luxury of time, and programs like MacPaint to play around with while they waited. Was there a more pleasant definition of purgatory in existence? Businesses weren't so willing to sit

---

* The so-called 'Fat Mac' with 512K of built-in memory was released in September 1984.

around twiddling their thumbs until the right software packages were designed, or a more powerful upgrade of the system was released. After all, very few managers asked for a spreadsheet to be left on their desk only as soon as their employee's computer was able to run one.

□ □ □

Apple regrouped. Out went the bleak dystopia of Ridley Scott's imagined future, in favour of something altogether more traditional. In that November's election issue of *Newsweek* magazine, the company took the unprecedented step of buying up all forty pages of available advertising space, at a cost of $2.5 million. 'It's unclear whether Apple has an advertising insert in *Newsweek* or whether *Newsweek* has an insert in an Apple brochure,' John Sculley quipped. To snag new customers, the Macintosh needed a new metaphor. They found one in the image of the humble automobile. In one regard, the idea made perfect sense. Apple has always had more in common with luxury-car manufacturers than many of its computer-manufacturing contemporaries. Cars stood out as the ultimate American consumer product; the status symbol that spoke of good, old-fashioned engineering, as well as safe transportation to wherever one needed to go. A thirty-second television spot was commissioned, showing what appeared to be the outline of a high-end sports car covered by a dust sheet. 'It's powerful,' purred the narrator, as the camera prowled over the shape's curves and smooth edges with fetishistic admiration. 'It's responsive. It handles beautifully. And it'll blow the doors off anything else in its class.' With that the cover was whipped away, revealing not the sleek chassis of a Ferrari or Mercedes, but, rather, a computer mouse connected to the Mac. This was ground that John Sculley felt altogether far safer on. 'Automobiles were bought for emotional reasons as much as any practical reason,' he says. There was also the unspoken bonus that nobody thought twice about upgrading their car every few years if they could afford it – whether there was anything wrong with it or not.

To go along with the advertisement, a new publicity campaign was arranged with dealers, entitled 'Test Drive a Macintosh'. The promotional strategy advised people in possession of a credit card to drop into their local retailer and 'borrow' a Macintosh for a period of twenty-four hours. Apple was sure that this was all the time that it would take to create a lasting bond, the likes of which it would be impossible to undo. The first hit would be free. The next, the user would pay for. 'The idea of test-driving a car itself is not a controversial idea,' says Mike Murray. 'It's very acceptable behaviour, whether you might buy the car or not. A person can go into a dealership, whether it's for a Ferrari or a Ford, and say they'd like to test-drive a particular car. The salesperson would say, "sure, let's go", and you'd hop into the car and go for a ride.' Psychologically the campaign was designed to tap into that same sense of no-cost fun of taking a car out for a spin. 'If we hadn't used that phrase and instead had created a campaign called "Take a Mac Home for a Night", no one would have done it,' Murray says. 'It wouldn't have made any sense at all for somebody to go into a computer store and say, "I want to take that computer home for a night." There would be no emotional memories or persuasion or feelings. We had to draw on past cognitive experiences, and the test drive of a car was the obvious metaphor.'

There was one inescapable irony which became apparent if one followed the car metaphor to its logical conclusion. Whereas prior to the Macintosh personal computers followed the hobbyist model of – in car parlance – encouraging one to 'lift the hood' to tinker with its internal workings, here was a computer that did just the opposite: closing the lid to that group of would-be engineers. If a purchaser even so much as unscrewed the case to look at the beautifully designed Macintosh innards, his or her warranty would instantly become null and void. The do-it-yourself transparency that the seventies hobbyists had seen as key to democratising technology was replaced by a transparency of use, which would prove indicative of the direction in which consumer technologies were headed. From this point on, when users claimed that a technology was transparent, they meant that they could pick it up and make it work immediately, not that they knew *how* it worked. It was a postmodernist spin to the medium that emphasised simulation over the 'real', and playfulness over seriousness. The Mac was a computer that begged to be taken at interface value.

Predictably reactions were split. Some found the approach liberating; others terrifying. Unsurprisingly for a machine that cast itself in such grand narrative terms, many users were tempted to discuss the opacity of the computer's interface in social or political terms – seeing it as alternately empowering and disempowering.* None spoke louder than some of the original hackers; men such as Richard Stallman, the software freedom activist who that year founded the Free Software Foundation. '[The Macintosh was] shocking in how restrictive it was in terms of what you could add to it,' Stallman says. 'The hardware was designed in such a way that you couldn't get other peripherals and put them in. As a result people wanting to do that wouldn't use the Macintosh, and maybe this was marketing genius, but I don't admire marketing genius; I don't admire genius that has selfish, greedy pursuits.' He may have been part of a vocal minority, but the accusations that Apple was betraying its libertarian roots would come back to haunt the company.

---

* The grand narrative or 'master narrative' is a term introduced by Jean-François Lyotard in *The Postmodern Condition: A Report on Knowledge*, to describe the narrativising of history in alignment with the predominant institutional and ideological forces of a particular time.

□ □ □

Computer dealers *hated* the 'Test Drive a Macintosh' campaign. The computers were in short supply as it was, and the idea of lending out machines that they could otherwise be selling seemed like a recipe for disaster. Then there was also the fact that many of those who did take a Macintosh home for a day would return it in an unmistakeably worse condition than that in which they had borrowed it. And this without even mentioning the additional paperwork it caused! In an effort, perhaps, to get dealers back on their side, Apple launched another promotion, promising a brand new Porsche 944 to the salesman in the United States who could rack up the most Macintosh sales. The winner was a 22-year-old aspiring vet from the Midwest, named Craig Elliott. Elliott had recently finished a course in animal science and microbiology at Iowa State University. He took a temporary job in sales while he saved money for veterinary school. He saw the competition announced, posted off his total number of sales and then promptly forgot about it. Not long after, he received a letter postmarked Cupertino, asking him to come and have dinner in San Francisco with Steve Jobs and the Mac team. Elliott had been an avid Apple user since the late seventies and was understandably excited about having the chance to meet the group behind the machine – not to mention America's favourite young entrepreneur. He flew to California for the first time in his life, rented a car and drove to the venue, a luxury four-star hotel in San Francisco, with beautiful views of the bay. When Elliott arrived Jobs was nowhere to be seen. He introduced himself to the members of the Macintosh team. When Jobs made his entrance, he did so in a manner that was sure to make the maximum impact. He appeared on the horizon, riding a BMW motorcycle, decked out in white shirt – cuffs flying – blue jeans, and penny loafers with no socks. Jobs brought the bike to a screeching halt, dismounted and surveyed the group. Every person there, Elliott included, was wearing a suit.

'Oh, should I have a tie on?' Jobs asked.

It was the kind of testing question that nobody knew exactly how to respond to.

'It doesn't matter, Steve,' someone ventured. 'Whatever you want.'

Jobs thought about it for a second, then walked over to Mike Murray, reached up, undid his tie, and fastened it around his own neck.

'Now we can eat,' he said.

Dinner was a daunting experience for Elliott. The restaurant at the Stanford Court Hotel was the most upmarket place he had ever eaten in, and Jobs was not the kind of dinner companion to make for a relaxing culinary experience. Tucking into his vegetarian main course, Jobs fired staccato questions at his new prize salesman, quizzing him about every aspect of the Macintosh. *What did customers like? What didn't they like? Which features most convinced them to part with their money?* Throughout the meal Jobs demonstrated his disconcerting habit of staring, unblinking, into the eyes of the person he was speaking to. It was an intense conversational experience, in which Elliot got the sense that not only was everything he said being intently listened to, but also closely analysed at the same time. 'As a guy from an agricultural part of the Midwest, I didn't believe in things like charisma,' he recalls. 'After meeting Steve Jobs ... he was the kind of guy that could lead you right off a cliff. At that moment I understood what it meant when people talked about having charisma. I was just hypnotised.' Eventually the question came up of how many Macintoshes Elliott had had to sell to win the competition. Feeling back on familiar territory, Elliot shot out his winning number.

'That's it?' Jobs snapped back. 'That's pathetic!'

He launched into an impromptu rant, saying that it was ridiculous to go to as much effort as the Macintosh team had if salesmen were only able to sell the kind of number Elliott had mentioned.

'Steve,' Elliott said, looking directly at him. 'I'm your best guy.'

Jobs returned the stare, considered the response, and then said, 'Okay.'

After the bill was settled, Jobs took Elliott outside to present him with his brand-new Porsche. Elliott had heard one of the managers at the table mention that he had an undergraduate degree from Harvard and an MBA from Stanford University. He was confused, though. *Hadn't Steve Jobs – that man's boss – been a college dropout, officially completing only six months of higher education at Reed College?*

'You didn't get any of those degrees?' Elliot asked.

'Exactly,' Jobs responded.

'What do you mean?'

'You don't hire people like yourself. It's always a trap. Apple's strength is bringing together people with different perspectives.'

The two men posed together for a photograph in front of the new car, in which Jobs can be seen directing animatedly from in front of the camera; Silicon Valley's answer to Cecil B. DeMille. He was still wearing Murray's tie, and its scrappy appearance clearly evidences the fact that it was put on in a hurry. After the picture had been taken, Jobs got back on his BMW motorbike. Before he drove off, he took a second to autograph Elliott's Apple II user manual from 1978, which the trainee vet had brought along just in case. 'We share a moment of history,' Jobs wrote and signed his name underneath. Elliott had made a good impression. Not long afterwards he received a second call from Apple, this time offering him a job in the sales team if he would consider dropping his studies to move full-time to Cupertino. He did, and went on to enjoy a lengthy, highly successful career with the company. He is one of the many people whose lives changed course courtesy of the Apple juggernaut.

If there was one market sector where the Macintosh flourished it was, perhaps unsurprisingly, the same arena in which youth protests and anti-computer rhetoric had so effectively spread just twenty years earlier. Universities became hotbeds for Apple product sales, something that continues to this day. From Apple's perspective university students ticked all the boxes when it came to potential Macintosh users. They were young and therefore not yet indoctrinated into the ways of IBM; plus, they had money and a compelling reason to own a computer. Previously Apple had taken a more grassroots approach to selling to students, but with the Macintosh it put in place an overarching strategy. Apple would turn the universities themselves into dealerships, bringing them on board as partners (or, as Apple phrased it, 'co-developers') to sell directly to the students on campus. Enter the Apple University Consortium.

With the Macintosh the strategy called for Apple to sell the computers to universities at the flat price of $1,000 each – less than half the retail price. There was a slight catch: to take up on this offer, the university would need to commit to buying $2 million worth of machines to sell to students over a three-year period. They also wouldn't be able to return them. Jobs' salesman approach was typically direct. Following a lunch with Donald Kennedy, the president of Stanford University, he brought in a Macintosh obscured by a sheet, whipped the cover away and asked, 'How would you like to buy a thousand of these for a million dollars?' There were no free lunches in the Apple University Consortium.

Universities fretted over the decision. They liked to be non-partisan with their support and worried that they might upset other companies, namely IBM, by adopting the Macintosh. IBM, in fact, seemed the logical company to go with. In the early 1980s they were attempting to push a stripped-down machine, called the IBM Junior, on to computer departments. IBM shot themselves in the foot, however. Their contracts for time-sharing computers had been long and complex; stipulating, among other things, that they be used only for university work and not for recreation or enterprise. With a personal computer, monitoring this would prove next to impossible. Step forward Apple with its far simpler contract, written in conjunction with the various university departmental heads.

Those inside Apple still knew that it would prove an uphill battle, but came up with several clever manoeuvres. Knowing that prestigious universities kept an eye on what each other was doing, it was determined that the best way to reach individual establishments was first to get their rival schools on board. On the East Coast, a vivacious Apple product rep named Stacey Bressler proved particularly difficult to shake off. If she was denied a meeting, she would simply turn up at one of the Ivy League schools' computer centres, set up a Macintosh in the lobby and start drawing pictures on it using MacPaint. Before long a crowd would gather and, invariably, a member of staff would arrive. At Dartmouth College, William Arms, the director of the computer centre, was immediately struck by the dream of students owning their own personal computers (Macs, of course!) which could be linked to a central university network. Arms was struck by just how intuitive to use the Macintosh was. After Apple's rep put the machine down on a table, he went off to fetch a cup of coffee. By the time he returned Arms had plugged in the keyboard and mouse, and was moving windows about on the desktop. That was the learning curve on a Mac. Arms kept the coffee mug as a souvenir. Before long every Ivy League school was signed up to the Consortium.

As it turned out universities needn't have worried about being stuck with vast quantities of computers that they couldn't sell. As Apple had predicted, students proved an ideal audience. At Stanford $2.5 million worth of Macs sold within the first two weeks of their going on sale, necessitating a lottery being conducted to see which students would receive their computers first. In the end it took Apple six months to fill orders of those who had paid their $1,000 upfront. On campuses around the country the Macintosh vastly outsold the IBM PC, in some places at a 40:1 ratio. The other computer companies were flummoxed. Feeling outmanoeuvred, IBM set up an around-the-clock strategy team to figure out how to respond. Their conclusion was simple: free IBM PCs or free money. Jobs wasn't perturbed. 'In some cases, they used IBM grant money to buy Macintoshes,' he said.

At California Polytechnic State University in San Luis Obispo, one hundred students protested at the president's office, bringing a petition signed by almost 1,000 fellow students demanding that the institution renegotiate with Apple to gain admittance to the Consortium. Baffled administrators had to explain that Cal Poly had turned down Apple's offer due to state laws stipulating that any such large contracts be open for competitive bidding. Meanwhile, some enterprising students even discovered that, for the difference between the reduced price they had bought their Macintoshes for and the price they could get for selling them to people not in education, a tidy profit could be made. Educators were forced to step in and play the role of big business. 'A student who sells his or her Macintosh is committing academic suicide,' said Drexel's associate vice-president for academic affairs. Universities, it seemed, had been pulled into the Apple nexus after all.

□ □ □

On 24 February 1985 Steve Jobs turned thirty. For anybody this is a milestone, but for a member of the generation taught not to trust anyone over that age it was a particularly potent blow. To all intents and purposes, Jobs was no longer computing's *enfant terrible*. After years of railing against it, he was now the adult supervision. He tried to get Bob Dylan to perform at his birthday party, but was declined, and had to make do with Ella Fitzgerald instead. Several Apple employees put together a 'greatest hits' video showing Jobs over the years. It ended with a Hindu proverb, stating, 'For the first thirty years of your life, you make your habits. For the last thirty years of your life, your habits make you.' The video was set to 'My Back Pages' by Bob Dylan, a song from the artist's 1964 album, *Another Side of Bob Dylan*. Commenting on the song's refrain, in which Dylan describes how he was 'so much older then' and is younger than that now, music critic Robert Shelton has described it as 'an internal dialogue between what [Dylan] once accepted and now doubts'. Jobs, too, was having doubts. At the ripe old age of thirty, was he a firmly established member of the technocracy? 'People get stuck as they get older,' he told one interviewer with more than a hint of veiled sadness. 'Our minds are sort of electrochemical computers. Your thoughts construct patterns like scaffolding in your mind. You are really etching chemical patterns. In most cases, people get stuck in those patterns, just like grooves in a record, and they never get out of them.' He concluded that it was rare that an artist in his thirties or forties was able to achieve anything of any real artistic significance.

One year on from the Macintosh's release, sales were a fraction of what they should have been. The computer had had its early adopters but new customers tailed off as it became apparent just how little software there was available for the machine. The plan to right that wrong was supposed to rely on the so-called Macintosh Office. Macintosh Office essentially boiled down to the idea of being able to connect a group of

Macs to a LaserWriter printer, as well as allowing them to share information using a device called a file server. It wasn't anywhere near as exciting as the launch of the Macintosh the previous January, but Apple had seen such success with its '1984' commercial that the pressure was on to provide a follow-up. Unfortunately, this proved as ill conceived as Macintosh Office ultimately was. With the '1984' commercial the viewer might have picked up on the irony that the television screen in the ad appears as the enemy, when they, too, were sitting zombie-like in front of a television set. At the same time, the spot was framed as a rebellious call to arms that they were being invited to join. Conversely the commercial which ran at the 1985 Super Bowl showed blindfolded businessmen falling to their doom like so many lemmings. Eventually one man removes his blindfold and looks around. He's alone in a world with the Macintosh Office.

'That was clearly a mistake, showing all of your potential customers jumping off a cliff,' says Alan Oppenheimer, an Apple engineer with the company at the time. He was right. So was Debi Coleman, Apple's controller of the Macintosh division, who was so offended by the spot that she quit the company there and then, typing up her resignation on her Macintosh and printing it out on the LaserWriter. 'The day after the Super Bowl, our telephone lines were overloaded with calls from irate people claiming they would never buy an Apple product again,' John Sculley recalled years later. 'They believed the commercial insulted the very people we were trying to court as customers in corporate America.' At one point an Auschwitz survivor phoned Apple to accuse them of using Holocaust imagery to sell personal computers. It was a bad day for all involved.

In March, former PARC employee Ted Kaehler joined Apple as a researcher. His wife, Carol, who sadly passed away in 1991

as the result of complications related to diabetes, had been at Apple for several years already, working on the Macintosh team, writing its official manual. From her, Kaehler felt that he had been adequately prepped to deal with Steve Jobs, a man with a reputation for being both inspiring and impossible to deal with in equal measure. He was wrong. Kaehler had arrived at the company during a period of great upheaval, like the soldier who enlists in peacetime, only to have war break out within his first weeks. Cracks were fast appearing in the John Sculley/Steve Jobs relationship and it was obvious even to newcomers that they were on a major collision course. The initial signs that here was a company in crisis were ominous. At PARC, Kaehler had had his own well-equipped private office, complete with a state-of-the-art Xerox Dorado computer for him to work on. When he started at Apple the Monday morning after leaving PARC, he was shown to a tiny workstation ('a cubicle in a sea of cubicles') with absolutely nothing in it. No computer. No phone. No desk. Not even a chair. By the time he finished work that night, Kaehler had somehow managed to get his hands on a Macintosh XL – a souped-up Macintosh which ran emulated Mac software on Lisa hardware: a last-ditch effort to sell Lisas – but still had nothing to put it on. He set it up on the floor. 'I was really discouraged by the end of the first day,' he says. 'I was thinking, "boy, what have I done?"'

The next shock came when Kaehler's boss – the man who had interviewed him for the post – was given the sack within three weeks of Kaehler arriving. A prickly individual to begin with, Jobs became increasingly abrasive in his final months at Apple as he sensed whatever power he had previously had within the company slipping through his fingers. He would have regular public showdowns with Apple's division managers, which often ended with him expelling from the company whoever had offended him. One of his trademarks was to arrange a meeting and then have his secretary come into the room within the first five minutes, carrying a phone message slip upon which there was clearly nothing written. Jobs would take the memo from her and pretend to read it. If he was interested in the subject of the meeting he would file the message away in a drawer, saying, 'I'll deal with this later'. If the meeting was boring him, he would claim that it was an urgent call to action which required an immediate response. With that he would get up and leave the room – not returning until the people who had bored him had departed. Shortly after Kaehler started at Apple, he and his boss arranged a meeting with Jobs to discuss a proposed machine called the Jonathan, a desktop computer capable of running the Mac OS, DOS and an early version of Microsoft's Windows. 'Apparently the first sentence or two was not interesting enough, because the phone message came in and Steve Jobs got up and left,' Kaehler says. 'That was the end of that.' He was stunned, unsure how to respond.

'Should we wait?' he asked.

It was clear that they shouldn't. The verdict was in: the Jonathan project bored Steve Jobs. The next day Kaehler's boss was fired.

□ □ □

The events that led to Jobs' ousting from the company he had co-founded read like the pages of a John Grisham thriller. For months Jobs sniped to friends and colleagues about Sculley's alleged incompetence, claiming that the CEO had no understanding of technology (failing to see that it was marketing, not technology, that Sculley had been hired for). When Sculley finally caught wind of what was being said he was incensed. He confronted Jobs at a meeting, telling him that this was no way to behave and that he should not pursue this course of action any further. Jobs failed to heed the warning. The next development came in a board meeting on 11 April 1985, when investor and board member Arthur Rock decided that he had heard enough utopian talk about how the sales of the Macintosh were going to turn around at any moment. 'Where is the evidence?' he asked. As it stood, sales were still a good 10 per cent lower than forecast. Rock made two decisions. The first was to cancel the Macintosh XL product line, thus ending the Lisa project for good. The second, designed to give Sculley more operating power within the company, was to remove Jobs as head of the Macintosh division.

The division would eventually be turned over to Jean-Louis Gassée, whose previous experience was successfully running Apple France. Like Jobs, Gassée was a product-marketing guy more than he was an engineer. He was a charismatic presence and fancied himself as that most Gallic of conflations: the superstar philosopher. He had appeared in the pages of *Vogue*, popped up in advertisements for mineral water, and even penned a tome, *The Third Apple*, which in his words was not so much a book as 'an invitation to voyage into a region of the mind where technology and poetry exist side by side, feeding each other'. Gassée could also be something of a renegade. During the Macintosh Office debacle, he had voiced his disapproval by loudly referring to it as the 'Macintosh Orifice' to whoever would listen.

Extricating Jobs from power within Apple was a task that required a degree of precision akin to that required of a brain surgeon. Thanks to the efforts of Regis McKenna, there was no one man more synonymous with personal computing, and Apple in particular, than Steve Jobs. Even with the Macintosh having delivered fewer sales than promised, and Jobs bad-mouthing the CEO at every opportunity, there were still those within the company who believed that he should be kept on. 'People started to panic, and when people start to panic what do they do? They fire the manager,' says Alan Oppenheimer. 'In baseball you fire the manager. In football you fire the coach. It was a traditional response by the Apple board. There was no question that Steve could be unthinking at times. He was hard to work for, he was overbearing and overfocused. He was a good scapegoat.'

Certainly this marked the end of an era at Apple. Not only was Jobs no longer running the Macintosh division, but a number of those who had worked on the project along the way began to drift off into the sunset. More than three years of non-stop work had left many people physically and emotionally drained. 'There were sweatshirts that Burrell Smith had printed up which read, "working 90 hours a week and loving it",' Jerry Manock remembers. 'You can only work that hard for a certain amount of time before you get burned out ... People were tired. I was tired.' He arranged a leave of absence to recharge his batteries, and took his wife and children on an extended trip to Scandinavia. By the time he returned Jobs had lost much of his power within Apple. Manock handed in his resignation and moved with his family to Burlington, Vermont, where house prices were cheaper and people were more likely to be discussing maple-syrup production than denting the universe with a personal computer.

Andy Hertzfeld was also done. 'I have an authority problem,' he says bluntly. 'I don't like people telling me what to do. I regard everyone I work with as my collaborator, not my superior or my inferior. But we got this manager called Bob Belleville, who gave me a poor review. Managers' biggest leverage is once or twice a year they get to write performance reviews for their employees, and he wrote me a very negative review for the period of time I did the best work of my life, creating the Macintosh operating system. He gave me a bad review not because of my technical work, which he said was fine, but because he said I was insubordinate ... He disillusioned me and I decided that I couldn't work for someone like that. I received the bad review in February of 1983. I couldn't quit then because I was too committed to creating the Macintosh, but I told myself that as soon as the Macintosh shipped I would leave. And that's what I did.' Like Manock, Hertzfeld took a leave of absence from Apple. When he was due to return he went to see Steve Jobs, who was still running the Macintosh team at the time, but perhaps beginning to sense that the tide might be turning. 'You don't matter as much as you think you do,' Jobs told him cruelly, projecting some of his personal trauma. Hertzfeld left. So did Bruce Horn, Burrell Smith, Joanna Hoffman and a slew of others.

Meanwhile, Jobs was put in charge of new-product development; finding the next mythical Macintosh in whatever garage it might be holed up in. The idea of keeping his head down and nose clean for a while would have been good enough for most people, but Jobs was not most people. He was mortified at losing power within the company and became alternately abusive and distraught. One evening he phoned up Jay Elliot, the vice-president of human resources. Elliot was at home, hosting a dinner party, when the telephone rang. He let it go through to the answerphone rather than pick it up. In front of all of his dinner guests, Jobs' voice came through sobbing loudly over the speaker. 'I've lost the support of the board, Jay,' he said. 'Help me win them back.' With Mike Murray's aid, Jobs formulated a scheme to buy up 1,000 Apple stores around the country. It was clear that he was looking for something, *anything,* that would give him back a measure of operating power on a par with Sculley's. But then Murray was stripped of his duties as head of Macintosh marketing. On Saturday 11 May 1985, Murray walked into his office to find Gassée there. When Murray asked him what he was doing, Gassée looked baffled. 'There must be some kind of mistake here,' he said. 'I'm taking over from you as head of Macintosh marketing. I start on Monday.' Murray felt like his head was spinning. He rushed to a chair to sit down. Worse was that Jobs had known about this, and had just 'forgotten' to tell his friend what was happening. When Jobs managed to get Murray back on his side, he announced that he was planning to stage a boardroom coup in late May while Sculley was away on a business trip in China, signing an agreement to export Apple computers. However Jobs made the error of revealing his plan to the newly appointed Gassée, who duly passed the information back to Sculley. The business trip to China was cancelled and Sculley – no stranger to boardroom politics – made Apple's board of directors choose between him and Jobs. If they chose Jobs, Sculley said that he would leave. They chose Sculley. Jobs stormed out of the room.

Not long afterwards Jobs was moved out of his office at Apple. The edict had come down from on high; an order from a level of management that no longer communicated regularly with him. One day Jobs' secretary came over to deliver a message. 'They want you to get out in two weeks,' she said, awkwardly. He was posted to a distant office, across the road from the now-vacant Macintosh building. 'I feel like somebody just punched me in the stomach and knocked all my wind out,' he would later recall. 'I'm only thirty years old and I want to have a chance to continue creating things. I know I've got at least one more great computer in me. And Apple is not going to give me a chance to do that.' Few people visited him. Relations with Sculley and the board of directors were frosty.

Jobs bitterly nicknamed his new office 'Siberia'.

PART II

THE NEXT BEST THiNG

9

THiS AiN'T THE SUMMER OF LOVE

'The Summer of Love was plainly drawing
to an end ... The spare-change panhandlers
disappeared from Haight Street along with
the tourists, and it was no longer a struggle
to get down the sidewalks. Fewer grimy,
lost-looking teenagers huddled in doorways
clutching lost-looking puppies or kittens. The
I/Thou Coffee House, whose lines had reached
out the door, was half empty. Haight Street
had the tawdry, exhausted air of a beach town
at the end of summer.'

Charles Perry,
The Haight-Ashbury: A History (1984)

Steve Jobs was miserable in Siberia. There might be one or
two phone calls to make in the morning, perhaps some post
to read, and then ... nothing. For a few weeks he kept himself
in the loop by reading the corporate management reports that
were sent around to everyone of a certain level within the
company, but then someone cottoned on to the fact that he
was receiving them, and even these petered out. After a while

he simply stopped turning up to work. Who was going to call him on it? Aside from a secretary and one security guard, he was all alone in the building. 'There was nobody really there to miss me,' he said. Still, Jobs was too close for comfort for some of the board members. At one point that summer there was serious consideration about sending him into outer space. NASA was approached about including Jobs, a 'global visionary', for its civilian astronaut programme, which was due to launch the following year with the voyage of the *Challenger*. But NASA turned Jobs down, selecting schoolteacher Christa McAuliffe instead, as the first participant in Ronald Reagan's Teacher in Space Project.* Several members of the Macintosh team joked that Jobs would not have liked space travel anyway. He wouldn't have been able to fly first class.

Next Jobs considered the prospect of moving to the Soviet Union and starting up an educational computer company. He travelled to Moscow with Apple's senior vice-president, Al Eisenstat, but became paranoid that a television repairman in his hotel room was actually a Soviet spy and decided to head home. Finally, he settled for taking a holiday and supervising renovations on his house.

Jobs' fall from grace was incredibly public. When he had become a self-made multimillionaire aged twenty-five, more than one journalist had likened him to the wunderkind talent of Orson Welles, the visionary filmmaker who directed his masterpiece, *Citizen Kane*, at that same age. Now the flip side of that coin was all too apparent. After *Kane*, Welles was never again given the freedom to pursue his own vision. In the eyes of many, his career went only in one direction from there: straight

---

* McAuliffe was one of the seven crew members to die in the 28 January 1986 Space Shuttle *Challenger* disaster, which saw the spacecraft disintegrate seventy-three seconds into its flight.

downwards. Would the same happen to Jobs? He had enough money to retire at thirty, as Mike Markkula had planned to do, or he could have been content relegated to the sidelines like Steve Wozniak. But that wasn't Jobs' style. 'I personally, man, want to build things,' he told a journalist from *Newsweek*, sounding once again like the idealistic barefoot hippie who had attended meetings at the Homebrew Club. 'I'm thirty. I'm not ready to be an industry pundit. I got three offers to be a professor during the summer and I told all of the universities that I thought I would be an awful professor. What I'm best at doing is finding a group of talented people and making things with them. I respect the direction that Apple is going in. But for me personally, you know, I want to make things. And if there's no place for me to make things [at Apple], then I'll do what I did twice before. I'll make my own place.'

Over the summer Jobs had come up with an idea. Some of his soul-searching days had been spent wandering the deserted Stanford University campus. The Macintosh was still selling at a rate of knots among the university crowd, and education had always been an area that Jobs was interested in. In early September he had lunch with the Stanford biochemist Paul Berg. Such was the world of the Apple co-founder that, in his own words, all he had to do was phone up the Nobel Laureate and say, 'You remember me, I'm ignorant about this stuff, but I've got a bunch of questions about how it works, and I'd love to have lunch with you.' Berg talked Jobs through the basics of the gene-repairing work that he was engaged in. Jobs began to get excited. *What if he could create the kind of powerful computer that would allow the experiments that had been taking two, three weeks to carry out in a wet laboratory, to be simulated in a fraction of the time?* The way Jobs told the tale, at this point the 59-year-old Nobel Laureate's eyes 'lit up'. The marginalised Apple co-founder left the lunch feeling elated.

Now all he needed was a team to help him.

As he had done with the Macintosh, Jobs found his cast of players among the brilliant but disenfranchised employees at Apple. First of all he spoke with Dan'l Lewin, a strapping six-foot-two, square-jawed marketer who had been a competitive swimmer before joining the company as head of the Apple University Consortium. When Jobs phoned Lewin, he found the normally amiable marketer griping about Apple. A recent company reorganisation had just about killed the University Consortium, and now Lewin had been taken out of college-specific sales and put in charge of marketing for the whole of education. He was feeling burned out and ready for a change.

For hardware, Jobs spoke with Rich Page and George Crow. Page was a systems wizard who had been instrumental in working on the Lisa project. At that time he had been at odds with Jobs, accusing him of trying to destroy the company by putting out the Macintosh so soon after the Lisa. Three years on, Page and Jobs found common ground. Page had been working on the so-called 'Big Mac' project. The Big Mac would be Apple's version of the mythical '3M' computer, offering a million-pixel display, a million bytes of memory, and be capable of running a million instructions per second. Page's prototype was almost ready to ship when Jean-Louis Gassée took over the Macintosh division.

Gassée disliked the Big Mac, which, despite being considerably more powerful than most other personal computers at the time, would remain essentially closed in the same way as the original Macintosh was. Gassée was a believer in 'slots', the expansion ports that would allow peripherals to plug into the machine, thus opening it up. Jobs had considered slots messy and overly complex, but Gassée vehemently disagreed. That was the official reason for his dislike of the Big Mac. The unofficial reason was that Gassée was busy establishing himself as the new Macintosh overlord. When a new alpha-male bear takes over a sleuth, the first thing it does is to cull those cubs which bear the genetic traits of its predecessor. In Gassée's case, this meant cancelling various projects. 'Jean-Louis associated my prototype with me and with Steve,' Page recalls. 'And since Gassée had just taken over the division he didn't want anything going through the door within six to twelve months that looked like it came from Steve. So he killed it.' Page was incensed at the death of the Big Mac, and more than happy to speak with Jobs about his start-up idea.

Crow had also been less than impressed by Gassée. 'I liked Jean-Louis personally, but he was the exact opposite of Steve,' Crow says. 'With Steve you knew on a minute-by-minute basis just where you stood with him, whereas Jean-Louis played his cards close to his chest.' Crow was working on a separate project which was being buffeted between teams with each company shake-up. When Crow went to see Gassée, he found that Gassée wouldn't speak with him. Nor did he phone him later that evening. 'After about six months of this, I called Steve and asked him whether he was going to do anything, because I was really ready to go and do something else. He said, "As a matter of fact, yes."'

The last two people Jobs spoke with were both veterans of the Macintosh project: user-interface expert Bud Tribble who had returned to the company after leaving, and financial controller Susan Barnes.

The group met for dinner at Jobs' luxurious – if slightly dilapidated – Woodside mansion, which he had bought the previous year from the Giannini family, the founders of the Bank of America. It was a unique property, made even more so by the fact that its current tenant had next to no furniture. Nor did Jobs heat the house, which stayed a chilly 15°C all year round. 'Until [I arrived and] saw the people in the room I had no idea who else Steve had been talking to,' Crow recalls. The group spent an evening eating vegetarian pizza, drinking Californian wine and compiling a laundry list of complaints about Apple. *What a mess*, everyone agreed. Jobs talked the five through his idea for a new computer company. It would be a risk, of course, but more than one of them had already talked privately to him about tendering their resignation. This would be an opportunity for a fresh start, in an area of business that had already proven to be successful. They just had to trust him. Then they had to tune in, turn on, and drop out of Apple.

□ □ □

On 25 June 1985, Bill Gates and Microsoft's marketing-applications director Jeff Raikes had sent a confidential memo to Apple. Gates and Jobs were long-time frenemies: sometimes friends, sometimes enemies, always rivals. Both had been born the same year, both were college dropouts, and both had been hurled into the world of computing following the January 1975 *Popular Electronics* cover story announcing the Altair 8800. Later on both men had seen the squandered value inherent in Xerox PARC, which Gates described as being akin to a rich neighbour whose house Gates broke into to steal the television set, only to find that Jobs had taken it already. In the same way that Apple had recruited numerous PARC alumni to create the Lisa, so Gates hired PARC's brilliant Hungarian computer scientist Charles Simonyi, the man responsible for Bravo, the WYSIWYG document editor which later became Microsoft Word. Despite Jobs' comments that Gates would 'be a broader guy if he had dropped acid once, or gone off to an ashram when he was younger', even this wasn't a notable difference. While Gates certainly didn't spend 1973 trekking around India, he wasn't immune to the charms of the counterculture. The Bill Gates on acid story – unconfirmed in an interview with *Playboy*, but not denied either – tells of him staring at a table, thinking that the corner was going to plunge into his eye.

Had Gates sent his memo six months earlier it would have been addressed, in part, to Jobs. As it was, Jobs was then exiled in Siberia and Gates's letter was marked for the attention of John Sculley and Jean-Louis Gassée. Both Gates and Jeff Raikes were more than familiar with Apple. Gates had developed for the Macintosh, while Raikes had actually been employed by the Cupertino company in 1980, before he was lured away to work at Microsoft. Gates was a firm believer in the Macintosh operating system. 'The next generation of interesting software will be done on Macintosh, not the IBM PC,' he told *BusinessWeek* in November 1984. As far as he was concerned, however, the future of personal computers lay in software, not hardware. If only he could get in at the ground floor with an operating system that would become the standard across the board, there could be no stopping him.

'Apple's stated position in personal computers is innovative technology leader,' Gates noted in his letter. 'This position implies that Apple must create a standard on new, advanced technology ... Apple must make Macintosh a standard. But no personal computer company, not even IBM, can create a standard without independent support ... The industry has reached the point where it is now impossible for Apple to create a standard out of their innovative technology without support from, and the resulting credibility of other personal computer manufacturers. Thus, Apple must open the Macintosh architecture to have the independent support required to gain momentum and establish a standard.'

When five weeks had passed and Gates had yet to hear anything back from a representative of Apple he wrote again. This time he mentioned the names of three companies which would be receptive to embracing the Macintosh operating system, as well as reinforcing his eagerness to be part of the deal in whatever capacity. 'I want to help in any way I can with the licensing,' he said. 'Please give me a call.' Nobody did. Gassée felt that allowing others to create their own versions of the Macintosh would just water down the exciting work that he himself was doing. As for Sculley; 'I saw no way that we could license the Macintosh without destroying the company,' he says. Gates was beside himself. He couldn't believe that Apple was willing to miss out on what was surely going to be the future of the computer industry. That November he released his own artless take on a graphical user interface-based operating system. Lacking a better name he called it 'Windows'.

'Board meetings at Apple were supposed to be held on the second Friday of every third month,' Rich Page recalls. 'If you had applied that formula to this board meeting it would have fallen on a Friday 13th. Now I don't know who it was, but someone decided to move the board meeting to Thursday 12th instead.' On 12 December 1985, Steve Jobs announced that he was planning to start up a new company. 'Don't worry,' he said. 'This isn't going to be anything to compete with Apple. I'm looking to take a few low-level people to build a high-end machine targeted at the education market.' The board was surprisingly happy with the suggestion. Jobs would be out of their hair, and all Apple would lose were a couple of entry-level coders the company could likely live without. At the time, the subject of who these 'low-level' people were wasn't broached. 'I don't know whether he didn't say or they didn't ask who he was taking, but it seems kind of naive to assume that they wouldn't be fairly senior people, right?' Page says. 'You could have looked at a list of people who were offended by the whole transition happening at the time and figured out who they probably were if you had half a brain.' The next day, Jobs wrote a note to John Sculley, listing the group he was referring to. That morning Lewin, Page, Crow, Tribble and Barnes drove in to work and tendered their resignations. The response wasn't good. It was anything but. 'Apple went berserk,' Crow recalls. 'These are not low-level people,' Sculley raged, and insisted that the five defectors be escorted off Apple property.

By the next week the company had filed a lawsuit, casting Jobs as a moustache-twiddling villain, engaged in a 'nefarious scheme'. This was something of an overreaction. How could it be that Jobs – considered by the board a generally useless manager while at Apple – was now perceived as invaluable? In short, it was a knee-jerk reaction; the response of a jilted lover more than that of a billion-dollar company with a genuine grievance. Jobs, too, responded with a startling lack of foresight. He sold off all his Apple stock immediately for more than $100 million, save for one share which would entitle him to the annual shareholder's report. (Given that the stock was depressed at the time – trading at just $18 a share – and kept rising in his absence, Jobs left almost $700 million on the table, which is what he would have made had he just hung on to it through 1987.) While the lawsuit progressed, Jobs and Dan'l Lewin used their time to travel up and down the country, visiting different universities, staying at some for up to a week. 'What are you looking for in your ideal computer?' they would ask. Some of the answers were very specific. Others could be maddeningly vague: 'I need to make a program for my class next Tuesday and to write that program on any other computer will take me four to five months and I don't have four to five months. And I don't have the money to pay somebody for four to five months. So I need it by next Tuesday and a grad student and I have to do it and we can invest about two days in it.' Okay, Jobs and Lewin would extrapolate, so you need a computer that's *fast* and *powerful*.

Before long, Apple withdrew its suit, realising that suing one of the company's co-founders looked to the rest of the world like high-tech patricide. What proved altogether more difficult was calling off Apple's lawyer. The company had hired a firm called Brown & Bain, which specialised in computer-industry lawsuits. Its principal partner was a 58-year-old veteran litigator named Jack Brown, who Silicon Valley legend had it, had never lost a case. When Apple said they were no longer suing Jobs, Brown had already smelled blood and was all limbered up and ready to rip his adversary to pieces. Stopping him was easier said than done. 'Jack was like a bulldog,' recalls Rich Page, who had also been named in the lawsuit. 'He was going to tear this thing to the ground like a pork chop and kill it ... He wanted his record to be forty-seven wins and no losses.' Eventually Brown was convinced to lay off, and Jobs agreed to a six-month moratorium on hiring any more Apple staff – senior or otherwise. At last Jobs was finally free to start his company.

In his book *Crossing the Chasm: Marketing and Selling High-Tech Products to Mainstream Customers*, Silicon Valley consultant Geoffrey Moore refers to the bell curve of technology adoption. Essentially this is a graph that rises and falls like a hill. It is divided into five different categories: innovators, early adopters, early majority, late majority and laggards. Between each category of potential customer is a small gap where, should fatal mistakes be made, consumers might fall away altogether and never return. By far the biggest of these gaps is between the early adopters and the early majority. It represents the difference between having a product taken up by a select few intrepid pioneers and one which, at least in baby steps, begins to enter the mainstream. By the end of 1984, the Macintosh had its early adopters. As noted, many of these had bought the Macintosh on little more than the promise that it represented. But promise is not enough to get the majority of people interested. Once this first wave of early adopters had received their Macs and were happily using them at home, sales declined. As had been the case with the Apple II and VisiCalc, what was desperately needed was that killer app so important that a certain targeted niche of people simply could not live without it. Apple found what it was looking for in a desktop publishing program called PageMaker.

PageMaker was the dream of a man named Paul Brainerd. From the look of him, Brainerd seemed to be no businessman. A gangly man with unruly blond hair, he appeared the classic example of the computer geek who stumbles across a world-changing idea and implements it before anyone else has the chance. Brainerd had previously been a newspaper editor, before joining Atex, a company which produced the high-end workstations that allowed suitably trained professionals to digitally lay out display adverts for newspapers and magazines. At the time this was revolutionary. Before companies like Atex, pages for print had to be laboriously pieced together manually; headlines cut out with scissors and stuck to paper using hot wax to keep them in place. Brainerd, however, saw that this process could be taken one step further, if only someone could write the appropriate bit of software. The month the Macintosh launched, he happened to find himself out of a job. Kodak had bought out Atex and subsequently shut down its Seattle outfit. Brainerd recruited four of his former colleagues and the group set about working in a new field which Brainerd dubbed 'desktop publishing'. They called themselves Aldus, after the fifteenth-century Venetian printer Aldus Manutius, who had invented the italic typeface and the semicolon. Brainerd put in $100,000 of his own money to help get them started and the four engineers agreed to work for half their previous salaries. The plan was to spend five years building the company up, sell for big money, and then quit the rat race. Everyone at Aldus would get rich.

With its WYSIWYG capabilities, the Macintosh seemed to be the perfect fit for desktop publishing. What made it even more so was the existence of the Mac's LaserWriter, a laser printer which, at $7,000 was more than $20,000 cheaper than its closest competition. Housing a logic board designed by none other than Burrell Smith, the LaserWriter had been a product of Steve Jobs' last burst of creativity at Apple before he was stripped of power. Brainerd's desktop-publishing program was the ideal supplement to Apple's printer. For the first time, users at home would have quick and easy access to the kind of technology that had previously only been available to publishing's elite. PageMaker's metaphor was the simple pasteboard used by editors around the world, only without the inherent messiness of real cut and paste. Experimentation was the name of the game. Without the cost and time delays of traditional paste-ups, users could – if they so desired – publish as many virtual layouts as they wanted before printing a single one.

The first version of PageMaker shipped in 1985, initially available only for the Mac. Its phenomenal success reinforced the computer's image of being the only logical machine for the graphics professional. Brainerd got rich, too, although Aldus didn't prove to be the 'all for one' partnership he had initially pitched it as. When the group met at the courthouse to sign the company's incorporation papers, the engineers discovered that Brainerd had given himself one million shares of stock to their 27,000. They refused to sign until their holdings were doubled. Brainerd got 90 per cent of the company. Years later he sold it to Adobe.

□ □ □

It should have been easier this time. Jobs was no longer the scruffy 21-year-old college dropout, but a seasoned thirty-year-old

Silicon Valley veteran, who had graced the cover of almost every major publication in the world. Instead of the garage of his parents' house, he was now working out of his 17,250-square-foot, fourteen-bedroom mansion. Jobs' surroundings had changed, but he was as iconoclastic as ever in his desire to have his team be taken seriously as Silicon Valley's newest rebel faction in the face of the monolithic corporation. 'We will never have as many ad dollars as the next person,' he said. 'We'll never have as big of an R&D budget. We'll never have as many salespersons. But we can *outthink* them. We can have the best products and we can have the best manufacturing and we can have the best strategy. Because that doesn't depend on scale, it depends on people, commitment and hard work.'

Only the unfeeling corporation had now changed. Instead of IBM, Jobs' anger was targeted at one company, and one person, in particular. 'Screw John Sculley,' he told anyone who would listen. 'Screw the Apple board!' Just months earlier he had been sadly accepting the fact that his day in the sun was over; that he was to be consigned to the Silicon Valley scrapheap alongside all those old Apple IIIs and other assorted bits of computing history that no one ever talked about any more. Now he was back in the proverbial garage – and it was tougher than he had remembered.

'Boy, I had forgotten how much work it actually is to start a company,' Jobs complained. 'It's a lot of work! You've got to do everything. You've got to come up with a name. You've got to come up with a logo ... You've got to open bank accounts. You've got to set up charts and general ledgers. You've got to set up management information systems. You've got to get a little kitchen set up – get a coffee maker. All this stuff!' Although he already had one multimillion-dollar company under his belt, on the last go-around Jobs had had Mike Markkula to ask for advice, and first Mike Scott and then John Sculley to actually run things. This time he was flying solo, learning on the job. Of course, being Steve Jobs meant that although he might have been doing a lot by himself, he didn't have to do things in the same way that everyone else did. When it came time to open a business bank account, for example, he simply picked up the phone, dialled the Bank of America and asked to speak with its president. Few other people could have got away with such behaviour, but few other people were the closest thing that Silicon Valley had yet produced to a Hollywood superstar. Sure enough, Jobs was invited in for an appointment. For a company name he had settled on Next, a moniker pregnant with anticipation. *What would Steve Jobs do ...?*

□ □ □

John Sculley was brilliant when it came to the marketing side of Apple. At the launch of the Apple IIc, the fourth model in the Apple II product line and the company's first truly portable computer, he debuted the computer in an audacious marketing stunt that would have impressed even a perfectionist like Steve Jobs. 'There were about two hundred of us in this auditorium in San Francisco,' remembers Bruce Tognazzini. 'We were seated in every third seat, with members of the press seated amongst us, unaware of who we were. Sculley unveiled the Apple IIc – which was about twice the size of a modern laptop – and held it up. He said, "This is the Apple IIc" and then looked around and asked, "Can everybody see it?" When the answer was no, he said, "Can anybody help me out?" and at that moment we reached under our seats, pulled out a bag containing the Apple IIc and handed it to the person next to us. There was just this audible gasp in the auditorium.'

Few on Wall Street had been sad to see Steve Jobs go. Although the stock remained down thanks to Apple's disastrous performance the previous year, it rose by $1 with the news that the disruptive Jobs was no longer with the company. Inside

Apple it proved a different matter. Sculley didn't need to catch sight of one of the surreptitious protest t-shirts ('WE WANT OUR JOBS BACK') to be aware of the whispered accusations about the East Coast executive, who didn't really understand technology, who had come in and booted out the company's visionary co-founder. 'I never had any difficulty when Steve was there because it was quite clear we were partners,' he says, of maintaining order within Apple in the days post-Jobs. 'It was really difficult after he left because there were a lot of bad feelings that there had been a showdown between us, when actually that wasn't the reality. The result was that I needed some way to inspire the people at Apple that we were still going to be an innovative company.'*

At a three-day Apple World conference in early 1986, Sculley laid out his manifesto for how Apple would move forward. He started by speaking about Apple's lack of a historical relationship with corporate America, about how Apple was a company with a soul, and how new start-ups eager to follow the Apple example should never be afraid, in words echoing Don Quixote in the musical *Man of La Mancha*, to 'dream the impossible dream'. In particular he cited the story of Apple's factory workers in Carrollton, Texas, where the two-millionth Apple II had rolled off the production line not much more than a year earlier. The factory had been shut down, as Apple's manufacturing was relocated to Singapore, but according to Sculley the employees had been so touched simply to have had the chance to work on the product that they turned up to work one last time wearing 'Apple II Forever' t-shirts. *Now, that was dedication!* 'My goal is to have Apple be the most exciting corporation in the world,'

---

* The truth is that there were bad feelings – at least on Steve Jobs' part. Sculley would remain cordial over the years, but Jobs never forgave him for driving him out of Apple.

Sculley said. 'Why is "exciting" so important? It's important because the things that Apple wants to do require passion – not just someone to come in and get a paycheque and perform forty hours of work. It requires a genuine passion to want to change the world.'

Sculley couldn't change the world on his own. He knew marketing like the back of his hand, but he relied on others when it came to technological advice. With Jobs out of the picture, he turned to Alan Kay, who had jumped ship to Apple the previous year after a three-year stint working for Atari. Kay had been made an Apple Fellow, the highest award the company could bestow on one of its scientists. One day he came to Sculley with a warning.

'Next time we won't have Xerox,' he said.

'What do you mean?' Sculley asked.

Kay talked Sculley through how much of Apple's technology had originally come from Xerox PARC.

'What can we do about it?' Sculley said.

Kay explained his personal theory that every idea takes around two decades to filter through from invention to commercialisation. 'The next great ideas are out there percolating right now in laboratories all over the world,' he said.

'He was my don, my guide,' recalls Sculley. 'He took me around and we got into labs and saw things that were remarkable. From that, Alan started to define what a personal computer might look like in twenty years and I came up with the idea to give it a name. I called it Knowledge Navigator.' The Knowledge Navigator was, essentially, a conflation of all the best ideas Sculley had seen. Its purpose was not to create a product that would make its way to the market within even the next several years, but, rather, to offer a tantalising glimpse of the long-term prospects of computing. After all, if people thought whatever Jobs was doing over at Next was impressive, wait until they could mentally process the notion of a miniaturised portable computer, as flat as a tea tray, which users could speak to in order to retrieve whatever information they so desired. An arbitrary date was decided when all this might come to fruition. The sixteenth of September 2011 seemed as good as any. To illustrate the concept, Sculley authorised a $60,000 budget and a six-week production schedule for the shooting of a short science-fiction video. 'I said "I wonder how we can inspire people with something that doesn't exist yet, so I went to George Lucas and asked if there was a way we could build a special-effects video that would create the illusion of reality",' Sculley recalls. To inspire them, the filmmaking team assigned the job of creating the Knowledge Navigator video read William Gibson's *Neuromancer*, along with Stewart Brand's newest book, *The Media Lab*, in which Brand discussed his hopes for a future filled with personalised newspapers and life-sized holograms. For production purposes, the Knowledge Navigator itself was made out of painted wood. 'It ended up on the cover of *Fortune* and various other places,' Sculley says. 'It looks remarkably like a product called the iPad. It wasn't that I knew anything about how to build an iPad, it was that I was using my experience of marketing as a way to create a vision that would [demonstrate] ... that Apple could still be an innovative company, even after the brilliant co-founder had left. And it worked pretty well.'*

☐ ☐ ☐

What said 'real business' more than a good company logo? Never mind that the iconic Apple logo had been created by a

---

* If the form factor itself was considered the future of computing within the cutting-edge developer community, Sculley proved remarkably accurate with his projected date when all of this might come to fruition. When, on 4 October 2011, Apple announced that Siri would be a key component of iOS 5 and the iPhone 4S, Sculley's 24-year-old prediction for a natural-language voice assistant that would be built into a touchscreen Apple device was less than a month off.

junior art director on a day rate, this time Jobs was dead set on doing business properly. Several top graphic designers came in to present the concepts they had come up with. 'Out of the fifteen that we saw there were a couple that were good and would have cost, maybe, a couple of thousand dollars,' recalls Rich Page. Not good enough for Jobs, it turned out. Jobs didn't just want a top graphic designer. He wanted to hire the best graphic designer in the world. He asked around and several people mentioned Paul Rand. Rand was a 72-year-old Yale professor, who had created the corporate identities for companies including ABC, UBS and, ironically enough, IBM. Susan Kare and David Kelley (the latter of whom had split from Dean Hovey and was now running an outfit called David Kelley Design) worked to convince Rand that Jobs was a client worth taking. 'Rand didn't care about computers at all, but he was intrigued by Steve's character and how he wanted to do everything to the hilt,' Kelley recalls. 'It was like two strong forces meeting and he liked that challenge.'

Rand might have enjoyed his dealings with Jobs, but he also carried the somewhat steep asking price of $100,000, which was to be paid in advance and was for one design only. If Jobs didn't like it, too bad – find someone else. Since Rand was working as a consultant for IBM, he told Jobs that he would have to seek permission from the East Coast giant if he was to avail himself of Rand's services. Had the shoe been on the other foot, Jobs would never have let one of his enemies use the same graphic designer as he did. Nonetheless, he put in a call to IBM and found ... that they accepted. It was the beginning of a somewhat unexpected partnership with the company Jobs had previously denigrated. Jobs also agreed to Rand's terms. The designer went away to his studio and worked for two weeks, coming up with the image of a black cube, tilted at a 28-degree angle, with coloured letters in orange, yellow, green and purple spelling out the company's name. As a graphical flourish, Rand had written the word in capitals, with the exception of a consciously lower-case 'e'. This, he explained in an accompanying pamphlet, could stand for '$e$ducation, $e$xcellence, $e$xpertise, $e$xceptional, $e$xcitement, e=mc².' Naturally, '$e$xpensive' was not one of the options cited. 'Don't get scared, this is not the design,' Rand joked as he handed out copies of the pamphlet, with its plain white cover, to members of the NeXT team at the company's first retreat. 'I think that's what floored Steve when he saw it and figured, "Jeez, a hundred thousand bucks down the drain".' Nonetheless NeXT employees were appalled at the wanton expenditure. From this point on – when Jobs wasn't around to overhear – they jokingly adopted 'milli-logos' as units of cost, so that when George Crow, for example, bought an oscilloscope he might report it as costing 800 milli-logos.

Whether intentional or otherwise, Rand's design bore a striking resemble to Robert Indiana's 1964 image 'LOVE' (which similarly features two stacked letters and an artistic usage of jaunty angles), which was later put on to the eight-cent US postage stamp in 1973. This was a company that was all about getting back to Jobs' roots. To do this he decided to make the company as flat as possible in structure, avoiding any form of unnecessary hierarchy. Business cards, for instance, would make no mention of specific job titles; just areas of the business, such as engineering or marketing. The idea was to create a commune-like environment in which NeXT *members* (not employees) would work together for the betterment of the *community* (not company). This notion extended to salaries, which were divided simply into 'founders', 'senior staff' and 'everyone else'. Founders received a starting salary of $100,000. Senior staff would make $75,000. Everyone else got $50,000. This being Silicon Valley the figures were, naturally, compounded by stock options and a bevy of other perks. NeXT employees received free health-club membership, counselling services, emergency loans of up to $5,000, low-interest loans for buying houses in the area, and health insurance covering both married and unmarried couples. Jobs also insisted that he would tell every company employee exactly what was going on behind the scenes, just so long as this was not shared with outside sources. Employees were able to access the financial records for salaries, so that they could check up on what others were making. It was going to be a regime free from the kind of petty squabbles that normally plagued companies. If there was any doubt as to whose regime it was, however, those were answered at the company retreat when Jobs addressed the five co-founders, standing in front of a whiteboard upon which was written 'The Sayings of Chairman Jobs'. This was the role that Jobs saw himself playing. Like Chairman Mao, he was a revolutionary, a guerrilla-warfare strategist, a philosopher and the man who was going to drag his empire – in this case, the personal-computer industry – kicking and screaming into the new century. Whether people liked it or not.

With $100,000 logos as the order of the day, NeXT proved to have a fearsome burn rate. The cash sailed out of Jobs' bank account like there was no tomorrow. By the end of 1986 NeXT was out of money. Attempts at cost-cutting didn't go well. On one occasion Susan Barnes, the company's financial controller, insisted that Jobs and the five other NeXT co-founders fly standard rather than first class to a national higher-education conference. 'Aren't you embarrassed to serve such shitty food?' Jobs loudly complained to the flight attendant as he sat in his cramped seat. Later one of the cabin crew mentioned to Dan'l Lewin that they had heard that the famous Steve Jobs was on the plane, and asked whether he was able to point him out. Lewin did; prompting the response, 'That's what we were afraid of.'

Jobs could, of course, have ploughed more of his own money into the company, but what sense did that make? Early on he had resisted the temptation to involve venture capitalists in the enterprise, but now he revised that stance and agreed to sell a certain share in NeXT in exchange for a significant investment. First of all he needed to discover what the company was worth. With only a handful of employees and no product on the market, this was easier said than done. Although he had only spent $7 million on the venture so far, Jobs somehow magicked up a valuation of $30 million. This had only the most tenuous connection to reality, although it carried with it the ego-boosting property of making NeXT the company with the highest valuation prior to going public. Still, the offer Jobs made to would-be investors was distinctly unappealing. To get only a 10 per cent share, a potential investor would have to put in almost half the money spent thus far. *Why would anyone do that?* As it turned out, very few were willing to.

Help came from an unlikely source. At the end of 1986 a documentary had aired nationally entitled *The Entrepreneurs*, talking about the plight of Steve Jobs as he grew a new computer company from the ground up. Indeed the first appearance of Jobs in the film showed him literally rooting around in a garden; a figurative return to the orchard commune in Oregon which he had left to start Apple. Symbolically this time Jobs wasn't picking apples, but pulling up carrots. *The Entrepreneurs* was well produced, but seemed to do NeXT few favours. Watching the team sit around talking about a computer which never gets built made it resemble something of a high-tech *Waiting for Godot*. The *New York Times* was stoically unimpressed. 'Its heroes of American capitalism no longer build better mousetraps; they talk and talk and hold meetings,' read its review of the show. At least one viewer was interested, however, and he proved to be the right one. Ross Perot was a 56-year-old businessman from Texarkana, Texas. After leaving the navy as a young man in 1957, Perot had joined IBM, quickly rising to the position of top salesman by filling his year's quota in just two weeks. From there he started his own Dallas-based data-processing company, Electronic Data Systems, which quickly picked up some lucrative government contracts for computerising Medicare records. When EDS went public in 1968, *Fortune* magazine named Perot the 'fastest, richest Texan ever'. Like Jobs, Perot continued to act like a scrappy underdog in a world of corporate giants long after he had made it big, referring to Electronic Data Systems as 'ratty little EDS' even when it carried a billion-dollar valuation. As an antigovernment crusader who had made his fortune on government contracts, he was also a similarly Jobsian mish-mash of contradictions.*

But Perot had money, and plenty of it. He watched *The Entrepreneurs* and the following day called the NeXT offices, telling Jobs, 'If you ever need an investor, call me.' Jobs waited a full week and then phoned him back. Perot duly sent three of his best men out to report back on what NeXT was doing. Jobs was canny. He had approached EDS to become an investor in Apple several years earlier and never heard back from them. This time he refused to speak with anyone except Perot. There was good reason for this: any amount of due diligence or scrutiny of its financial records would show that NeXT wasn't worth close to what Jobs was claiming. Perot could have balked, but he liked Jobs' *mano a mano* attitude to business and agreed to come out to California to meet him. Jobs gave him a personal tour of the automated factory in Fremont where he was planning to put the

---

* Despite EDS receiving $36 million from the federal government between 1966 and 1971, the self-made myth surrounding Perot subsists to this day. (By comparison, EDS's closest competitor, Applied Systems Development Corporation, received just $275,000 during this same period.) Even Perot's profile on Forbes.com describes the source of his wealth – an estimated net worth of $3.4 billion – as 'computer, real estate, *self-made* [emphasis added]'.

NeXT Computer into production. 'The entire factory looked like something out of a science-fiction movie,' says NeXT's former director of materials, Matt Medeiros. 'It was all done to a Steve Jobs spec and not to an equipment manufacturer's spec.' At one point Jobs seemed about to blow the deal when he chose this as the moment to begin screaming at one of the NeXT employees over some minor transgression. Perot looked on, strangely unfazed, then turned to the person next to him and said, 'I used to be like that when I was his age, but then I learned you catch more flies with honey.' The tour proceeded.

There was something else that Jobs needed to mention. NeXT's valuation had mysteriously jumped up again. Had he previously said $30 million? *Well, there was that juice maker and that new whiteboard and ...* The point was it was now worth $100 million. Perot didn't care. He liked Jobs and presented him with what amounted to a blank cheque. 'I pick the jockeys, and the jockeys pick the horses and ride 'em,' he told him. 'You guys are the ones I'm betting on, so you figure it out.' It was almost too good to be true. In the event, the jockeys got $20 million, in exchange for Perot owning 16 per cent of the company.

'Boy, if we've got this much from Ross, maybe if things go well we can get more later on,' said one NeXT employee, turning to Jobs.

As far as Perot was concerned, however, NeXT was a sure thing. He had missed out on a golden opportunity to purchase Microsoft from a young Bill Gates back in 1979. Gates recalls Perot offering him between $6 million and $15 million; Perot claims that Gates wanted $40 million to $60 million. Either way, he had missed out on the bargain of a lifetime. When Microsoft went public in 1986 its market valuation quickly passed the $1 billion mark. Investing in NeXT was like turning back the clock. If anyone was going to outdo Gates it was Steve Jobs, and Perot trusted him implicitly. 'In terms of a start-up company, it's the one that carries the least risk of any I've seen in twenty-five years in the computer industry,' he claimed at the time.*

As had been the case with John Sculley, Perot was beyond impressed by Steve Jobs. Addressing the National Press Club several years later he painted the NeXT founder not as the acid-dropping Silicon Valley bad boy, but as the personification of All-American youth that owed more to the white picket-fenced vision of America in the 1950s. He described him as 'a young man, so bright they let him sit in the engineering classes at Stanford in high school, so poor he couldn't afford to go to college – working in his garage at night, playing with computers chips, which was his hobby, his dad came in one day – his dad looks like a character out of a Norman Rockwell painting – and said, "Steve, either make something you can sell or go get a job." Sixty days later, in a wooden box that his dad made for him, the first Apple computer was created. And this high-school graduate literally changed the world.' It was a remarkably cleaned-up version of the Apple Creation Myth (so cleaned up, in fact, that Wozniak's part in it had been entirely scrubbed from existence), but also demonstrated just how totally Perot had bought into the Steve Jobs mythos. Not long before he had also reiterated the Macintosh ideology, now firmly centred on Jobs, during an interview for *Life* magazine. 'This young man went up against IBM in a capital-intensive business and ate 'em alive,' he said, every inch the proud Texan father.

Perot's attachment brought another benefit for NeXT. With

---

* An amusing thing happened the day after news broke about Perot's involvement with NeXT. On that particular day several NeXT employees, vice-president Rich Page among them, arrived at work before the receptionist turned up. The phones were already ringing. Page answered to find a reporter from the *Detroit Free Press*. 'Is it true that Ross Perot has invested $20 million in NeXT?' was their first question. Page confirmed that it was indeed true. 'And who is Steve Jobs?' was their second. No one dared to tell Jobs that there was a media outlet unfamiliar with him, but the employees got a kick out of it.

some adult supervision now firmly on display, two of the larger education establishments felt safe to climb aboard. Carnegie Mellon and Stanford University each put in $660,000 for a respective 0.5 per cent stake in NeXT. It was a tiny investment for a company that had guzzled $7 million in one year alone, but it also suggested faith from the Establishment. For the first time since its founding, NeXT appeared to be a company built on firm ground. And Jobs owned 63 per cent of it.

□ □ □

On 6 February 1987 Steve Wozniak ended his full-time employment with Apple. After promoting the Us Festivals he had rejoined the company as an engineer with the aim of continuing to work on the Apple II product line, but found that, despite the fact that the division continued to bring in money, it was treated as an afterthought. 'Apple's direction has been horrendously wrong for five years,' he was quoted as saying by the *Wall Street Journal*. Wozniak's new plan was to develop a universal remote control which would be able to operate everything from the VCR to the hi-fi system. He rented a small office above a Swenson's ice-cream parlour and starting working with an engineer friend named Joe Ennis. 'It felt just like the early Apple days,' he enthused. Wozniak couldn't have been happier. He called his new company CL 9, standing for Cloud Nine. By this time he rarely spoke to Steve Jobs. In 1985, when Wozniak had taken his previous leave of absence, his former friend learned about it in the newspaper. When Jobs discovered that Wozniak was using the same design firm as him to create some of the prototypes for CL 9, he raged about it and forced them to drop Wozniak as a client, citing it as a conflict of interest. To make sure that the message was received loud and clear, Jobs picked up Wozniak's universal remote, which was lying on a table, and hurled it against a wall. He then calmly picked up the broken device and handed it to a shocked frog designer. 'Send it back to him,' he said. As it transpired, Jobs' and Wozniak's interests were not entirely dissimilar – although the scale they were enacted upon did a good job of summing up both men.

While Jobs travelled around the country, wining and dining the nation's top education officials to try and persuade them to buy his NeXT Computer, Wozniak instead settled for helping kit out local schools with Macintosh computers. 'I've always been computers, computers, computers,' he said. 'But as far as big dreams that make any sense to me, now it's education.' He also became well known for throwing lavish parties, often inviting more than a thousand people at a time to descend upon his ranch-style home in the hills of Los Gatos. On one such occasion he had the house decorated with 100,000 helium balloons, tied together to form giant rainbows. For entertainment he hired thirty different acts, ranging from magicians, clowns and a puppet show to popular comedians and musicians. Fearing that his abode wasn't impressive enough to entertain that number of guests, he spent $4 million hiring an architectural firm to remodel the property, fitting it out with meditation gardens, koi ponds, oak trees (flown in via helicopter) and a series of oddball adornments that suited his offbeat sense of humour. One of these involved adorning the garden with special motion sensors rigged up to trigger a chorus of gorilla grunts, lion roars and other assorted jungle sounds, which were transmitted by way of hidden speakers from a nearby CD player. He also converted a space that had once been the laundry room into a futuristic arcade, complete with twinkling lights in the ceiling to simulate the night sky. Without a doubt the most unusual feature was a man-made, 800-square-foot cave, which Wozniak had excavated into the hillside at the rear of his property, hidden from view behind a cascading waterfall feature. Inside were fake dinosaur footprints, sealife fossils, prehistoric human skulls, petroglyphs and sparkling amethyst and quartz formations – all scientifically accurate according to experts brought in from the nearby California Academy of Sciences. 'I try to have a simple, peaceful life right now,' Wozniak told one visiting journalist, exhibiting monk-like austerity.

□ □ □

A decade after Apple became incorporated, there was little doubting that Steve Jobs had drifted away from being the weird, raggedy-looking hippie who dropped acid and wandered around Reed College with no shoes on. 'I was really out of touch with him by that time,' says his old friend Daniel Kottke. 'But if you look at the photos, Steve was wearing a suit while promoting NeXT. He had to tone down his countercultural influences.' Those who still refused to believe that Jobs had changed all that much received confirmation of this when NeXT announced its new ally. It was Big Blue himself, IBM. 'A funny thing has happened since we started working with IBM,' Jobs told *Fortune*. 'We like them and they like us.' Given how the personal-computer revolution had started, on the surface there were no stranger bedfellows. The meeting of minds had occurred that June, when Jobs was invited to attend a glittering seventieth-birthday gala for *Washington Post* publisher Katharine Graham. Graham had famously overseen the running of the newspaper when it broke the news of the Watergate scandal, leading to the resignation of President Nixon. She was among the most influential people in Washington and the fact that Jobs had even been invited to her birthday spoke volumes about the circles he was moving in. Among the other invitees was then-current President Reagan, no doubt hoping that he was not indicted in any similar scandals.

'There's one word that brings us all together here tonight,' said humourist Art Buchwald, standing up at the party and raising a glass of champagne. 'And that word is "fear".'

The audience erupted with laughter. There was another word that had brought Jobs to the party, however, and that word was 'networking'. Through Malcolm Forbes, publisher of *Forbes* magazine, he was introduced to the 52-year-old IBM chairman, John Akers. Amidst the polite small talk, Jobs casually let slip that he thought IBM was taking one hell of a gamble by betting its software strategy on Bill Gates and Microsoft.

'Why is it a gamble?' Akers asked, genuinely interested.

'I just don't think their software's any good,' Jobs responded. He continued with his spiel, remarking that Microsoft was simply grafting second-rate Macintosh features on to a less than sturdy software framework. As far as he was concerned it was too little, too late.

'How would you like to help us instead?' Akers asked.

Jobs couldn't believe his luck. He said that he would like that very much.

Several weeks later he and Bud Tribble travelled to IBM headquarters in Armonk, New York. Jobs noted how much the situation reminded him of the visit to Xerox PARC. But this time it was them in possession of the futuristic technology and IBM who could be the proverbial foxes in the henhouse. 'We asked ourselves how much we should show them and decided, hey, these are senior executives,' Jobs recalled. 'They're not going to run back to their computers and start programming.' He decided to pull out all the stops. What Jobs and Tribble sold IBM on was something called NeXTSTEP. NeXTSTEP was the advanced, object-oriented layer of software that sat between NeXT's operating system and the application programs. IBM was suitably impressed. *Jobs was in!* Or, at least, that was what he thought. Half of him enjoyed playing a big shot with the largest data-processing company in the world – then doing a storming $60 billion per year in business – but the other half remained terrified that IBM could crush him at will. When it came time to commit, Jobs dragged out proceedings, running hot and cold about the deal. In meetings he would explode in rage at seemingly trivial issues and stalk out of the conference room yelling, 'It's not worth it', before returning minutes later to resume his place at the table. Once, when IBM lawyers arrived in California wielding a contract that ran well over a hundred pages, Jobs greeted them, flicked through the thick document, and then dropped it straight into the waste-paper bin. The IBM representative wasn't entirely sure how to react. Nonetheless, Jobs held out and was rewarded by a $60-something million offer from IBM to use the NeXTSTEP software. 'IBM got *boned* on that one,' Paul Vais, the NeXT marketing manager, later told journalist and author Alan Deutschman.

Meanwhile, development continued on the NeXT Computer itself. Right from the start Jobs was clear that he didn't just want to build a personal computer. He didn't even want to build another insanely great computer like the Macintosh. This time he wanted to build *the* personal computer. Early on in the process he had gone to visit David Kelley in his offices in Palo Alto.

'David, I'm interested in your opinion,' Jobs said. 'As a designer what is the simplest shape you can think of?'

Kelley considered for a moment.

'A sphere,' he said.

'You're wrong,' Jobs said. 'It's a *cube*.'

Jobs wanted to create something that would be iconic in shape; computing's answer to the geodesic dome. Rather than start with the technological innards and build the packaging around it, he wanted to select his form factor and then let the engineers work out how to squeeze in the technical components. Such was Jobs' desire to do something totally different from every other computer in existence that for a while he toyed with the idea of hiring the Japanese fashion designer Issey Miyake to come up with a distinctive look. Finally he realised that there was only one person for the task. That person was Hartmut Esslinger.

❏ ❏ ❏

An outspoken German designer from the country's Black Forest region, Esslinger was eleven years older than Jobs – and every bit as much a product of the sixties protest culture as the Apple co-founder, albeit with more of a Germanic spin. 'Because I grew up in war-torn Germany, a country still coming to grips with the atrocities of the Third Reich, my aesthetic and my ideologies could not help but be informed by the changing world around me,' Esslinger would recall in his book *A Fine Line*. 'Too young to really understand the concept of war, we children always tried to listen to adults when they spoke in hushed tones about "the bombings" and "Stalingrad", but none of it made much sense to us. No one spoke of the Nazi terror or the Third Reich – not even my extended family, which, as I learned many years later, had lost seven of its own members to the concentrations camps.' When the fathers of the tiny village he grew up in – Johannes Heinrich Esslinger among them – returned from the camps they were virtual strangers; sad, brooding men with explosive tempers that could be triggered by the slightest provocation. Hartmut Esslinger wasn't an easy child, constantly being kicked out of class for misbehaving. After graduating from high school he joined the army, which enabled him to study engineering. While at the University of Stuttgart in 1966 he came across his first computer. His time at Stuttgart was cut short, however. One day Esslinger's mechanics professor came over to speak with him. 'Your drawings are too nice,' he said. 'You won't make it as an engineer. Why don't you look at doing design instead?' Esslinger had no idea there was any such profession. Nonetheless he looked into it and switched to a design course in Schwäbisch Gmünd, a picturesque town in eastern Germany, in the state of Baden-Württemberg. 'My mum fainted when she heard [what

I had done] – my dad wasn't excited either – but I knew this was what I wanted to do,' he remembered. Esslinger's time as a design student was initially frustrating. He was used to electrical engineering and found the work his contemporaries were doing unspeakably dull. 'I was shocked that designers were still doing tableware and stupid furniture. I told the professor that I wanted to do high-tech, and he said, "do it". So I started designing consumer electronics: electronic synthesisers and fancy radios.' Following a vicious argument over a radical clock-radio design he had dreamed up, Esslinger said 'screw it' and struck out on his own. In 1969, at the age of twenty-five, he started up his own company in a garage.* 'It was tough in winter with gloves, but we made it. We were baking models in the oven so there could be no cake, no roasts, no turkeys anymore, but it was fun.' Before long Esslinger began to attract some major clients.

In 1982 he was approached by Steve Jobs about doing some design work for Apple. At the time he was less than impressed by the designs the company was coming out with. 'The Apple II was successful because it was an innovative product, but the design and the peripherals looked like absolute crap,' he says. 'The Apple III was a little better, but it was too expensive. The Lisa was also incredibly ugly.' Still, he liked Jobs, who shared his fondness for driving fast cars, sought perfection in everything he did, and was the only high-level person working in computing to wear a t-shirt more 'old and worn' than Esslinger's own. Jobs' initial pitch was unsatisfactory. He wanted Esslinger to supply the designs for products that would then be pieced together, according to his specifications, by Apple's industrial-design

---

* Originally the less-than-inspired Esslinger Design, the firm would later be renamed frogdesign, then frog design, then finally frog. Even its name – an acronym for the (F)ederal (R)epublic (O)f (G)ermany – conveys something of a rebel yell. Explains Esslinger, 'frog ... is always printed in lower case, a rebellion against German grammatical rules that, forty years later, other companies are beginning to adopt'. Today the company employs more than 1,600 people around the world.

group – then run by Jerry Manock. Esslinger demurred.

'That's not going to work,' he said curtly. 'The way you do it, you'll destroy my designs in the first second.'

Jobs had previously been thinking about incorporating Sony-style design concepts into the visual language of Apple products. Esslinger was well known for working with Sony, but he explained to Jobs that Sony was a Japanese company and its products reflected this. What Apple's computers needed to do was to look American – more specifically Californian. Esslinger had always loved American pop culture, right back to his first time seeing James Dean in *Rebel Without a Cause* at the age of twelve. 'I saw it twice, on a weekend,' he recalled. 'It was so incredible [that] we got a bus on Monday afternoon to drive to the American Army headquarters in Stuttgart to get t-shirts.' Esslinger's company was given the contract with Apple and the designer moved to California and set up shop. The first challenge he faced was refiguring Apple in such a way that the autonomy of the design department was heightened. 'At the time the Apple designers were reporting to engineers, who were four layers under the CEO,' Esslinger recalls. 'I explained to Steve that design has to be top-down. Whenever design isn't at the top, it doesn't work. He believed that.' Making the changes to the structure proved an uphill battle. 'The engineers complained because previously they had told the designers what to do, and now that situation changed,' he says. Nonetheless, Esslinger's team began work on a project to find a new design language for Apple: one that would look, for want of a better word, *Californian*. 'Everyone was designing East Coast and I told Steve that we had to design West Coast,' Esslinger says. 'I looked at the art of the Native Americans, the Navajo. They did paintings in sand and I thought that was interesting because that same sand went into the silicon that made the computer chips.' When Jobs left Apple, Esslinger went with him. It was, after all, the chance of a lifetime.

□ □ □

One week before the biggest launch of his life, Steve Jobs was doing his best to stay calm. He was standing in a high-school gym in Berkeley, rehearsing for the forthcoming 12 October 1988 launch of the NeXT Computer. The gym wasn't going to be the site of the grand unveiling – that would be San Francisco's Davies Symphony Hall – but it was available for rehearsals that day, while the Symphony Hall wasn't. Jobs wore blue denim jeans and a red flannel shirt, and repeated his carefully scripted lines into a wireless microphone. As he rattled them off, trying to sound as though each one were a spontaneous thought plucked off the top of his head, the NeXT Computer threw up image after image, and sound cue after sound cue, to correspond with Jobs' words with pinpoint accuracy. At the eleventh hour the pressure was really on, but at this level of business when *wasn't* it? Jobs was faintly aware of a gag doing the rounds among employees of Silicon Valley tech companies, likening him to the charismatic cult leader Jim Jones. If you set a cup of grape Kool-Aid in every seat at this NeXT launch, the joke said, the audience would drink every last drop. *Well, that was what the NeXT team was hoping for at least.*

The stage set up for the launch was going to be almost spartan in its simplicity: just one man, a microphone, a vase of flowers and the NeXT Computer. The only later additions to this cosy party would be a physics professor from Reed College, to demonstrate an application called Mathematica, and a violinist from the San Francisco Symphony who would come on stage as the grand finale, to play a duet of Bach's A Minor Violin Concerto with the machine in order to show off its superior audio capabilities. To help bring this vision to life Jobs had called in the services of George Coates, a countercultural San Francisco-based performance artist who had first moved to the area in 1969, and whose interests revolved around injecting a little Frank Zappa into his Information Age surroundings. 'I like to create worlds that exist outside the artificial corporate commercial culture,' Coates once claimed. The emphasis on aesthetics was more than Jobs being Jobs. In part it was a piece of conjuring misdirection, designed to distract people from the fact that the launch was taking place at least a year after it was promised. It was the Macintosh all over again, except that where the Mac had been a top-secret project within Apple – Cupertino's answer to the Manhattan Project – the NeXT Computer had been a revolution brewing in full view of the public, with gossip columnists all too eager to question whether Jobs was really a one-hit wonder.

Several people at Apple had followed suit, with a few wags quipping that in three years all NeXT had so far managed to ship was a t-shirt. One person even took the extra step of creating a badge with a take-off of the infamous $100,000 Paul Rand logo. Complete with lower case 'e', it read 'NeVER'. Feeling harried from all directions, Jobs' answer might have been to rush something out before it was ready, but that wouldn't have meshed with his perfectionist tendencies. 'When you manage a project there are three things that you can try and control,' says Rich Page. 'One is the content, i.e. what is the project? Another is cost, which is generally measured by the number of people you have working on it. And the third component is the schedule. At best, you can only control two of these components, and you have to give up on the other one.' At NeXT the scale of the project had never diminished. As it was, the schedule had only been reached by throwing more and more money at the project. Besides, delays were a temporary annoyance. Six months down the line people don't remember whether a product has shipped on time. They remember whether it worked or not.

Back in the Berkeley high-school gym, all was going well

with rehearsals until suddenly a software error made the NeXT Computer hang. The screen froze. So, too, did the NeXT employees standing in the wings, sure that a classic Jobs outburst was about to be aimed their way. Only it didn't happen.

'We'll fix that,' Jobs said calmly. 'No problem.'

There was an almost palpable sigh of relief. The NeXT Computer might still have been suffering from software problems, but at least the machine itself looked beautiful. Hartmut Esslinger had outdone himself with the industrial design. Taking the form of a pitch-black magnesium cube – measuring exactly a foot square – the NeXT Computer looked like something from another planet. The cast magnesium not only had a pleasing solidity to it, giving the impression that the NeXT Computer could crush any number of beige boxes underfoot, but also provided a pleasingly aesthetic solution to the problem of radio frequency interference. It brought with it as many problems as it did answers, however. For one thing, magnesium solidifies much faster than plastic during the injection-moulding process, being far more likely to result in air bubbles or other defects. For another, it was highly reactive. 'One of the factories that we had … blew up because of the magnesium dust getting in the water,' says Matt Medeiros, NeXT's director of materials. It was also prohibitively expensive. The cost of the moulds alone had come to $650,000 and necessitated a nationwide search to discover a metal shop capable of creating them. (One was finally discovered in Chicago.)

Yet another obstacle was the colour. Jobs spent hours agonising over exactly what would be the right shade of low-gloss black for the computer, eventually finding one which met his exacting requirements on the tone-arm of a stereo turntable. But this shade proved almost impossible to achieve. Ross Perot introduced Jobs to the paint experts at General Motors, who advised him on the difficulties of what he was suggesting. Black, they said, was the colour least forgiving when it came to hiding flaws in body defects. Jobs ignored them. *No problem*, he figured, *we'll just make sure that we don't have any flaws.* He did, but it came at a hefty price. In the end a single finished case came in at ten times the amount the team had budgeted for. Although universities had said they would be happy to pay no more than $3,000 (ideally $2,000), the NeXT Computer was now priced at $6,500.

The day of the launch finally dawned. Four thousand five hundred invitations had been sent out, inviting friends, well-wishers and industry opinion-makers to come and pay homage. The expectant crowd was dotted with revisionist hippies and unrepentant yuppies, wearing a mixture of tuxedos, business suits, cardigan sweaters and corduroy jackets – as if those attending weren't entirely sure whether they were turning up to a night at the theatre, a christening, a funeral, a hobbyist meeting or the business happening of the year. In a sense the answer was a little bit of all five. 'I couldn't sleep at all last night,' said one anonymous attendee. 'It's like I'm six years old, it's Christmas Eve and I know I'm getting a train set in the morning. But until I open the box and actually see it, it's not real.' Even those who worked in the media weren't immune to the hyperbole. Rick Smolan, a highly respected photographer who has worked for *Time*, *Life* and *National Geographic* magazines, likened the arrival of the NeXT Computer to the chance to see Thomas Edison unveil the phonograph. Turning up to the launch wasn't yet something he was doing for himself; it was something that he *owed* to all those who would come after him. 'I don't want to tell my grandchildren I was invited but didn't go,' he said, explaining why he had taken the red-eye from the East Coast in order to attend.

One of Jobs' real strengths was product demos, something he would demonstrate time and again following his later return to Apple. Even taking out the public-speaking element of engaging an audience, demos can be nerve-racking occasions. The software is, quite often, still buggy or incomplete, and a glitch or crash at this late stage can prove catastrophic. Since the NeXT software was still not finished, a second system was set up behind a curtain in San Francisco's Symphony Hall. Jobs, in front of the crowd, ran System A, while out of sight a feverish group of engineers copied his every move on System B. Both were linked to a video screen which could be switched to show either system in the event that an application crashed. The audience would be none the wiser. 'I think I speak for everyone at NeXT, saying it's great to be back,' Jobs started, striding on to the stage to a thunderous round of applause, his hands clasped in front of him in what looked like prayer. He ran through the NeXT Computer's abilities, showed footage of the automated factory in Fremont where the machines racing off the production line were 'completely untouched by human hands', and proudly talked about how NeXT had made the decision to 'risk our company' by including a read-write optical disk capable of storing 250 megabytes. Jobs demonstrated the machine's audio playback facilities by playing John F. Kennedy's 'Ask not what your country can do for you ...' inaugural address and Martin Luther King's 'I have a dream', both from the first half of the 1960s. The NeXT Computer, he concluded, wasn't just a tool; it was more like a 'partner in thought'.

It was a bravura three-hour performance. Afterwards the journalists trotted down the hallway for a news conference. Jobs was again on top of his game, whipping out one sound-bite friendly comment after another.

'How many NeXT Computers do you hope to produce?' asked one reporter.

'We can make a lot, but probably not enough,' Jobs answered coolly, to a chorus of laughter.

'What do you tell those who want one of your computers but aren't in higher education?'

'Enrol,' he said.

Negative or critical comments were roundly ignored, and not even the Jobs-approved lunch served up to the press – cream cheese and sprout croissants with mineral water – could dampen the high spirits.

'This machine will replace sex,' wrote one magazine editor.

□ □ □

In 1989, Apple, or more specifically Jean-Louis Gassée, unveiled the Macintosh SE/30, the ninth incarnation of the Macintosh. Given the comment about the NeXT Computer replacing fornication, it is somewhat ironic to note that if Apple had continued with the naming process started with Gassée's previous machine, the Macintosh IIx, this would have been titled the Macintosh SEx. The Macintosh SE/30 once again offered significant improvements over its predecessor, being four times faster than the previous-generation Macintosh SE, although at equally significant price. This was Gassée's business model for the Macintosh, which he saw as occupying a 'high right' quadrant of the market; meaning that in a chart showing performance against price, the Mac would lead the way in both. Inside Apple Gassée's mantra was 'fifty-five or die', referring to his goal of an astounding 55 per cent profit on the Macintosh.

'We don't want to castrate our computers to make them inexpensive,' he told a journalist for *InfoWorld* who quizzed him about the SE/30's $4,369 price tag. 'We make Hondas, we don't make Yugos.' On the record he claimed that Apple was considering introducing a sub-$1,000 computer at some point, 'although we can't do it in the next two or three years.' Away from the press, he was more defiant. In a meeting with employees from Apple's R&D department, the question was again raised about whether the company should start developing products aimed at lower-end consumers, as the likes of Dell and Gateway were doing with some success.

Gassée practically bristled at the suggestion.

'We don't make products for techno-peasants,' he allegedly proclaimed. 'Fuck them!'

The engineers in the room, most of whom preferred working on high-end computers anyway, broke out into a spontaneous round of applause. *Yeah! Right on! No techno-peasants! Fuck them!* Tobey Fitch, one of the senior managers in charge of the organisational and leadership-development team, was horrified. 'I was in the back of the room going, "Isn't 90 per cent of the market a techno-peasant? Aren't they spending money with these other people? Isn't that part of our problem?"' he recalls.

But Gassée had spoken. In order to liberate the world from ugly, cheap hardware, a certain amount of elitism was necessary for people's own good. There was another change with the Macintosh SE product line, too; a feature that struck people as highly un-Jobsian. It wasn't quite as immediately apparent as the price, but it whirred along in the background all the same. The SE/30 came with an in-built cooling fan.

If there was ever a clash of the billion-dollar giants in countercultural clothing, it came when Apple Computer was sued by Apple Corps, the multimedia conglomerate started by the Beatles.* Apple Corps, whose headquarters were in London, had been set up in 1968, almost a decade before Apple. The initial face-off between the two companies came in 1978 when Apple Corps sued for trademark

---

* The Beatles' Apple Corps did as good a job as any corporation of helping its founders appear to successfully negotiate the murky waters between retaining a hippie sensibility and looking like cold, hard businessmen. Asked about it during a 1968 appearance on *The Tonight Show* by host Johnny Carson, John Lennon replied, 'Our accountant came up and said, "We've got this amount of money. Do you want to give it to the government, or do something with it?" So we decided to play businessmen for a bit because we've got to run our own affairs now. So we've got this thing called "Apple" which is going to be records, films, and electronics – which all tie up.'

infringement. That suit was settled in 1981, with Apple shelling out $80,000 to Apple Corps, as well as signing an agreement promising to stay out of the music business. That seemed reasonable enough. By 1989, however, Apple had made the decision to include a Musical Instrument Digital Interface (MIDI) Manager with its new version of the Macintosh, which would support musical instruments being plugged into the machine as peripherals. This triggered the second Apple Corps lawsuit. At the time, Jim Reekes was Apple's senior software engineer in charge of the audio and system sounds. It was a largely thankless task. 'No one understood the audio system on the Mac,' he says. 'It was basically an orphan. No one cared about it and no one wanted to do anything with it, so I took over that job. I was doing that for the next ten years.' Due to the 1981 agreement, Apple was (perhaps understandably) hesitant about venturing too deeply into this area of research. When Reekes later suggested that Apple develop an application to allow users to download music, organise it into tradeable playlists and burn their own mix CDs, his idea was given the pocket veto. Reekes's job was stressful, but his interaction with upper management and high-level company politics was, thankfully, limited. Now, as Apple's go-to guy for audio, he was getting subpoenaed as part of the Apple Corps suit, with lawyers constantly arriving in his office with orders to make copies of his files or to go through his emails. A part of English law states that written documentation is inadmissible, and somebody – possibly Reekes himself – would have to take the stand to read out whatever material they required in court. 'The whole thing was ridiculously stupid and I was completely fed up with it,' he recalls.

One night the System 7 engineers were working late, as happened with some regularity. It was in the early hours and fuses, burning the midnight oil, were dangerously short. Reekes was in the middle of working on the sound control panel, revamping the OS by creating new audio cues to incorporate into the user interface. 'I was trying to expand what the Mac could do sonically,' he says. 'The lawyers end up telling me that the beep sound I had made for the start-up, which I called "chime", would have to be renamed because that word sounded "too musical". That was the final straw, and now this thing was just beyond ridiculous.' Exasperated, Reekes read out the message to his colleagues.

'What are you going to do?' they asked.

'Rename it, I guess.'

'To what?'

Reekes thought about that for a second.

'Let it Beep,' he said, entirely straight-faced.

'You can't do that! If you use a name that sounds like one of the Beatles' songs, this whole thing is going to blow up completely. You'll be looking at a lawsuit yourself.'

'So sue me,' Reekes retorted.

Another flash of inspiration hit. He liked the way that phrase sounded. *So sue me. So-sue-me.*

'I've got it,' he said. 'I'm going to call it "Sosumi", and we'll tell the lawyers that it's a Japanese word that doesn't mean anything musical. In fact, we'll tell them it means "a complete absence of musicality".'

The name stuck. So did the sound: a version of the short xylophone sample has been featured on every subsequent Mac OS.

□ □ □

In March 1989, NeXT announced that it had signed a deal with Businessland, America's largest computer retailer, to sell computers to the non-education market. To many this reeked of dishonesty. Jobs had talked about creating a revolutionary machine for education and now he was turning around and selling it to large corporations? The simple fact, however, was that NeXT simply was not sustainable selling computers to the higher-education market alone. Nor were software developers all that keen to develop for a market sector in which piracy was so widely tolerated, accepted even. Institutions like Carnegie Mellon and Stanford University may have disagreed, but they had invested in NeXT and were now having to think like the big businesses which, in fact, they were. The Businessland deal benefited both NeXT and Businessland in equal measure. For NeXT, it meant that the company didn't need to train its own dedicated sales force, technical-services department or worry about focusing promotion solely on businesses. For Businessland, as VisiCalc's Dan Bricklin pointed out, having NeXT as a client was 'like being able to put a Maserati on the showroom floor'.

Businessland had impeccable credentials for the task it was being asked to do. Having been founded in 1982, by 1989 it had expanded from one store to more than a hundred dedicated outlets nationwide. Its sales had followed a similar trajectory: $44 million in 1983; $157 million in 1984; $351 million in 1985, and rapidly upwards – passing the $1 billion mark in 1988. While it had initially catered for small and medium businesses, Businessland's CEO, David Norman, had been instrumental in pushing towards favouring the high-volume Fortune 500 company market. Projecting sales for NeXT Computers, Norman claimed that they should easily do $150 million. At a base price of $9,995 per unit – with no discounts – this meant selling between 10,000 and 15,000 machines. Jobs told him that this was no big deal; the NeXT factory was gearing up to produce ten times that number. On the day the agreement was signed, Jobs invited the senior executives from Businessland to his house for dinner. As he clinked glasses with Norman, Jobs toasted, 'Let's go kick the shit out of some people.'

Meanwhile, there was further reason to be cheerful. Jobs, Page and NeXT's director of materials, Matt Medeiros, flew out to Tokyo to meet executives from Canon. The company, which produced the optical disk drive for the NeXT Computer, along with a key component for its printer, was suitably impressed and agreed to invest $100 million in exchange for a 16.67 per cent stake. By this time NeXT had an astonishing market value of $600 million: unbelievable for a business with just two hundred employees. To celebrate, the company moved to expanded premises along the San Francisco Bay in Redwood City. Although the building was new, Jobs set about redecorating it from top to bottom. 'Get rid of the elevator,' he said, referring to the plain but functional lift which stood in the main lobby. There were concerns about wheelchair access, but these were quickly overruled. I. M. Pei, the noted Chinese-American architect whom Jobs admired immensely, was brought in to design a special freestanding staircase in its place. Nor was that the extent of the refurb. Everything at NeXT had to conform to the highest possible standards. In the employee bathroom Jobs specified that the floor be made up of dark green tiles with a light-grey grout, going so far as to pick out the exact colours to be used. After a team of tilers had been working for several days, he went in to inspect the finished job.

'The grout's the wrong colour,' he immediately snapped.

The lead contractor insisted that he had used the right colour, Jobs accused him of lying and NeXT employees admitted that they really couldn't tell the difference. The upshot of it all was that Jobs ordered the whole tiled floor be torn up and replaced, at a cost of $30,000.

Even more extreme was the saga of the conference-room table. Jobs insisted that this be made of granite and be installed in one piece. When he was told by numerous companies that the size of table made this impossible, but that they could piece together the two parts so carefully that no one would ever be able to spot the join, Jobs paid for a long-closed quarry to be reopened in order that a piece of granite of the exact specifications could be excavated.

The rest of the NeXT offices followed along these lines. There were $2,200 chairs in the common areas and a $10,000 sofa in the lounge. What was the most expensive phone you could find, with a special high-capacity T1 phone line? 'Get one of those for every room,' Jobs said. 'No, forget that – for every desk.' He even wanted a line installed in his house, so that he could patch into work from the comfort of his home if need be. NeXT's manager for telecommunications looked into it and realised that it would be unfeasibly expensive to run. Jobs wasn't going to be happy. He ducked into his boss's office to break the bad news.

'It's going to cost $660 per month,' he started.

Jobs didn't look up from his computer screen.

'I don't care,' he replied. 'I'm rich.'

Looking around at the luxurious decor, the NeXT offices seemed like the headquarters of a company that had really, finally, *profoundly* made it. Dan'l Lewin proudly announced to the press that already a hundred of the companies in the Fortune 500 had NeXT Computers installed. But the truth was something altogether different. Lewin's statement was factually accurate, but deliberately misleading. In reality, while the numbers he stated were correct, all these machines were loaners, sent over for evaluation free of charge. Outside of the William Morris Agency – the Hollywood talent agents who had committed to buying 270 machines by 1991 – not a single company had bought a NeXT Computer. As the 1980s came to a close, Businessland had sold just 360 of Steve Jobs' machines. The sheen of freshly minted success that existed at NeXT's headquarters was all smoke and mirrors. And freestanding staircases.

10

SEX, DRUGS AND PiXAR ANiMATiON

'It seems to me quite possible that the 1960s
represented the last burst of the human being
before he was extinguished. And that this
is the beginning of the rest of the future,
and that from now on there will simply be
all these robots walking around, feeling
nothing, thinking nothing. And there will be
nobody left almost to remind them that there
was once a species called a human being, with
feelings and thoughts. And that history and
memory are right now being erased, and that
soon no one will really remember that life
existed on the planet.'

Wallace Shawn and Andre Gregory,
My Dinner with Andre (1981)

Less than six months after leaving Apple and at about the same
time that he founded NeXT, Steve Jobs quietly invested in
another business. Or as quietly as he knew how, which meant
headlines in the *Wall Street Journal*, *Los Angeles Times* and *Variety*.
The company in which he opted to buy a majority stake was a

spinoff from George Lucas's cutting-edge Lucasfilm computer-graphics division. It had an odd-sounding name: one that looked like it described some high-tech gadget, but rolled pleasingly off the tongue in the way of a Spanish verb. It was called Pixar and it was the company that would eventually transform Steve Jobs into a real, no-fooling billionaire. Not that it looked that way at the time, of course. Once the impact of those initial February 1986 newspaper headlines ('Jobs Acquires Lucasfilm's Graphics Unit') had sufficiently faded from the memory of casual readers, Pixar would be all but forgotten next to industry heavyweights like Silicon Graphics and Sun Microsystems. *Rich guy buys new company shocker.*

For those in the tech industry, or those who knew Jobs personally, the company would come to be seen as a rich man's folly; an overpriced hobby with a direct line of credit to Jobs' bank account which was fast bleeding the Apple co-founder dry. Those in Hollywood were just as baffled. Was this simply an abortive attempt to get Jobs a foot in the door of one of the most glamorous professions in the world? Pundits even came up with a tongue-in-cheek name for the unlikely convergence of Silicon Valley technology and Hollywood moviemaking. They called it 'Sillywood'.

But Jobs knew something that few of them did. At every level he realised that the dream of Pixar's high-tech artisans echoed the one that he himself had set out to lead at Apple. Although their chosen mediums differed, both were part of the same revolution; concerned with the creation of a utopian world in which graceless, number-crunching machines were reconfigured as tools with the ability to create something beautiful. Pixar's later *Toy Story*, for example, would present a potted, family-friendly version of the countercultural ideology of the techno-geeks. The film's cowboy character, Woody, representing the fifties archetype of overland frontiersman, stands for the simple charm of hand-drawn animation, while Buzz Lightyear, the sixties archetype of the space frontiersman, carries the alluring sheen of high-tech. Only when the two work together can real magic start to happen. More than a decade later, in 2008, Pixar released *Wall-E*. This time the film's storyline depicted an earth so ruined by the despicable actions of an evil mega-corporation, Buy 'n' Large, that it had been left uninhabitable by humans. Just to drive home the point that Apple was nothing like said corporation, Apple's head of industrial design, Jonathan Ive, would be dispatched to the Pixar offices to design the film's female robot character, Eve, who acts as Wall-E's love interest. Not altogether unlike a toned-down, fleshed-out version of the '1984' Macintosh commercial for the under-tens, in the film only the free-spirited Wall-E and Eve (the latter closely resembling an Apple product; the former with at least sense enough to desire one) are able to hurl the metaphorical hammer and go on to rebuild earth's shattered ecosystem.* Stewart Brand would certainly have approved.

Although Jobs entered the Pixar story a decade after its founders' journey began, he would also have appreciated their efforts as they clawed and scratched to see their particular revolution kick off. It was a situation that he could certainly relate to.

❑ ❑ ❑

Alvy Ray Smith grew up a Silicon Valley nerd, with the distinct geographical disadvantage that he didn't live in the Valley. He was a naive kid from New Mexico with a crew cut and cowboy boots. Then he arrived in California at the height of the 1960s,

---

* The film includes numerous nods to Apple products. Wall-E watches his favourite film each night on an iPod screen; he makes the same 'boot up' sound that Macintosh computers have made since 1996; and even the villainous AUTO, the autopilot of the spaceship *Axiom*, is voiced by Apple's text-to-speech system, MacinTalk.

ostensibly to study computer science at Stanford University, and found himself up to his eyeballs in the countercultural revolution. 'In other words, I walked smack-dab into the middle of it all,' he says. 'I had no idea what was going on at first. Then I realised, "oh my God – I can wear anything I want." I let my hair get long, I grew a beard. Art was finally being celebrated, instead of being this thing that was written off. I got right into it and was a full-on hippie within a year of arriving. I was taking drugs in Golden Gate Park and hanging out with all the famous musicians, and going to the ballrooms in San Francisco to hear the Grateful Dead, Janis Joplin and Jefferson Airplane. It was the full sixties experience.' Amidst all the craziness Smith somehow had the wherewithal to complete a Ph.D. For his dissertation he wrote a study of cellular automata, the concept of self-replicating machines. Not long afterwards he discovered an article on the subject in *Scientific American*, written by the famous mathematician and science writer Martin Gardner. Smith was a big fan of Gardner's work, which also included a famous annotated study of the mathematical concepts in Lewis Carroll's *Alice's Adventures in Wonderland*. He sent Gardner what amounted to a fan letter, and the mathematician was sufficiently impressed to write back and ask if he could come and visit Smith on campus. When he arrived, he was shocked beyond belief by the unkempt appearance of his correspondent – and even more so when he learned that his books were not, in fact, being read by the thoughtful middle-aged academics that he thought they were. '*The Annotated Alice* was *de rigueur* reading for every acidhead in America,' Smith laughs. 'He didn't know that. It was clear that he had no idea whatsoever about the drug culture. He was surprised at all the long hair. I didn't say anything, because I was so amazed that this man, who was so smart, was unaware of the culture and of who was buying his books.'

After graduation Smith headed to the East Coast, where he became an associate professor at New York University. But he found that he couldn't easily forget what he had seen in the Bay Area and nor could he escape the nagging feeling that he was meant to be doing something more with his life than he was. When he had a skiing accident in New England, Smith was forced to wear a full-body cast for three months. It gave him a chance to evaluate the direction his life was heading in. He had always been a painter, but with his thirtieth birthday rapidly approaching he was worried that he was letting that talent slip away. 'I said, "You know, Alvy, you're not doing anything about your art",' Smith recalls. '"You're basically feeding the Vietnam war-machine by teaching these kids computer science. Why don't you drop out? Go to California and find something that involves computation and art." So that's what I did.' As soon as he had healed sufficiently, Smith resigned his academic position and travelled back west in his Ford Torino, based on nothing more than a hunch that something good was about to happen there. After a slow first year – primarily spent bumming around as a penniless hippie – Smith bumped into Dick Shoup, an old friend from New York who was now working at Xerox PARC.

'What are you doing there?' Smith asked, mentally writing off any job that involved research and development for a big East Coast corporation.

'I'm building a computer for artists,' Shoup replied. 'Why don't you come over and take a look?'

Alvy Smith was puzzled, but interested. 'To tell you the truth, I couldn't figure out what he was talking about,' he says. Nonetheless he drove from Berkeley to Palo Alto, across San Francisco Bay, to see what his comrade was engaged in. Shoup worked on the floor below the other engineers at PARC. While they were busy thrilling over a still image of *Sesame Street*'s Cookie Monster on the prototype Xerox Alto, he connected up a black and white video camera to one of the lab's minicomputers and set about recording the world's first ever digital video. From there, he expanded the so-called 'frame buffer' to allow it to handle colour and import footage from a variety of different sources. He named this revolutionary new program SuperPaint. When Smith saw SuperPaint for the first time he was absolutely floored. 'I looked at it and said, "This is it. This is art and computation combined",' he remembers. From that moment on he was desperate to get a job at PARC. To do this he knew that he needed an angle. Xerox's research laboratory had some of the smartest computer scientists in the world. *Why would they want old Alvy Ray Smith unless he brought something new to the table?* He considered the dilemma and decided that he would sell himself to PARC management as an animation expert. Since he had no real background in the subject, he went out to the local bookstore and picked up a $1.50 copy of Preston Blair's famous *How to Animate*, through which he picked up some of the basics.

Because there were no job openings available, Shoup hired Smith by writing out a purchase order, in the same manner that one might buy a new batch of clipboards or an office chair. Smith didn't care in the slightest. He just wanted time to spend with SuperPaint. He and Shoup threw themselves headlong into experimentation, constructing elaborate video sequences which would be utterly impossible to create in any other medium. One day they took an image of Shoup's girlfriend's face and transformed it into an otherworldly kaleidoscope effect. On another they created a gecko's-hide texture which changed from red to yellow to green to blue like a psychedelic rainbow. On yet another they loaded in old episodes of *Star Trek* and modified the colours to make it look as if Captain Kirk and the *Starship Enterprise* away team had landed on a planet perpetually tripping on LSD. *It's Haight-Ashbury, Jim, but not as we know it!* Smith couldn't believe his luck. He was having a whale of a time at Xerox PARC, breaking ground and blowing minds. As far as he was concerned it was one giant party. And there was no end in sight.

□ □ □

As the oldest of five siblings in a strict Mormon family in Salt Lake City, Utah, Edwin Earl Catmull – today the president of both Walt Disney and Pixar Animation Studios – dreamed of working as an animator for the movies. He spent hours in front of the television and fashioned crude flick-books in which the main character was a hybrid of a man and a unicycle. When asked who his heroes were, he cited Albert Einstein, Pinocchio and Peter Pan. Not necessarily in that order. For all the movies and cartoons that Catmull watched, and all the high-school art classes that he took, however, he harboured a dark secret: deep down he knew that he wasn't a talented enough artist to make the grade as a professional animator. By the time college came around, he decided to cut his losses and signed up instead to study physics and computer science at the University of Utah. Utah was the nation's leading centre for computer-graphics research. Its computer-science department had been started in 1965 by Dave Evans, a former professor from Berkeley and an elder in the Church of Jesus Christ of Latter-day Saints. The 1960s were an exciting time to be involved with high-tech graphics work. The field was heavily sponsored by the Pentagon's Advanced Research Projects Agency, who saw potential military application for the research. The influx of funding saw one ground-breaking development after another. One of Catmull's professors was Ivan Sutherland, the man who had invented Sketchpad, a revolutionary program which allowed users to draw images directly into a computer using a light pen and then to move, rotate or resize them while retaining the image's basic properties. Several of Catmull's peers were equally impressive. Teaching one of his programming classes was a young graduate student named Alan Kay, who was carrying out his own work based on language paradigms and models of systems interactions. Then there was Jim Clark, who later went on to start Netscape and Silicon Graphics; John Warnock, who created Adobe; and Nolan Bushnell, best known for founding Atari and giving a young, unkempt Steve Jobs his first break in the business. 'It was quite a creative stew,' Catmull would recall. Computer graphics was a field that attracted a colourful bunch of characters. Alan Kay funded his studies by moonlighting as a jazz guitarist, which often resulted in him oversleeping and turning up late for lectures. Jim Clark, meanwhile, had been a childhood prankster, suspended from school for igniting a smoke bomb on a bus and, on another occasion, smuggling a skunk into a formal dance.

Catmull was obviously intelligent, but his interests were so wide-ranging that he never seemed capable of knuckling down to focus on just one area of research. One moment he was interested in Z-buffering and methods for the display of bicubic surfaces, the next he was writing papers on texture-mapping algorithms and geometrical modelling. 'I guess I didn't know any better,' he said. As a graduate student in 1972, Catmull and a friend collaborated on an experimental, one-minute film, which took viewers on a journey around – *and eventually inside* – a digital recreation of Catmull's left hand. The effect was created by encasing his hand in plaster of Paris to make a mould (sadly forgetting to shave off the hairs first), then covering this in latex and drawing around 350 tiny polygons over its surface. The distances between these polygons were then fed into a computer using a Teletype keyboard, which in turn created a fully rotatable 3-D model. *A Computer Animated Hand* proved to be an early step in not only mapping textures to a shaped, three-dimensional surface, but also demonstrating that this could be done without leaving messy, jagged edges. Catmull's supervisor, Ivan Sutherland, was impressed. He wanted to send his young protégé to Disney, feeling that if Catmull could wax lyrical about the potential of computer graphics, it might prove a good way of building up links between the two establishments. Catmull went, but his visit didn't turn out well. Following a meeting with Disney's studio head he was introduced to Frank Thomas, one of the company's legendary animators, then in his fifties. In Thomas's office Catmull noticed a creaky vintage typewriter and some other antiques, which he took to be a reminder of the studio's early days.

'Those are some great props,' he said, trying to strike up a conversation. 'It's amazing that you've managed to hold on to all this old stuff after all these years.'

'They're not props,' spluttered Thomas. 'That's my office equipment.'

Perhaps Disney wasn't quite ready for computer animation yet, Catmull thought. His day ended early when the studio was hurriedly evacuated. Someone had phoned in a bomb threat. Still, he refused to be discouraged. By the time he left the University of Utah with a Ph.D. under his belt, Catmull felt confident that he had discovered the intersection between computation and art. He now knew what he wanted to do with his life. 'When I graduated my goal was not as lofty as emulating all of reality; it was to make an animated film,' he said. With a wife and young family to support, he began looking around for work, but found that few places had the inclination or resources to throw money at computer-graphics research. 'No studio was interested,' he lamented. 'Neither were research sites. The problem was graphics technology at the time was fairly crude, and most people would look at it and judge it by what they saw. They couldn't imagine what it would become.' Catmull took a job with a CAD firm in Boston named Applicon. He was there for just one month before being contacted by an eccentric entrepreneur named Dr Alexander Schure, who ran an oddball East Coast operation called the New York Institute of Technology. He had an offer that Catmull would find difficult to turn down.

□ □ □

Alex Schure had founded the New York Institute of Technology in 1955. It was a moneymaking scheme, a way of capitalising on the wave of new students studying under the government's GI Bill, which guaranteed education to returning war veterans. Nobody was entirely clear about where Alex Schure's money originated, although it was obvious that he had an abundance of it. Rumour had it that his grandfather had made a fortune in the fur trade, and this had since been multiplied thanks to some handy connections in the military. Initially New York Tech (as it was known) functioned simply as a third-rate correspondence school, handing out technical certificates to would-be engineers. Even in doing this it stood out as unusual. A person signing up for one of Schure's courses would find themselves inundated with a mess of electronic parts through the post, supposedly taken from Soviet warehouses during the Second World War. Still, the gimmick proved successful and before long Schure was able to expand his outfit to include three campuses in upmarket parts of Manhattan, Long Island and Old Westbury. At the same time he had a second ambition. Walt Disney had died in 1966, leaving a void in the world of high-quality animation. Even with no knowledge of the subject, Schure saw no reason why he shouldn't be Disney's natural successor. He began assembling a team capable of putting together the world's first feature-length film completely animated using a computer. He also travelled around the country, visiting research sites like the University of Utah, and buying up all the tools he could lay his hands on.

'What equipment do you have?' Catmull asked, when Schure first approached him about running the new Computer Graphics Lab at New York Tech.

'I don't know,' Schure said. 'I just bought one of everything.'

Schure's Computer Graphics Laboratory was based in New York Tech's Westbury Campus. It was a magical backdrop, one which appeared to have come straight out of an F. Scott Fitzgerald novel. Set on the wooded hillsides of Long Island's North Shore, the Westbury Campus was comprised of four sprawling estates joined together, the former homes of wealthy New York families like the Rockefellers and the Vanderbilts. Dotted throughout were quaint cottages and servants' quarters, while the manor houses themselves were retrofitted into laboratories and classrooms. The Computer Graphics Lab was located in the garage of a building called the Gerry Mansion. Those who worked there liked to joke that it was, quite literally, a garage operation – although unlike a company such as Apple, which had been started out of the garage of Steve Jobs' parents, this was a four-car, two-storey garage. Later on, when enough people had joined the team, the Computer Graphics Lab was moved into a nearby stately home called Holloway House, which featured fine Corinthian columns, eight bedrooms and a bear pit in its back garden. Catmull was over the moon. 'We just started programming our brains out,' he said.

◻ ◻ ◻

Back at Xerox PARC, Alvy Smith was having a ball. 'The personal computer was being invented around me,' he recalls. 'The first personal computer, the first laser printer, the first mouse use, the Ethernet – and Dick and I were doing the first colour graphics. It was the scene as we now know it. I was vaguely aware that the revolution was in our hands. You look back and realise how awesome it was. Of course, we didn't understand that fully at the time.' Then one day it all suddenly came to a halt. Smith discovered that his purchase order had been cancelled. Assuming that there must have been some form of clerical error he went in to speak with one of his supervisors.

'We've decided not to go with colour,' he was told.

'Not to *go with* it?' he said.

'We're not going to do it.'

Smith recoiled. 'That's nuts,' he exclaimed. 'The future is colour and you guys own it completely.'

'That may be true,' Smith's supervisor continued, 'but it's a corporate decision to go with black and white.'

Smith was stunned. The PARC scientists may have been ivory-towered researchers, but they knew a valuable discovery when they saw one. Xerox was throwing it all away. Following his firing, Smith found himself at a loose end. He was bummed at having lost a steady paycheque, but more so about no longer having access to the frame buffer. At the time, even a powerful computer was barely capable of throwing a single monochromatic image up on screen, let alone trying to coax movement out of it. What was he going to do? Keeping his ear to the ground, Smith started to hear about another frame buffer – the second one in the country – being housed at the University of Utah. Determined to finish what he had started, he packed up his belongings and headed east. Accompanying him on the journey was David DiFrancesco, a hippie-artist friend who was heavily into the motorbiking scene and had also worshipped at the SuperPaint altar. The two arrived in Salt Lake City in Smith's Ford Torino and found themselves in the middle of squaresville. Thanks to its high level of government funding, the University of Utah was one of the straighter education institutions in the country – as notable for its military-regulation haircuts as for its pioneering computer graphics. 'A couple of unruly long-haired artistic hippies weren't consistent with the defense department, let's put it that way,' Smith says. He and DiFrancesco were hurried through, but not before Smith received another lead. It transpired that there was another frame buffer – *a third!* –

in existence. In fact, they had just missed it. Alex Schure had recently visited Utah, and a frame buffer was among the items he had purchased and taken back to New York Tech with him.

'Who should I look for there?' Smith asked.

'Get hold of Ed Catmull,' he was told. 'He's a former University of Utah student.'

Smith and DiFrancesco spent the last of their money on plane tickets and jumped on a flight to Long Island. They arrived at Schure's Westbury Campus in the middle of a heavy snowstorm and sought out the Computer Graphics Laboratory. Walking in, they came face to face with a person who fitted Catmull's description.

'Ed?' Smith began.

This was actually Malcolm Blanchard, another New York Tech employee. From behind Blanchard another figure emerged. He walked up to the two visitors. 'I'm Ed Catmull,' said the man, who wore glasses, a neatly clipped beard, and had his shirt done up to the top button. 'Can I help you?'

To the casual observer, Alvy Ray Smith and Edwin Catmull were about as different as two people of the same age could be. Where Smith was an exemplar of the counterculture, with long hair and a taste for dope-smoking and LSD, Catmull was extremely straight, with a Mormon wife and two young Mormon boys at home. However, they did have something in common. Catmull, like Smith, was staunchly opposed to the Vietnam War. While Smith had been able to dodge the draft on account of his studies, Catmull had instead become a conscientious objector. The results were dire. Not only was he forced to undergo a year-long legal battle, but he had fallen out with both his church and his family. The Mormon Church vehemently backed the government's decision to send troops to Vietnam, while Catmull's father was a Second World War veteran and deeply embarrassed by his son's antiwar views. Sensing a unified cause, Catmull was all too happy to hire Smith and David DiFrancesco to work in the Computer Graphics Lab. All involved realised that they had struck gold by stumbling upon New York Tech. They entered into a honeymoon period during which it was all they could do to tear themselves away for the time it took to eat lunch and dinner. 'We stayed awake as long as possible and worked as hard as we could every day,' says Smith. 'When we fell asleep, we'd do so for the minimum amount of time possible and then get right back to it. It was incredibly exciting!' As would prove to be the case with Pixar, the atmosphere in the Computer Graphics Lab was collegiate to a fault. They had no set direction whatsoever – save advancing computer graphics in some positive way. By comparison, life at New York Tech made the lack of supervision at Xerox PARC seem like a military boot camp. There was also no shortage of equipment. Dr Schure would do whatever was in his power to ensure that New York Tech remained the leading computer-graphics centre in the world.

Smith and Catmull decided that their patron was one of the most fascinating men either had ever encountered. 'Our vision will speed up time, eventually deleting it,' were the first words Schure said to Smith. He didn't converse in the manner in which normal human beings interacted, but, rather, in a long stream-of-consciousness diatribe with no discernible direction or pattern. The only negative about working for Schure was his habit of declaring that traditional animators were going to be put out of a job by the forthcoming computer-graphics revolution. He didn't mean it quite as maliciously as it came across, but it did nothing to endear the Computer Graphics Lab to the traditional animators that they would have liked to have hired. 'Why would we want to help you put us out of business?' the animators figured. Smith and the others would beg Schure not to say these things. 'Alex, it doesn't work that way,' Smith pleaded. 'Computers can't take away the art. You can get rid of the groundwork, but you can't take away the art. Come on – don't say that. That's horrible for us.'

☐ ☐ ☐

After several years of innovative work, Smith and Catmull began to grow restless. Although the setup at New York Tech was idyllic, they were becoming increasingly certain that Schure was heading nowhere fast. A good case in point was *Tubby the Tuba*. Realising that computers graphics were going to take some time to perfect, Schure had established a separate, hand-drawn animation unit to create the feature film that would announce the existence of New York Tech to the world. With customary aplomb, he launched into it fully, hiring a team of veteran animators and the highest-profile voice actors he could find. The film's titular tuba was voiced by the comedian Dick Van Dyke. There was one slight problem: despite his complete lack of experience, Schure himself had insisted on directing. The results were catastrophic. The animation was shoddy, the artwork looked flat and the plot was boring. 'Anything that could be wrong with animation was wrong with it,' Smith says. Catmull agreed. 'It was awful, it was terrible; half the audience fell asleep at the screening,' he recalled. 'We walked out of the room thinking, "Thank God we didn't have anything to do with it, that computers were not used for anything in that movie!"'

By now both men were disillusioned with the New York Institute of Technology and with Alexander Schure. They wanted out. 'We realised that no matter how wonderful it was there on the north shore of Long Island this guy hasn't got it, he can't take us into the future like we thought he could,' Smith says. In the meantime both he and Catmull had started making yearly (secret) pilgrimages to the Walt Disney Studios in

Burbank, California. 'Why on earth isn't Disney doing computer graphics?' they asked one another. Because of its name value, Disney had its pick of animation talent, along with a great deal of money, but was sadly lacking in ambition.

Thankfully, in 1979 the West Coast finally came calling. More accurately, the movie director George Lucas did, and more specifically Lucas's real estate agent did. If anyone was going to understand Smith and Catmull, it was George Lucas. He, too, had come of age in the sixties, as part of a generation dead set on shaking up their chosen field with a much needed creative revolution. '[The Hollywood studio system] died when the corporations took over and the studio heads suddenly became agents and lawyers and accountants,' the filmmaker once said. 'The power is with the people now. The workers have the means of production.' After making the movie *American Graffiti*, Lucas had scored big with his sci-fi blockbuster *Star Wars*. With the proceeds, he set about expanding Industrial Light and Magic (ILM), the dedicated special-effects unit he had installed in an old warehouse on Valjean in Van Nuys. It was clear that computers were going to be an increasingly important part of the film-making process and Lucas was eager to have a team capable of handling it. His technical knowledge was limited, however. John Dykstra, the special-effects whiz who had created a computer-controlled camera for Stanley Kubrick's *2001: A Space Odyssey* before joining ILM, once complained that if Lucas could have made *Star Wars* by hanging up a black sheet in his garage and waving model space ships around on broomsticks to simulate flight, he would. Through some recommendations, Lucas had been pointed in the direction of Schure's Computer Graphics Laboratory. When Smith and Catmull heard that the filmmaker was interested in speaking with them, they wasted no time in flying to Marin County, California, for a meeting. Lucas himself didn't show. Instead he sent his representatives – or 'minions' as

Alvy Smith called it. The three items that Lucas wanted, his staff relayed, included a digital video-editing system, a digital audio synthesiser and a digital film printer which could transfer images from film to computer, manipulate them and then transfer them back again. Could Smith and Catmull create those? 'You bet we can,' they said.

□ □ □

Lucas hired Smith and Catmull, along with some other members of the New York Institute of Technology, and installed them in an office in San Anselmo. Early on, the pair did a rough, back-of-an-envelope calculation to try and work out how much it would cost to create a feature-length film on computer. If they had previously thought of themselves as ahead of their time, then these numbers confirmed it. With the available technology a ninety-minute computer-animated movie might cost in the region of ... *$75 billion*. This at a time when the average big-budget Hollywood film was, maybe, $15 million. 'You did the computation and figured out that if you needed *this* many instructions to compute a movie, and it would cost *this* much for a million instructions per second, how much would it cost to draw the instructions per second that we thought we would need?' Smith says. 'It was up in impossible numbers.' Until Moore's Law caught up with the work they were doing, Smith and Catmull figured that they would sit tight, keep honing their techniques and help George Lucas out wherever possible with his own projects. That is, if he ever called.

By now production of the *Star Wars* sequel, *The Empire Strikes Back*, was underway, but there was no request whatsoever that the Graphics Group get involved. In fact, the word that got back to Smith and Catmull was that the film would contain no computer graphics whatsoever. The realisation of what had happened hit Smith like a ton of bricks. *George just wants us to build machines for him*, he realised. To Smith and Catmull, hardware was nothing more than tools. What Lucas was doing was the equivalent of hiring a team of virtuoso painters and then having them spend their time sorting paintbrushes into size order. All they wanted – what they had assumed George Lucas wanted – was to create animated feature films on computer. There had evidently been a grave misunderstanding. 'It took us about a year to realise that George didn't get us any more than Alex Schure did,' Smith says.

It was from Paramount Pictures that Lucas's computer division finally got its big break. Paramount was in production on its own science-fiction epic, *Star Trek II: The Wrath of Khan*. The producers had approached ILM about creating a short sequence using computer graphics for one of the movie's big set pieces. What they were after, they said, was a sixty-second sequence to show off the plot's so-called Genesis Device, a machine which could spur an otherwise barren planet to life, creating primitive life forms such as vegetation. Smith saw this as more than a chance to showcase to the world the work that the Graphics Group was doing; it was also an opportunity to win over their as yet unconvinced boss. In his own words the aim was to create something that would 'knock George's socks off'. This wasn't just about the visual content. If Smith was really going to impress, his sequence would have to include the kind the acrobatic camera movements that would be impossible to carry out in any other way. The shot would start as a wide view of a barren planetoid from space, before journeying down to the surface, riding a wave of fire, travelling across – and through – newly formed mountain ranges, and ultimately pulling back to where it had started; now revealing an earth-like planet teeming with life. It was a technological *tour de force*. When Lucas saw the finished product he was a convert. 'Great camera move,' he told Smith, ducking into the office on a rare visit. From then on, Lucas's films relied increasingly on computer graphics.

During this time, Smith and Catmull continued making their annual pilgrimages to Disney. On one of these trips they met a young animator by the name of John Lasseter, who was both enthusiastic about computer graphics and unafraid that they were going to make him creatively redundant. 'He wasn't frightened of us,' Smith recalls, still amazed. The more the two men discovered about the 26-year-old upstart, the more impressed they became. He might just have been the most naturally talented animator either man had ever seen.

❏ ❏ ❏

Born in 1957, Lasseter's boyhood was the familiar story of the child who grew up wanting to work as an animator in the movies. Although he would go on to become one of Hollywood's more gregarious personalities, as a youngster he was shy and more than a little weird. At the age of five he won first prize in an art competition sponsored by the local grocery store, for a picture he had drawn of the ghoulish Headless Horseman. He loved reading the funnies that ran in the newspaper, but free time was mostly reserved for whichever animated shows he could watch on television. 'I loved cartoons more than anything else,' he said years later. 'My parents could not get me out of bed on a school morning, but come Saturday morning, boy, 6.30 – boom – I was out of bed, bowl of sugar-frosted flakes, six inches from the TV and waiting for the cartoons to start ... It was just my time.' After graduating in 1975, Lasseter applied to study at the California Institute of the Arts. Widely known as Cal Arts, this was an art college based in Valencia, Los Angeles County, which had been founded by Walt Disney himself more than a decade earlier. He was in the first year able to take a course in Character Animation, which taught traditional Disney-style rendering. His classmates included future directors such as *Edward Scissorhands'* Tim Burton, *Aladdin*'s John Musker, *The*

Nightmare Before Christmas's Henry Selick and *The Incredibles*' Brad Bird.

Following Cal Arts, Lasseter picked up a job working as an animator at the Disney Studios proper. At the time that he joined, Disney's animation studio was in a state of flux. The year before, Disney's chief animator Don Bluth had left the company, along with seven other animators and four assistants, to create animated feature films for Steven Spielberg. The animation department itself had shrunk from an all-time high of 650 to just 200. Nor were the films they were working on all that impressive. Due to budgetary cuts the newer projects on the production slate looked decidedly worse than those created forty years earlier. Many of Disney's greatest innovations had long since been consigned to the scrapheap. Chief among these was the multiplane camera, which was capable of producing a stunning three-dimensional animation effect by filming multiple pieces of artwork moving at differing speeds and distances from the camera. The Cal Arts animators had grown up watching the classic Disney films of the 1940s and 1950s and were keen to bring many of these techniques back.

'This would be a great multiplane shot,' Brad Bird said of one scene they were shooting for *The Fox and the Hound*. 'Let's show audiences that we can still do that stuff.' The Disney accountants went away and worked out a cost, returning with the gloomy estimate of $200,000. 'It's totally unworkable,' they said. Bird realised that the numbers had been predicated on exhuming the old camera from storage and then hiring four union workers to operate the hand cranks, as if this were still 1940.

'Why don't we just use a computer?' he said, bewildered.

Soon afterwards Bird was hauled into the office for a dressing-down. 'They were basically saying that if I'd stop complaining about quality, I could hold on to my job,' he recalled. 'I said, "I'm complaining about stuff your master animators taught me to complain about. So either I'm getting fired, or I'm selling out everything you guys supposedly stand for.'" They fired him.

In addition to cartoons, the new generation of would-be animators had grown up on a steady diet of high-octane Hollywood blockbusters. John Lasseter had been in his sophomore year at Cal Arts when *Star Wars* was released. He went to the cinema to check it out. The experience left him literally shaking in his cinema seat. 'We wanted to do *Star Wars*-level entertainment with animation,' he said years later. 'We were so excited, but the creative leadership didn't know how to handle us and worked really hard to keep us under control.' The feeling that Disney's animation department perhaps wasn't quite the magic kingdom it had been made out to be was further intensified when the 1982 movie *Tron* was screened as a special presentation to the young animators. The film had been co-written by Alan Kay's wife, Bonnie MacBird, and proved to be a ground-breaking advance in computer graphics. When Lasseter got out of the screening theatre he struck up a conversation with Glen Keane, a fellow Cal Arts student who had graduated three years ahead of him. They had both been blown away by what they had seen, but faintly depressed at what it suggested about the work they were doing. 'I felt like [the people who made *Tron*] were doing what was really exciting and we were doing dumb, flat drawings,' Keane later said. 'There the computer was moving in three dimensions ... It was all so real, and motorcycles were moving around and it was just – wow! It was eye-candy!'

The two animators decided to team up to create a film that would combine computer-animated backdrops with traditional hand-drawn characters. They settled on using Maurice Sendak's *Where the Wild Things Are* as a jumping-on point and set about creating a short test film based on it; calling in the services of the New York-based Mathematics Application Group, Inc. (MAGI) who had created many of the special effects for *Tron*. While

MAGI put together the three-dimensional computer-generated background, Keane took charge of the hand-drawn character animation, which was then composited together under Lasseter's exacting direction.

It was at this time that Lasseter first met Alvy Ray Smith and Edwin Catmull. As a junior animator, one of Lasseter's jobs involved working in the 'morgue', the basement archive where the drawings for every Disney film ever made were stored. Showing off his Disney credentials, he offered to give a tour of this to his two new acquaintances.

'What do you guys want to see?' Lasseter asked.

Before Catmull could say anything, Smith spoke up.

'Do you have Preston Blair's dancing hippos from *Fantasia*?' he said.

Lasseter walked over to a ledger, looked up Preston Blair and shepherded the duo over to a filing cabinet from which he withdrew a manila folder. Inside were the original images from *Fantasia*, then forty-three years old. Smith especially was ecstatic. The film was a personal favourite, bringing back all kinds of memories. '*Fantasia* was quite a famous movie among certain members of the sixties generation,' Smith laughs. Like many hippies, on occasion he would drop acid and then go and sit in a darkened cinema, spending ninety minutes – with tears pouring from his eyes – tripping out to a film like *The Jungle Book* or *The Aristocats*. Walt Disney, of course, could not have foreseen a generation of long-haired LSD lovers forming a substantial part of his target audience, but there is also no doubt that he was not himself immune to the allures of psychedelia. In the case of *Fantasia*, Disney had employed the services of a mescaline subject by the name of Kurt Beringer – previously best known for his 1927 monograph *The Mescaline Inebriation* – as his chief visualist for the project. The result, at least according to the movie's promotional material, is an animated film in which the viewer can 'see the music [and] hear the pictures'.

All this went through Alvy Smith's mind as he flicked through the drawings. At the end of the day Catmull made his excuses, said how great it had been to meet John Lasseter, thanked him for showing them the archives, but pointed out that he really ought to get back home to his family. Smith and Lasseter were left alone in the basement.

'What else would you like to see?' Lasseter asked.

Smith pondered the question for a moment.

'How about the "Pink Elephants on Parade" scene from *Dumbo*?' he finally said.

Lasseter looked that one up, found it and the two men began flicking through those drawings, too. This was better still. If anything, *Dumbo* is an even trippier film than *Fantasia*. The insanity of the Pink Elephants sequence especially, in which the titular floppy-eared elephant sees a series of phantasmagorical 'things you know that ain't' after drinking from a barrel of water spiked with alcohol, remains arguably the most psychedelic piece of animation ever created. It can be read as a Disneyesque ode to the subject of LSD-fuelled consciousness expansion. Dumbo discovers his ability to fly not through any obeisance to the status quo, but, rather, in the throes of a psychotropic visionary state.

Smith came away from the meeting feeling elated. He wanted to hire Lasseter immediately, but the young animator was under contract with Disney. Lasseter, meanwhile, was having his own problems. Based on the success of the *Where the Wild Things Are* test, he was given permission to start work on developing a longer film, based on a story he had read and liked in *The Magazine of Fantasy and Science Fiction*. The next six months were spent in pre-production, developing storyboards for the film that would be his directorial debut. One day he got an alarming message: he was to give an immediate presentation on the film to Ron Miller, the head of Disney Studios, and Ed Hansen, the manager of the animation department. This would normally have been nothing out of the ordinary, but Lasseter had picked up on some coolness from Hansen about the direction he was heading in. Nevertheless he gathered up the storyboards and went to the meeting. His fears proved to be well founded. Neither Miller nor Hansen was smiling when Lasseter walked in. They listened to his pitch in silence and afterwards hit the wannabe director with a fateful question.

'How much is this going to cost?' Miller asked.

'No more than any of our other movies,' Lasseter responded. 'With computers we—'

'The same price?' Miller said.

'Yes, sir.'

'What's the point of doing this if it's going to cost us the same? I thought the whole point of animation on a computer was to do it cheaper and faster.'

With that, Miller got up and left the room. Lasseter headed back to his desk in stunned silence. Five minutes later his phone rang for a second time. It was Ed Hansen. 'Congratulations,' he said. 'You've completed your assignment. Your contract with Disney is now terminated.' Lasseter was fired.

□ □ □

In 1983 Ed Catmull was at a Siggraph computer-graphics event in Long Beach, California.* It was a somewhat unique setting for such an event, taking place on the retired *Queen Mary* ocean liner, which had been docked and converted into a convention centre. Catmull was on a business call to Smith back at Lucasfilm

---

* Short for 'Special Interest Group on GRAPHics and interactive techniques', Siggraph has been running since 1974 when the inaugural event was held for a crowd of six hundred in the city of Boulder, Colorado. Today it routinely draws in excess of 20,000.

when a familiar figure walked past his open door, offering a quick greeting as he did.

'Who was that?' Smith asked down the phone.

'That kid, John Lasseter, just walked by is all,' said Catmull.

'Is he still at Disney?'

'I don't think so. As far as I know, he's not working anywhere right now.'

That was all Smith needed to hear.

'Ed, what are you waiting for?' he said excitedly. 'Get off the phone and go hire him.'

Catmull did. Lasseter joined the Graphics Group back at Skywalker Ranch in Marin County. His initial job description was that of interface designer. 'We couldn't have called him an animator or George would have axed it,' Smith says. 'Nobody knew what [an interface designer] was but they didn't question it in budget meetings,' Lasseter, who was by no stretch a computer expert, later said. Even with the knowledge that a feature-length film was out of reach, the Graphics Group team had not been idle. Using photorealism as their benchmark, they set about creating computer-generated still images to perfect their various techniques. One depicted a road winding away into the distance, set against a sky in which a beautiful rainbow is shining. It was a striking image: one that, purposely or otherwise, echoed Stewart Brand's similar 'Stay hungry, Stay Foolish' photograph which had so inspired Steve Jobs. Another image showed pool balls on a green felt table. Created the year that the Macintosh went on sale, the numbers on the balls spelled out the date: *1-9-8-4*. Before long the team was ready to create its first film, a whopping two-minute short to be directed by Alvy Smith, animated by John Lasseter, and with technical contributions from everybody else. The title Smith selected was a riff on the art film *My Dinner with Andre*. It would tell the story of a child pursued by a giant bumblebee. As with Steve Wozniak at the Homebrew Club, the purpose of *The Adventures of Andre and Wally B* was not to make money. It was to showcase to other people in the industry and have them stand up and scream in delight at what had been done. The venue selected for the film's premiere was the following year's Siggraph conference. The team finished it with literally days to spare.

The great reception afforded *The Adventures of Andre and Wally B* should have been the triumph to end all triumphs. Sadly it was marred by other, real-world events. News had emerged at Lucasfilm that George Lucas and his wife, Marcia, were divorcing after fourteen years of marriage. Lucas was shattered. When the divorce was finalised it cost him in the region of $50 million. Suddenly money was tight. Even though Lucas had come around to the work being carried out by the Graphics Group – who had recently finished creating a dazzling sequence for the Steven Spielberg-produced *Young Sherlock Holmes* – the division was nowhere near turning a profit. Before they knew what had hit them, the team was being sold off as part of a firesale. Smith and Catmull discussed their options with one another. A conversation with Jim Clark provided a few pointers. Clark had known Catmull since their University of Utah days, and Smith since they had worked together at New York Tech. He had since gone on to found his own company called Silicon Graphics, which was then the hottest start-up in Silicon Valley. Clark spun them a line about how liberating it was to be involved in big business. He had distrusted authority ever since spending four bitter years in the navy, being badly mistreated by his superiors who thought of him as a delinquent no-hoper. Now he was rich, and there were no stuffed suits telling him what to do.

'What are you two still doing working for The Man?' he asked.

Smith and Catmull had no answer. After pursuing their dream for ten years they were beginning to think that nobody understood what it was they wanted to achieve. Alex Schure had the right idea, but didn't have the means to get things done. George Lucas, at least initially, had the means, but also the wrong idea. Maybe it was time to break out on their own? As he had done when he was learning about animation, Smith visited a bookstore and picked up a 'how to' business manual about starting one's own company. That night he sat down with Catmull and drew up a business plan. Since Moore's Law stated that they were still about ten years away from being able to create a feature-length computer-animated film, they would need a revenue source in the meantime. They decided that they could sell the powerful graphics computer that had been developed by Catmull and his team. Next up was choosing a name for the company. The machine in question was called the Pixar Image Computer, named by Smith and another employee of the Graphics Group called Loren Carpenter. Having grown up in New Mexico, Smith had wanted a word that sounded technical but was Spanish in tone.

'What about Pixer?' he said, throwing out a name.

Carpenter thought for a couple of seconds.

'It's good,' he answered, 'but how about Pix*ar*? You know, as in "radar".'

Smith liked it. *Pixar it was*. More than a name, what Pixar really needed was someone to fund it. The company had a fearsome burn rate of millions of dollars per year. Selling the Pixar Image Computer was one way to make money, but they needed major financial backing to keep afloat in the meantime. The venture capitalist that wound up investing was an unlikely one. He was thirty years old, two years older than John Lasseter and a decade younger than Smith and Catmull. He was no stranger to the situation in which Pixar's co-founders found themselves, having just come from an ugly divorce himself – in his case a split from the company that he had helped start. It was Steve Jobs.

□ □ □

Jobs had first heard about the work that Smith and Catmull were doing from Alan Kay, while both were working at Apple. 'You should really see what these guys are doing over at Lucasfilm,' Kay mentioned to Jobs one day while the two men were taking a walk. They chartered a limousine and drove over to Lucas's Skywalker Ranch, where they spent an afternoon touring the facility and chatting with Smith and Catmull. Smith had previously been introduced to Jobs at a design conference at Stanford several years earlier, but this was the first time he had had the chance to speak directly with the Apple co-founder one-on-one. He noted that Jobs was impressed by the work they were doing, but that he also came across as arrogant. Never one to react well to not being the centre of attention, Jobs commandeered a whiteboard and, in Smith's words, drew 'a curve [showing] the worth of Apple from nothing to a large number. I guess we didn't express enough awe because Alan Kay coaxed us to see what an amazing achievement it had been.' It would prove the start of an often bristly relationship between Steve Jobs and Smith.

According to Jobs he went back to Apple and tried to convince them to buy the company. Then CEO John Sculley recalls no conversations about Apple acquiring Pixar, despite the fact that 'Alan Kay and I talked a lot about what George [Lucas] was doing'. Nonetheless, Jobs invited Smith and Catmull to his home in Woodside, and had his chef serve them up a lunch of edible flowers while reiterating how interested he was in the work they were doing. There was one thing stopping him, however, and that was the sale price. In an effort to recoup some of his losses, Lucas was asking for $15 million for his stock in Pixar, while another $15 million would have to be guaranteed to fund the operation. Jobs thought this was too high. As far as he was concerned, Lucas's desperation was almost palpable. Give him a few months with no takers, Jobs figured, and that price would fall dramatically. He was correct in his assumption. Jobs was keen to buy Pixar outright, but both Smith and Catmull rejected the idea. They had no desire to lose control of their fate yet again. What they wanted was a VC willing to invest in the company for a later return. For a heated few days Catmull held out serious hope that Disney might come through with an offer. 'We need the money and this would be a perfect fit,' he enthused. Several employees, including Roy Disney, were keen, but Frank Wells, Disney's president and chief operating officer, quashed the deal. 'We're not an R&D company,' Wells argued. 'It's going to cost twice as much as they say, and I don't believe it will ever save us a dime.' Disney was out. Another source of interest came – ironically enough – from Ross Perot, who weighed in with an offer from Electronic Data Systems (which had been bought by General Motors in 1984 for the sum of $2.5 billion). Perot was offering an astronomical figure. Did Jobs have a counter-offer? 'How about $5 million?' he said. Smith and Catmull laughed the Apple co-founder out of town.

The EDS offer looked to be a promising one, until it too fell through when the news came down the wire that GM had forced Perot to step down from its board of directors. Finally there were no other options. It was time to call Steve.

'Make your move again,' Smith told Jobs in a phone conversation. 'Lucas is desperate and doesn't know what to do with us. He'll probably accept your offer.' Jobs did and Lucas, as predicted, agreed to the terms. On 3 February 1986 Jobs wrote out a cashier's cheque for $5 million for the purchase of the Pixar technology and guaranteed an additional $5 million by way of funding. In return he received what amounted to 70 per cent of the spinout company. The employees and management of Pixar, then numbering forty people, including Smith and Catmull, would own the other 30 per cent. On the founding documents, Smith, Catmull and Jobs were listed as company directors, while Jobs was also named chairman of the board.

Pixar had gone from one multimillionaire to another.

□ □ □

Hot on the heels of *The Adventures of Andre and Wally B*, the Pixar crew began developing another film to premiere at the following year's Siggraph. This was going to be the first film under the Pixar banner, so the pressure to perform was even greater than it had been previously. Lasseter came up with the idea of doing a short involving lamps, inspired by the Anglepoise Luxo lamp he kept on his desk. 'I had done some student films with them and they were kind of fun,' he recalled. The result was *Luxo Jr.*, a two-minute short about an adult and child Luxo lamp playing with an inflatable ball. The film was even better received by the crowds of Siggraph than the group's previous effort. Amidst the show reels and special-effects demos on display at the conference, Pixar's film shone out like a beacon. It had... *a story?* After the screening Jim Blinn, a pioneer in computer graphics, ran up to Lasseter.

'John, John, I have a question for you,' Blinn said.

Lasseter looked at him, wondering what he could possibly say about graphics to a man with a Ph.D. in computer science.

'Okay,' he responded.

'Was the parent lamp a mother or a father?' Blinn asked earnestly.

It was a moment of triumph for Lasseter. 'That simple statement made me realise that we [had] achieved something that computer animation hadn't before,' he recalled. 'What was interesting to people was the story and the characters, not the mere fact that it was made with a computer.' Steve Jobs was present for the Siggraph triumph, as he would be for that year's

Oscars when *Luxo Jr.* was nominated for an Academy Award for Best Animated Short Film. It didn't win, but its mere inclusion suggested that the tide was starting to turn with regard to the public's acceptance of computer animation. Jobs saw another, more immediate benefit, too. Just as *The Adventures of Andre and Wally B* had been a calling card for the work being done by the Graphics Group at Lucasfilm, so *Luxo Jr.* could be used as a two-minute advertisement for what it was possible to create using a Pixar Image Computer.

To the Pixar employees Jobs was every inch the wealthy jet-setting Silicon Valley tycoon. He was less hands-on than Alex Schure and, if possible, even less so than George Lucas. At least Lucas made films. There was certainly none of the mucking in that had characterised Jobs' early days at Apple. Smith still found him arrogant, and difficult to deal with. 'Nobody has much of a kinship with that guy,' he says. 'He doesn't allow it, as far as I can tell. He insists on a master–slave relationship and the notion of being kin with him just doesn't ring true.' On the subject of whether he felt any connection with Jobs based on their shared hippie background he was quick to rebuff this: 'I know all about his going off to India and I know about his vegetarianism, but I'd never seen anybody named Steve Jobs that matched the kind of person that I think of as counterculture.'

Jobs did bring one characteristic that went beyond being a money man. With his vision for spotting how technologies could be adapted for the mass market he saw that Pixar's 3-D graphics-rendering program had the potential to bring to the masses the kind of high-end graphics capabilities that had previously been the province of those with 32-bit VAX machines and Cray Supercomputers. This wound up becoming a mass-market product called RenderMan. 'We're talking software that three or four years ago was only available to Lucasfilm and Industrial Light and Magic people, and we're moving it down to microcomputers,' boasted RenderMan's product manager, Paul Yarmolich, at the time. But it quickly became evident that there simply wasn't a market for high-end graphics-rendering in the same way that there was for personal spreadsheets or desktop publishing. Pixar may have bragged about the abilities of its RenderMan software but, as a spokesperson for Silicon Graphics commented, 'right now, they can't point to anyone who's using it'.

In the meantime, the company had managed to find two major clients for its Pixar Image Computers. One was Disney, which had finally come around to the idea of computers being used in the animation process, even if it didn't publicly acknowledge it for fear of 'taking away the magic' of all of those hand-drawn images. Disney wound up using the Pixar Image Computers for CAPS (Computer Animation Production System) which would allow digitisation of the traditional, hand-drawn animation process.

The other new customer, strangely enough, was the CIA. Just as Smith was reconciling himself to the fact that his hard work liberating computers in order to create art was going to go straight back into the military-industrial complex, he had another bit of bad news. In order to sell computers to the government, he would have to undergo a background check. Suddenly Alvy Ray Smith, the protest-marching, dope-smoking, free-loving hippie, was going have his past gone over with a fine-tooth comb.

'Shoot, I'll never get a top-secret clearance,' he complained. 'They'll take one look at a picture of me back in the sixties and fail me right there. Steve's not going to get one either, because we were all dropping acid back then.'

Eventually one of Smith's colleagues calmed him down, telling him that the clearance process had been litigated so many times that as long as a person didn't outright lie on the questionnaire, the government was almost legally obliged not to turn him or her down. 'You can be a card-carrying homosexual communist, and, as long as you say so, they've got to pass you,' Smith laughs. 'The thing is that you just can't be a blackmail subject. As long as you're open and honest, you can't be blackmailed. That's the theory.' He agreed to give it a whirl. After writing his name, date of birth and social-security number, the first question on the form was about past illegal substance use. Smith noted with amusement that whichever square had written the questionnaire had misspelled 'marijuana' by including an 'h'. His amusement turned to fury when he looked further down the page and saw that heroin was included as another option. *Marijuana and heroin on the same list!* Worse, this wasn't a checklist, just a general 'yes or no' question. Smith gritted his teeth and ticked 'yes'. The next question was grimmer yet: "Have you ever bought, sold or grown any of the same list?" 'I realised that if I said yes to that one, I'm basically making myself out to be a heroin dealer,' Smith says. 'So I lied and checked "no". The rest of it was easy to fill out.'

Amazingly for Smith, he got his security clearance. He was in the same room as Jobs, when Jobs received a phone call from a government official regarding his own background check.

'Yes, sir, I promise I'll never take it again,' Jobs could be heard saying over the phone, giving a big pantomime wink to the Pixar employees as he did so. Smith claims that the moment was the closest thing to camaraderie the group ever achieved with Steve Jobs.

□ □ □

Like the NeXT Computer, the Pixar Image Computer wasn't selling in the numbers it should. In fact it wasn't selling, period. By 1988 the company had only sold 120 units, a dismal number. The main impediment was the cost. The machine was valued at $135,000, but in order to run it one also required a $35,000 workstation from Sun Microsystems or Silicon Graphics. Pixar began work on second-generation, lower-cost models, but even these sold for $30,000. Jobs was beside himself. What kind of company had he bought? 'Lucky for us, Jobs could not sustain the public embarrassment of the failure of his first investment after being pitched out of Apple,' Smith says. Literally the only thing keeping Pixar alive was Jobs' pride. The sales figures for NeXT weren't public knowledge, but it was still clear to any outsider that they weren't high. If Pixar went belly up, too, Jobs would be regarded as a one-hit wonder; a man who got lucky by meeting Steve Wozniak and was incapable of recreating that early success.

As they had at Lucasfilm, Smith and Catmull began to worry about their future. They drew up contingency plans and business proposals to try and sell themselves to another buyer in the event that Jobs dropped out. Only he never did. Each time that it seemed as if he was reaching his breaking point, Pixar would pull off some miraculous feat to earn it a stay of execution. In 1989, the year after *Luxo Jr.* had wowed the world, the team won an Academy Award for its five-minute short, *Tin Toy* making it the first computer-animated film to win an Oscar. But while the award boosted Jobs' ego, it did nothing for his bank balance. Funding Pixar continued to be like throwing money into a black hole. He told the team that they would have to come up with another way to monetise the company. Ralph Guggenheim, a video graphics designer who had been working with Smith and Catmull since the heady days of New York Tech, came up with a plan: the Pixar team could hire themselves out to work on television commercials. Jobs approved the new strategy, but instead of continuing to pour money into the company he decided to engage in a bit of corporate ju-jitsu. At the time Smith and Catmull each owned 4 per cent of Pixar. Jobs wasn't looking to share.

'I'll give you the money,' he said, 'but I want some of each of your percentage.'

The Pixar co-founders were so desperate to see their dream for a computer-animated film come to fruition that they acquiesced. Over the next few years Jobs systematically whittled away the Pixar employees' share in the company down from 30 per cent to ... zero. 'The end result was that Jobs had poured $50 million into the company and finally taken all the equity,' Smith recalls. As far as Jobs was concerned he was simply making the best of a bad situation. If any potential buyers had come forward with $50 million to spend he would have unburdened himself of Pixar in an instant. After all, when it came to hardware companies that weren't selling any hardware, he already had NeXT on his books. Why have two companies that were doing the same thing? Besides, NeXT continued to be Jobs' focus. Between 1986 and 1992 some estimates put Jobs at the Pixar offices no more than five times in total. That arrangement was just fine with Smith and Catmull. 'Ed and I quickly worked out that the best way to run Pixar was to keep Steve as far away as possible,' Smith says. 'Because when he showed up he would distort the world and we would have to spend a week or two putting it back into shape.' Instead of Jobs coming to them, Smith and Catmull would drive the two hours to Redwood City each month for board meetings to report on progress.

Finally, after chipping away at their dream for the best part of two decades, it seemed like there was some forward momentum. In late 1990, Peter Schneider, Disney's head of feature animation, called the group at Pixar and asked whether they would be interested in making a feature film. Within three months they would thrash out a deal to deliver three of them. John Lasseter put on his thinking cap and started coming up with ideas. He had always liked the kind of buddy movies he had grown up watching. What if they could pitch an animated update of a film like *Butch Cassidy and the Sundance Kid*, only with the two leads played by toys? And they could clash with one another over which one their owner liked more. And then they could wind up on a dramatic chase, and ultimately become friends. The ideas began to flow, and Pixar geared up for its biggest undertaking to date.

Only Alvy Ray Smith wouldn't be a part of it.

□ □ □

Anyone who has ever seen one of Steve Jobs' famous Stevenotes presentations will be aware of just how performative they could be. This was not simply a show for the buying public. In board meetings Jobs was every bit the showman he was when the entire world was watching. Manning the only whiteboard in the room he would constantly jump up and down to illustrate the points he was making as he spoke. He liked to be provocative. 'What he would typically do when he walked into a board meeting was to say something outrageous,' Smith recalls. 'We'd say, "Steve, that's boloney", and he would just back off – because it frequently *was* boloney.' As did the people who regularly worked with Jobs, the Pixar crew quickly came to realise that this was Jobs' way of grabbing the attention of the room. Seeing Alvy Smith as the strongest personality in the group, Jobs would often make a comment and then stare directly at him to see whether he was going to react or not. Most of the time this arrangement worked well for both men. Then one day things turned nasty. Jobs was complaining that one of the boards for the Pixar Image Computer was late.

'But, Steve, one of your boards is late for the NeXT machine,' Smith said.

All hell suddenly broke loose. 'It was a typical exchange, but his response was bizarre beyond belief,' Smith recalls. 'He turned on me as a full-on street bully and started insulting my accent. Shouting at me and bullying me. I had never been in that position before. Nobody had ever treated me like that. I responded just as irrationally, and both of us went into full-blown rage, shouting at full voice, right in each other's face, four inches apart. You step back and think, "I've got to lose this. It's crazy, this is the guy with all the power and the money", but we were both irrational.' Midway through the heated confrontation, Smith – losing on decibels – marched past Jobs and began drawing on the whiteboard. 'There was this unspoken rule that Steve sat in front of the whiteboard and was the only person who could use it,' Smith says. 'It was unspoken, but I hate unspoken rules.'

'You … you… you can't do that,' Jobs protested, almost incandescent with rage.

'What, write on a whiteboard?' Smith responded.

The whole display was utterly ridiculous – a boardroom farce of the highest order. Jobs considered what to do for a second, then stormed out of the room. 'I never liked him since, I wanted him out of my life after that,' Smith says. A year later, Smith left Pixar and set up another company called Altamira, which dealt with image editing software. Jobs purchased a 10 per cent share. Non-confrontational to a fault, Edwin Catmull was slotted in as Pixar's sole company director. Pixar remained a quiet backwater of the animation industry as the team worked on their quirky buddy movie.

John Lasseter had come up with a working title for it, which would be a good enough placeholder until a better idea came along.

He was calling it *Toy Story*.

11

ANOTHER BRiCK IN THE WALL

'Technique requires predictability and, no less, exactness of prediction. It is necessary, then, that technique prevail over the human being. For technique, this is a matter of life and death. Technique must reduce man to a technical animal, the king of the slaves of technique. Human caprice crumbles before this necessity; there can be no human autonomy in the face of technical autonomy. The individual must be fashioned by techniques, either negatively (by the techniques of understanding man) or positively (by the adaptation of man to the technical framework), in order to wipe out the blots his personal determination introduces into the perfect design of the organization.'

Jacques Ellul,
The Technological Society (1964)

As a young boy from Chingford in Essex, the world of California – with its ten-lane motorways and oversized fruit and vegetables – seemed almost impossibly exotic to Jonathan Ive. Ive (known as Jony, pronounced 'Johnny', to his friends) was born in

1967, the year in which hundreds of thousands of young people converged on the Haight-Ashbury neighbourhood in San Francisco to celebrate the Summer of Love, and Fairchild Camera and Instrument – the East Coast corporation that would later spin off Silicon Valley's Fairchild Semiconductor company – built a metal-oxide semiconductor product for data-processing applications, which went on to become the industry standard. At the time, neither of these developments appeared to have any direct implication to him, of course. Aside from a few retrospective articles in the newspapers and the odd late-night showing of *Easy Rider* on Channel 4, the American counterculture might as well have been another world. Nor was Ive a technology geek. Like the vast majority of the British public, he had only been using computers for a handful of years; most of them spent frustratedly tapping away at machines like Acorn's BBC Micro, which was standard issue in schools, and Clive Sinclair's ZX Spectrum, which came packaged with a rubber keyboard. Part of Ive's disinterest in computers stemmed from the same difficulty he had run into elsewhere at school, and the same impediment shared by a great many people who choose to go into design as a career: he was dyslexic.

Academics aside, however, Ive had been a normal enough student at school; a slightly chubby teenager who played prop forward for the rugby team and spent his evenings and weekends as the drummer for a local band called Whiteraven, performing cheap gigs in draughty church halls. Had it not been for his later success, his biggest claim to fame might well have been that he attended the same state secondary, Chingford Foundation School, as David Beckham – albeit eight years apart, meaning that they were never there at the same time. What interested Ive most was design, a passion inherited from his dad, Mike, a former silversmith who had gone on to work as an HM Inspector for craft, design and technology in secondary schools. 'I remember always being interested in made objects ...' Jony Ive has said. 'I remember taking apart whatever I could get my hands on. Later, this developed into more of an interest in how they were made, how they worked, their form and material.' Instinctually he learned the basics: the fundamentals of form factor, how shape and colour define perception of an object, the names of one or two famous designers whose work he liked. By his teenage years his mind was already made up. He would become a designer. But what to design and make? He had the impending feeling that at some point he would most likely have to choose an area to focus his attention on. After some consideration he settled on cars, which seemed as good as anything, and applied to study automotive design. The affair didn't last long. Ive found the other students on the course to be too weird; kooky artists masquerading as designers. 'They were making *vroom, vroom* noises as they did their drawings,' he recalled years later in an interview with the *Observer*. He looked around for other options, packed his bags and moved up to Newcastle, where he threw himself headlong into the Polytechnic's industrial-design course.

While Ive impressed his lecturers with his dedication, he was frustrated by his almost total lack of drawing ability. 'I started hand-drawing and had a horrible time,' he said. It was in this capacity that he hoped that his old nemesis, personal computers, may have been able to help him. Computer-aided design had made great leaps forward in the 1980s, but all the machines that Ive tried were too unwieldy and user-unfriendly. Then, in the final year of his degree course, he discovered the Macintosh. 'I remember coming across this remarkable product,' Ive said. 'It was easy to use, and you could do little things like change the noise when you made a mistake. It sounds small, but at the time that was remarkable. I went from feeling stupid to feeling empowered. I somehow connected to the people that made it.' Eager to uncover more, he began reading up on

Apple's history and discovered that he could relate to a lot of the company's ideals. 'The more I learned about this cheeky, almost rebellious company the more it appealed to me,' Ive recalled. 'It unapologetically pointed to an alternative in a complacent and creatively bankrupt industry. Apple stood for something and had a reason for being that wasn't just about making money.'

□ □ □

Helping pay Ive's way through his degree course was a London-based design agency called the Roberts Weaver Group. The general manager of Roberts Weaver, a man named Phil Gray, was a friend of Mike Ive. Several years earlier Mike Ive had asked Gray to take a look at some of his son's drawings. Gray was impressed. 'For someone in his late teens, he had incredible craftsmanship,' he says. 'He had a real acute eye for detail, a very precise young man.' He took the matter to some of the other people he worked with and they agreed to give Ive a small stipend to aid with his living costs while studying. It wasn't very much, but it helped out. Perhaps more importantly, they made him an offer that would get his foot in the door: they offered him a summer job.

□ □ □

Clive Grinyer was a senior designer at Roberts Weaver. He hadn't been there for more than a few months, having recently returned from Palo Alto, where he had been working for the company that today is known as IDEO, which had grown out of David Kelley Design. Grinyer had studied at the Central St Martins School of Art and Design. After graduating he was given the opportunity to go and work for Bill Moggridge in San Francisco. Moggridge was the man every design student in England wanted to work for. He had opened up his first consultancy in London at the height of the Swinging Sixties and his second in Silicon

Valley a decade later. Bringing in exciting young designers from Europe (this at a time when work visas were easier to obtain) he started a revolution in the way in which high-tech products were conceptualised, created and ultimately presented to customers. In Palo Alto, Grinyer had been working with Moggridge on several exciting projects, including the GRiD Compass, the world's first laptop computer, which opened and shut with the now familiar clamshell mechanism. Now Grinyer was back in London, his San Francisco tan having faded. When he walked into the Roberts Weaver office on this particular morning in 1987 he immediately noticed the new intern with the weird haircut. Jony Ive was eight years younger than Grinyer, and sporting a shoulder-length mullet, which was intensely backcombed so that the fringe rose up vertically and then cascaded.* 'I thought he looked like a hairbrush,' was Grinyer's first thought. His second was, *oh my God, he's working on everything.*

It quickly transpired that Jony Ive was the best intern the Roberts Weaver Group had ever had. For the most part interns were a generally hopeless lot; teenagers whose main area of expertise seemed to lie in getting coffee orders for the team wrong and sneaking off early. Ive was different. A big project in the studio at the time was a set of pens for Japan's Zebra Pen Corporation. Ive churned out drawing after drawing, eventually finishing up with an elegant pen with a tactile ball and clip mechanism on top. He was obsessed with what he called 'the fiddle factor' – the deliberate design flourishes added for no reason other than to give the user something to play about with. The team at Roberts Weaver took to referring to it as an

---

* Today Jony Ive sports a fully shaved head, resembling one part Zen Buddhist monk and one part skinhead punk. It's difficult altogether to miss the symbolism of the original '1984' Macintosh commercial in which a group of shaven-headed drones are given the chance to unleash their full creative potential thanks to the power of Apple computers.

object's unique 'Jony-ness' and watched, helpless, as the pen became their boss's new favourite plaything. 'His designs were incredibly simple and elegant,' says Grinyer. 'They were usually rather surprising but made a complete amount of sense once you saw them. You wondered why we had never seen a product that way before.' Grinyer and Ive hit it off immediately. They shared a similar sense of humour and Grinyer liked the young intern, who was softly spoken and self-effacing, but quietly confident. For his part, Ive was fascinated by Grinyer's stories of working in California. He asked every question he could think of about life in the United States.

After Ive's internship was up, he left the Roberts Weaver Group and headed back to Newcastle. But he stayed in touch with Grinyer and when it came time for his end-of-year degree show he phoned him up and invited him to come and take a look. Grinyer dutifully drove the five hours up the M1 and spent a night grabbing fitful snatches of sleep on the floor of Ive's crummy student flat in Gateshead. Sharing the sleeping space with him were heaps of foam models which Ive had constructed for his final project: a hearing device with an earpiece to allow teachers to interact with deaf students. (All in dazzling iPod-white, of course.) The sheer number of prototypes was breathtaking. Where most students would build six models and then go off to the pub, Ive had built more than a hundred. 'It was incredible that he had made so many and that each one was subtly different,' Grinyer recalls. 'I imagine Charles Darwin would have connected with them. It was like watching a piece of evolution really. Jonathan's desire for perfection meant that every single model had a tiny change and the only way he could understand if it was the right change or not was to make a physical model of it.'

◻ ◻ ◻

Once he had finished at Newcastle Polytechnic, Ive had the chance to go out to California. His hearing-device project had netted him an RSA Student Design Award, which came with a travel bursary to go and work overseas for six weeks. The company that was chosen – Pitney Bowes, a manufacturer of office equipment, based on America's East Coast – was hardly inspiring, but following the completion of his time there, he was able to go off travelling. With the Student Design Award as his door opener, one of the studios Ive called into was Lunar Design, a consultancy based in Palo Alto, run by a man named Bob Brunner. Brunner was a child of the computer industry, his father having developed the first Winchester disk drives for IBM back in the 1970s. The meeting with Ive was pleasantly cordial ('Nice to meet you; that's some impressive work you've done; if there's ever an opportunity in the future ...'), and then Ive flew back to the UK. It wouldn't be the last he would hear of Bob Brunner.

Clive Grinyer didn't stay at the Roberts Weaver Group for long. He moved to Cambridge and started working as a designer at its Science Park, which had long been eyed as the UK's answer to Silicon Valley. With its high-tech buildings and well-manicured lawns, this part of Cambridge *almost* looked like Palo Alto if one was to squint, he reasoned. At the same time he had other ideas he wanted to pursue; ideas that didn't necessarily involve working with three hundred 'nerdy, geeky' engineers for the next twenty-odd years. One weekend he was back in London, visiting an old friend named Martin Darbyshire. Grinyer and Darbyshire had known each other since they were students together at Central St Martins in the early eighties. Darbyshire was one year older than Grinyer, but had been set back an academic year after suffering a nasty motorbike accident. Over dinner in a Kentish Town curry house, Grinyer made his pitch. He wanted Darbyshire to join him in setting up their own design consultancy. Grinyer had recently been approached by Commtel, a phone company based in Peterborough, with an offer to go and design a range of telephones for them. They had earmarked £20,000 for the job, but Grinyer had managed to persuade them to let him set up his own firm to service them, rather than acting as their employee. In other words, he had the start-up capital ready to go and was just looking for a partner. Darbyshire agreed.

For their first eight months, the two men ran their design business from the front room of Darbyshire's home in Finsbury Park. 'My wife and I moved into the back bedroom and we converted our front bedroom into a makeshift studio,' Darbyshire recalls. To make it all seem like it was really happening, Grinyer went out and bought them a couple of drawing boards and a Macintosh computer. They decided to name themselves Landmark, because it had an air of upscale legitimacy to it and Darbyshire liked to stay in Landmark Trust houses when he went on holiday. 'We found it a fairly interesting name,' Grinyer says. 'In actuality, it wasn't. And it turned out that a Dutch company called Landmark had started up eight months earlier, who then rather aggressively tried to sue us. They seemed to think we had done this maliciously and it all got a bit crazy.' Not particularly enamoured by their first stab at the naming process, the pair went back to the curry house to think of another. Grinyer suggested Orange, the mobile-phone network still being five years away from coming into existence. It ticked a lot of boxes, but Darbyshire pointed out that it had connotations of Agent Orange, the toxic defoliant used by the United States military during the Vietnam War. Then Grinyer prodded about and discovered that a loose collective of designers in Denmark was already trading under the name. Orange was out. Since it was coming up to Christmas 1989, Darbyshire put forward Satsuma, which was then modified to Tangerine. Tangerine carried the suggestion of Tangerine Dream, the 1960s German electronica group of whom Grinyer was a fan. It also turned out to be a Chinese good-luck symbol. *Tangerine it was!*

□ □ □

Once they had done enough work to sound faintly respectable, the pair used some more of the start-up money to take out page adverts in design journals to announce themselves to the world. 'The first ad just said "who are Tangerine?" and then had a list of clients that we were working with,' Grinyer remembers. 'That caused a big splash. People said, "bloody hell, these people have come from nowhere and are already doing some interesting work". It was a recession at the time so most people just assumed we had either been fired or made redundant. The idea that we had done this voluntarily was quite surprising.' Next it was time to find a more respectable base of operations. A graphic designer they knew had recently set up shop in a large converted warehouse in Hoxton, north of the City. He had a couple of floors spare and the pair were able to scrape together enough money to rent half of one, which they shared with a slightly pompous furniture designer. It was at this point that Jony Ive turned up. He had just finished a stint with Roberts Weaver (contractually stipulated by their funding of him through Newcastle Poly) and was bringing with him a client in the form of Ideal Standard, a plumbing-fixture company. Grinyer introduced Ive to Darbyshire and he joined them as an equal partner in Tangerine. Darbyshire recalled: 'It was immediately obvious when you looked at his work that he was an immensely talented designer – and a really nice person to boot.' The group was growing. For a while Grinyer's sister joined them as a PA while she looked for work in London after finishing university.

With Tangerine seemingly going from strength to strength it was an exciting time for all involved, although, as Darbyshire points out, 'there was still a lot of pain, along with the pleasure'. In 1989 Hoxton was still among the rougher parts of London; almost a decade away from the urban regeneration that would see it become home to fashionistas in Ray-Bans and skinny jeans. 'The area had been the first "no-go" zone for police in the 1960s,' Darbyshire recalls. 'They never walked around there; they only ever went around in their patrol cars.' Grinyer lost track of the number of radios ripped out of his car, or the amount of times his tyres were let down parking there each day. For lunch he and Ive would go out to a local caff, which doubled as a boxing gym. 'You couldn't get a tuna fish sandwich. That was considered camp. It was all white rolls, ham and cheese.' Designers got funny looks as it was – not helped by the fact that Tangerine's office was immediately next door to an enormous disco with blacked-out windows, which they quickly learned through trial and error was the centre of Shoreditch's gay scene. Business proved to be a similar conflation of different elements. One day they might be designing for a major firm like Ford or Hitachi; the next they would be working for a small-to-medium-sized family-owned business based in Norfolk. The other market Tangerine proved popular with was among the local personal inventor community, who would try and get around the issue of payment by promising shares in whatever idea they had once they hit it big. 'They meant well,' says Darbyshire charitably. To keep business ticking along nicely was a matter of keeping multiple plates spinning simultaneously. Whenever a client came to visit they would load the studio with every foam prototype they had for that particular project, to demonstrate that the team was giving it their full, undivided attention. Once the meeting was over and the client had left, the other work could be brought back in and the myriad extra foam models shuffled out. Unsurprisingly it wasn't too long before they ran out of storage space.

Rather than spend money they didn't have to expand Tangerine's premises, Ive suggested that they hire a van for the day and drive some of the prototypes down to his parents' house in the village of Curry Mallet in Somerset. They did, and by the time they left to return to Hoxton his dad's garage was so full of cardboard toilets and baths that it was practically impossible to move without violently colliding with one.

If foam models took up space, however, what really took up time was working out business proposals. The vast majority of Tangerine's clients had never used a design agency before and had no real idea what to expect. Most assumed they would be shelling out about £1,000 for what amounted to a couple of sketches. When the full proposal subsequently arrived – calling for custom injection-moulding and carrying a daunting price tag of around £30,000 – some would-be customers would outright balk. 'This isn't quite what we were expecting,' they would stutter.

□ □ □

In March 1990, Ive drove to Hull to deliver a presentation to one of the managing directors of Ideal Standard about a ceramic bathroom-sink project he was working on. Right from the start of the day he was dreading the meeting. His first concept for the project had been strikingly bold. With a conventional sink the basin is screwed to the wall, while an understated plinth hides the pipes from view. Ive had taken the plinth and made it into a far more central aspect of the design; leaning, angular, against the wall and supporting a freestanding bowl. It was an original concept although, from Ideal Standard's perspective, completely unfeasible. They pointed out that it was too big, would be incredibly expensive and far too heavy for a plumber to install. Biting his tongue, Ive had modified his idea, coming up with a far more sensible plinth, which still retained some of the uniqueness of his original concept. It was this revision that Ive was now pitching.

From the moment he was greeted at reception, the signs were ominous. It was Comic Relief Day and the Ideal Standard director Ive was meeting was wearing a novelty red nose, which he refused to take off as they sat in a conference room to talk business. Ive gave his presentation, but seemed to come up against a wall of resistance. When he had finished speaking, the director – still wearing his red nose – began tearing into the work Ive had spent hours perfecting, pointing to all the 'useless' design features Ive had added in. This was not a situation Ive was completely unfamiliar with, but on this occasion it pushed him too far. When he got back to Hoxton later that day he was beyond furious. To all intents and purposes this was the moment when the UK lost one of its best young designers.

It was Ive who rebounded by harpooning the big Cupertino whale. A major company like Apple didn't come along every day. In the two years since Ive had toured California, Bob Brunner had been steadily moving up the Apple chain of command; first working on projects unofficially (frog design was still carrying out Apple's design work and the company was worried about upsetting them by bringing in another consultancy to do extra work), then full-time as Apple's new director of industrial design. Looking to shake things up a bit, Brunner decided that what Apple needed was a fresh pair of eyes. It was decided that the company would organise a talent search to find the world's hottest young designer. At this point he remembered Jony Ive. On a business trip to Germany, Brunner took a layover in London and drove to Tangerine's office in Hoxton Street. He dropped in for a meeting and introduced himself to the group. Tangerine at the time were working on a fizzy-drinks machine with a swing-door mechanism that could open without knocking over any cups. Brunner loved it.

'This is *exactly* the kind of creative thinking we're looking for,' he gushed.

Brunner explained that he was heading up a strategy project codenamed Juggernaut, whose *raison d'être* was to explore possible future design concepts for Apple computers. This, he said, was where Tangerine might come in. Out of his bag Brunner pulled the first PowerBook laptop that anyone at Tangerine had seen, still several months away from being released. He laid it on the table. 'These are the lines we've been progressing along so far,' he said.

Several subsequent conversations – and Tangerine's first modem – later, and the agency was given the contract. For the next six months, they worked on three different ideas. The first was for a small clutch bag that would open to reveal a miniature widescreen, palm rest, trackpad and keyboard. The second was for an enhanced desktop with a portrait screen and an angle-displaced keyboard. The third was a wireless computing concept, which featured a detachable keyboard and other elements. Tangerine had to work within the existing Apple design language – which meant sticking to the battleship-grey of that generation's product line – but were free to push boundaries. So long as what they were dreaming up had some root in reality, they were free to go for broke. Ive had heard wireless keyboards talked about, so why not add one, he figured? 'I remember one day seeing Jony with this foam model of the tablet on his desk,' says Peter Phillips, another former Roberts Weaver designer who had joined Tangerine by this juncture. 'He was sitting away from it with his knees up, typing away on this foam keyboard, and saying "this feels really good".'

Things very nearly went wrong. Tangerine was too small to have a machine shop on site, so the team would produce mock-ups and then pass these over to self-employed model makers around the country to create working prototypes. No matter how many hours Grinyer, Darbyshire and Ive spent designing a concept, it was with these model makers where the buck ultimately stopped. 'The products were quite inventive in the way they would fold up or the keyboards would fold out,' recalls Grinyer. 'We found a local model maker, who had done a lot of film and advertising work. When we got the models back, they looked fantastic but would break after you had used them once. Apple had this pile of broken models that they couldn't do anything with. That was all a bit of a disaster.' (Conversely, Darbyshire remembers them all working very well.)

□ □ □

Eventually it was time to go and present the models in Cupertino. Grinyer, Darbyshire and Ive flew over to San Francisco for the presentation. Grinyer and Ive loved being at Apple. Darbyshire wasn't so sure. 'Apple is a deep culture,' he says, choosing his words carefully. 'You have to want to belong to it. It's almost to the point of being a cult, and I have to say that I find that kind of spooky. There are fantastic sides to the culture as well: a great sense of freedom, constant encouragement to develop and look for something new to perfect, but also a slightly cliquey, weird side that I find too claustrophobic. It's almost a religion – and I can't deal with that.' Still, the presentation went exceptionally well. As they were about to leave, Brunner asked if he could speak with Ive in private for a few minutes.

When the two men were alone Brunner explained to Ive that doing small research projects while at Tangerine was one thing, but if he really wanted to create something radical he needed to think about joining Apple as a full-time employee. 'We've always suspected that the project was given to Tangerine to try and lure Jony to the sunny climes of California, to see if they could poach him over to the States,' says Peter Phillips. '[Jony] saw the sunshine, saw the work, did a little calculation on the paycheque and decided to join our team' is how Brunner phrased it. In fact, even with his Ideal Standard experience hanging over him, Ive characteristically agonised over the decision. Did he really want to leave Tangerine? Did he want to leave *Britain*? Would his wife, Heather, also the daughter of a schools' inspector, want to move to California?

Apple offered to fly Ive and Heather out to Cupertino for another weekend. When Ive came back to the Tangerine office he was still sweating the decision. Grinyer and Darbyshire decided to intervene. 'We all said, "It's a great opportunity, Jon, how can you not take it?"' Darbyshire recalls. Eventually he made up his mind. He would go to California. He said goodbye to his Tangerine colleagues and packed his bags.

In September 1992, Jony Ive signed his contract and became a full-time employee of Apple's industrial-design department.

12

FEAR AND SELF-LOATHING ON INFINITE LOOP

'We can't stop here. This is bat country!'

Hunter S. Thompson,
Fear and Loathing in Las Vegas (1971)

In January 1990, Steve Jobs gathered his best lieutenants around him at the NeXT offices in Redwood City and voiced the question that many felt he should have been asking years earlier: 'Are we making the right product?' The answer, of course, was that they were not. The core technology of the NeXT Computer may have been unparalleled, but its software was also lacking and its price point unfeasibly high.

'So what should we do about it?' Jobs asked.

The answer, it was decided, was to forget the detour that had been taken along the way and to return to the concept that had first been discussed in 1985, at those early company retreats in Pebble Beach. What they needed to do was to build a $3,000 computer.

Dan'l Lewin, however, wouldn't be around to see it come to fruition. Lashing out at someone to blame for the NeXT Computer debacle, Jobs had stripped away the operating responsibilities of the vice-president of marketing. In February, Lewin left the company, headed to another start-up that would offer the thrill of setting out on a new high-tech venture, but without the pain of doing so with Steve Jobs. Lewin wasn't the only one who found working directly for Jobs troublesome. George Crow, who was responsible for the analogue and mechanical aspects of the NeXT Computer, went to the doctor one day for a routine checkup and instead received something of a wakeup call.

'If you work for that man for another six months you might die,' his doctor allegedly told him. Although Crow didn't leave, the NeXT structure was reshuffled so that Crow would instead report to Rich Page, who would act as a Steve Jobs buffer.

Jobs decreed that there would be two new computers in the NeXT product line. One was to be the low-cost computer the company had promised from the start, which would be shaped like a pizza box to save on manufacturing costs. (Because rival company Sun Microsystems already produced a computer shaped like a pizza box, and Jobs didn't want to be seen as doing anything derivative, he insisted that the so-called NeXTstation be instead referred to in polite society as a 'slab'.) The other would be the higher-cost NeXTcube, which would retain Esslinger's striking magnesium-cube form factor, but be significantly faster than its predecessor and feature full colour graphics.* Being a perfectionist, it was this second machine that held Jobs' interest best. As it was, any attempt to get him to scale back his vision to cater for lower-end consumers was, in the words of materials director Matt Medeiros, 'a battle'. One example of this was Jobs' total insistence that, aside from the company logo, every component in a NeXT machine, down to the circuit boards, be entirely black. Rich Page and Medeiros argued back and forth with Jobs about the insanity of adding so much to production

---

* Despite its colourful logo, the first NeXT Computer, like the first Macintosh, had a monochromatic display.

costs for a part of the machine that the majority of users were never going to see.

Meanwhile, the press continued asking questions. As a private company, Jobs kept his sales figures secret; bamboozling reporters with vague terms like the suggestion that NeXT was in 'active dialogue' with major purchasers, or was 'starting to close some really seriously big deals'. But other parties were starting to speak up. 'A year is enough time to evaluate the system,' said the chairman of the computer-science division at the University of California, Berkeley, who had bought twenty NeXT Computers. 'It looks like it just hasn't hit home.' More damning were the quarterly reports from Businessland which revealed that, far from making money from the NeXT deal, Businessland was actually *spending* $10,000 in overheads per machine sold, due to the investment that it had been forced to make in creating a dedicated sales and marketing team. At a meeting with Businessland's regional vice-presidents Jobs pounded his fists on the table and screamed, 'If you can't do better than that, you shouldn't be in sales at all.' Inside NeXT, the utopian tenets the company had been founded on were breaking down. One vice-president was already out of the door and others were talking of leaving. To attract new superstar players to the company, or simply to keep people from throwing in the towel, raises and bonuses were handed out at a dizzying rate of knots. A junior accountant used to making around $25,000, joined NeXT on a $50,000 salary and was shunted up to $90,000 within a couple of years.

Jobs and his team were pinning all their hopes on the new product launch in September 1990. Entering a new decade, Jobs, now thirty-five, continued to paint himself as the convergence of high-tech and liberal arts: half hippie-shaman and part big-business poster boy. At a dinner held the night before the Unveiling II, Jobs sat with the unlikely pairing of conservative

Ross Perot and Graham Nash, the English folk-rock singer, who was a proud owner of the NeXT Computer. 'Anyone who can seat a veteran of Woodstock and an ultra-capitalist at the same table is all right with me,' quipped a reporter for *Computerworld*.

The most dramatic moment of the launch, which once again took place at the Davies Symphony Hall, came when Jobs showed a clip from *The Wizard of Oz* on the NeXTcube. At the time, the notion of a computer powerful enough to play a film was almost unthinkable. The scene Jobs had selected was the one in which Judy Garland's Dorothy opens the door to the Land of Oz, at which point the black and white film suddenly transitions to colour. The audience whooped appreciatively. The reaction was real. The technology wasn't. Because the especially designed video chip wasn't yet ready, Jobs had faked it. Rather than coming from the NeXTcube itself, the footage was being relayed from a LaserDisc player hidden behind the curtain. As metaphors go, Jobs couldn't have chosen a more suitable movie.

◻ ◻ ◻

By September 1990 Jean-Louis Gassée was out of Apple. The downfall of the charismatic Frenchman, who had once looked set to take over Apple after John Sculley departed, was a product called the Macintosh Portable. This had been a long-time dream for the Macintosh product line, dating back to Jobs' time in the company, but it was Gassée who finally moved ahead with it. The resultant Macintosh Portable contained a lot of neat features, but was hampered by an excessive price tag of $5,799 and a weight in excess of seven kilos. Feeling that the 'high right' business model was no longer working, Sculley relegated Gassée to the sidelines and stepped into the breach himself, acting as Apple's chief technology officer. Having lost power, Gassée left.

One of Sculley's favoured projects was a device called the Newton, which appeared to herald the arrival of the CEO's

Knowledge Navigator concept almost twenty years ahead of schedule. The Newton was the brainchild of an engineer named Steve Sakoman, who possessed an overriding passion for handheld computing. Sakoman had started his career at Hewlett-Packard, where he served as the project manager for the world's first battery-powered portable MS-DOS PC, the HP 110. In 1984 he was tempted over to Apple by Jobs, only to find himself dumped unceremoniously into the Macintosh division following Jobs' resignation, helping to create the Mac Plus, Mac SE and Mac II. By March 1987, Sakoman was bored and ready to quit. He wanted to return to handheld computing and talked with Mitch Kapor about bankrolling a new company to build them. He had been convinced to stay by Gassée, who arranged for him to start up a handheld-computing skunkworks project at Apple. Sakoman recruited a team of capable engineers, including Steve Capps, co-author of the original Macintosh Finder. Capps had left Apple in 1985 to start his own music-software company, through which he produced what he described as 'the ultimate air guitar'. The Jaminator ultimately failed to make waves, however, and Capps was convinced to return to Apple to join the Newton team. The group had moved into an abandoned warehouse on Bubb Road in Cupertino, half a mile from the main Apple headquarters, and got to work.

Chief among the Newton's features was its handwriting recognition, which could be inputted by writing directly on to the screen by way of a stylus. Apple wasn't the only company in Silicon Valley investigating pen-based computing at the time, but it was one of the few willing to do something more than simply graft a stylus on to existing computer architecture. For many it seemed that the pen represented the next iteration of a pointing device like the mouse. 'There is all this fine motor control in your hand that makes pens work that is completely wasted on a classical computer with a keyboard and a mouse,' says Walter Smith, an engineer who worked on the Newton team. 'If you talk to an artist, drawing with a mouse is like being asked to draw with a bar of soap. There are all these neurons that don't connect to a regular computer that are for driving your fingers in very complicated fine ways.' Taking up the metaphor of a pen and pad of paper, the Newton would allow interaction in a natural, intuitive manner. Editing a document, for example, would be completely transformed. Suddenly the various proof-reading marks that had been honed over the years – allowing for underlinings and crossing-outs – could be transferred to the computer screen. But handwriting recognition proved more challenging than anticipated, due to the difficulty of adapting to different writing styles. (The Newton team didn't make it any easier on themselves by insisting that the device recognise cursive, so that users wouldn't have to simplify their handwriting when entering sentences.) A bizarre breakthrough had occurred when Apple vice-president Al Eisenstat was on a business trip in Moscow. Hearing a frantic knocking on his door, he opened it to find a nervous programmer standing there. Without saying a word, and glancing over his shoulder to see if he was being followed, the man handed Eisenstat a floppy disk containing handwriting-recognition software, then quickly left. Upon his return to the States, Eisenstat handed the code over to the Newton team who, finding it remarkably accurate, built it into their device.

The project had grown quickly. Sakoman originally imagined the Newton as a small, lightweight computer that would cost the same as the original Macintosh. But with no adult supervision on display, the project began succumbing to the unique Silicon Valley ailment of 'feature creep'. By the time all was said and done the device had expanded in both physical size and scope of ambition to comprise a slate, measuring 8.5 by 11 inches, with a touch-sensitive screen, a stylus for handwriting recognition, a hard disk, a battery that would run for weeks at a time, and an infrared port that would allow the devices to communicate with one another. A quick totting-up suggested that this would add up to $8,000. Steve Sakoman left the project.

It was at this point that John Sculley discovered the Newton. When he did he threw his full support behind it. 'It was Sculley's Macintosh,' says Frank O'Mahoney, one of the marketing managers assigned to the project. 'It was Sculley's opportunity to do what Steve had done, but in his own category of product.' Ending its status as a research project, Sculley set a shipping date and a price for the device: a Newton handheld would be ready by 2 April 1992, costing less than $1,500. For a while the team explored different form factors for the tablet. Larry Tesler, who was placed in charge of the project, favoured a model named Senior, which measured 9 by 12 inches and would carry a price tag of $5,000. Tesler was concerned that releasing a smaller, underpowered Newton straight off the bat would do irreparable damage to the brand. Others argued with him, claiming that another high-end product line was the absolute last thing that Apple needed. This latter faction ultimately won and development blazed ahead on Junior, a device that measured 4.5 by 7 inches and would cost around $500. Apple partnered with Sharp for the release of the Junior (now retitled MessagePad).

Some reports suggest that the Newton was rushed to market despite clearly not being finished, because one of the higher-ups in the division had put up their expensive wine cellar as collateral in a wager that stated that the device would be released by a certain date. 'Nine months before it was launched it was just obvious that this thing was not ready,' recalls O'Mahoney. 'The PR people and the product people understood this, and as a result when we were previewing it we would never let reporters touch it. We knew that if we did and they got that little stylus in their hand and wrote their name it would come out completely wrong. So we completely controlled the access that reporters had.'

As it happened, one of the most damning indictments of the Newton came from a man who had never even used one. Gary B. Trudeau, the cartoonist behind the popular *Doonesbury* comic, released a series of cartoon strips that ran within a month of the MessagePad's release. Trudeau was an Apple fan, but his joke at the expense of the new technology proved all too prescient. In one of the strips the central character attempts to get the Newton (or a device that looks remarkably like it) to recognise his handwriting. 'I am writing a test sentence,' he inputs. 'Siam fighting atomic sentry,' the machine responds. He tries again. 'I am writing a test sentence.' 'Ian is riding a taste sensation.' One more time. 'I am writing a test sentence.' Finally the device understands what is being entered. 'Catching on?' he writes. 'Egg freckles?' says the machine.

□ □ □

For a computer company started by a couple of countercultural college dropouts, Apple wasn't doing badly, all things considered. In late 1992 it was the most profitable personal-computer company in the world, had sold more computers than any other vendor worldwide and held cash reserves exceeding $2 billion. But John Sculley still felt that it needed to either be sold to or merge with a larger company. 'I wanted to leave and go back to the East Coast,' he says. 'I'd had my time in Silicon Valley and I went to the board and talked to them about it in early 1993. They didn't see a logical successor in the company, and so we set out on a path whereby we hired Goldman Sachs to try to sell the company.' Goldman Sachs, the multinational investment-banking and securities firm, put out feelers to several companies, including AT&T. There were no immediate bites. Sculley was accused of abandoning the Macintosh platform to pursue other avenues of interest. One was politics, in which he was increasingly associated with Bill Clinton, to the point at which there were rumours that Sculley had only been eliminated as a potential running mate because he was twice divorced. The other was the Newton, which looked well on its way to being a disaster.

'Sculley was an impractical visionary,' says former Apple employee Alan Oppenheimer. 'Like Steve, he came up with these visions that most people said were impossible, but in his case they really *were* impossible. Maybe because he couldn't distort reality like Steve could, or maybe because they were just impossible based on present-day technology.' After Apple's earnings per share collapsed in 1993, the board of directors asked Sculley to step down. As with Steve Jobs in 1985, he would retain the title of chairman and was asked to talk about nothing publicly except for the Newton. 'During the board discussion ... I made it clear I was prepared to be full-time chairman of Apple,' Sculley noted afterwards. 'I am not walking away from Apple. I have no plans to go anywhere else.' By October, when Apple revealed a 97 per cent drop in its earnings for the year's fourth quarter, he resigned.

'In hindsight, it would have been much smarter if I had gone back to Steve Jobs and said, "hey Steve, this is still your company, you created it, let's figure out a way for you to come back and run Apple",' Sculley says. 'It would have avoided all this stupid stuff that happened after I was pushed out, where they went and licensed the technology and instead of building really cool products, built really crap products. They destroyed every principle that Steve had come up with, and that I tried to maintain while I was there, and almost bankrupted the company. All of that could have been avoided if I'd been smart enough to go and talk to Steve and see if he wanted to come back and head his company again. [That's] the [decision] I regret the most.'

As he notes, this is all with the benefit of hindsight. At the time, Apple continued to push forward on its own. Whatever problems might have existed under Sculley were about to get worse. Much worse.

□ □ □

Jobs' finances were in a sorry state. True, $25 million was a lot more than most people had in their bank accounts, but it was also $72 million less than he had had prior to starting NeXT and investing in Pixar. Meanwhile, the dominoes were coming dangerously close to falling. In February 1991, Susan Barnes, NeXT's chief financial officer, gave two months' notice and announced that she was leaving the company. Jobs responded by locking her out of her voicemail and email. In June, Ross Perot resigned from the board of directors, saying that he needed to dedicate more time to his company, Perot Systems. That same month Businessland – once the mightiest computer retailer in the nation but now wobbling on the brink of bankruptcy – finally gave up the ghost. Its pieces were sold for a pittance to a New York company which installed heating and ventilation equipment. Then Burrell Smith, who had been suffering from schizophrenia in the years since leaving Apple, showed up outside Jobs' home – which he now shared with his wife of several months, Laurene Powell-Jobs and their newborn son, Reed – and broke sixteen of his windows, along with his car windscreen. On top of this the NeXT machines were still not selling. By any stretch of the imagination, it was a trying period. Still Jobs refused to compromise.

Although his computers were woefully lacking in software, he balked at the idea of letting any old developer port any old program over to his beloved system. A perfectionist who would micromanage development down to the choice of plating on the interior screws of his computers, Jobs would rather have *no* software than the wrong software. When he did deem a company suitable, he felt that simply acknowledging this badge of honour would come across as the highest praise imaginable. 'We're thinking of having you develop for us,' he told an executive at Symantec, a large software company which developed for the Macintosh. 'Excuse me?' they responded, taken aback. In an industry where potential major developers were often wined and dined the approach left some people cold. Symantec declined the offer. Meanwhile, IBM was struggling to adapt the NeXTSTEP software for its own uses, and, despite having paid Jobs the money it had agreed to, eventually gave up. 'Somewhere along the line this diamond got dropped in the mud and now it's sitting on somebody's desk who thinks it's a dirt clod,' Jobs complained to *Fortune*. 'Inside that dirt clod is still a diamond, but they don't see it.'

At times NeXT showed hints of breakthrough success, often from unexpected sources, as when the company sold an undisclosed number of computers to the US defense department, or when an upmarket real-estate company (run by the wife of a former Pixar employee) bought 120 NeXT machines because it wanted to stand out from its IBM-using rivals. But the company's running costs were always too high. Looking to rectify the situation in some small way, NeXT's director of materials, Matt Medeiros, argued with Jobs over his insistence on mailing out every NeXT product, as with Apple products today, in a perfectly white box. 'Everything was shipped by FedEx or UPS and these boxes just looked terrible by the time they reached the customer,' Medeiros recalls. 'The only thing we could do was to put a protective sleeve over the white box. Now, when you add the cost of a white box, which is five times the cost of a normal box, then you have an outer-skin box – and you're in the process of trying to save money because you don't have any more funding coming your way ... It becomes an emotional issue when you're laying somebody off and yet you're still trying to preserve the white box.'

The remaining NeXT co-founders were starting to doubt that the ship was ever going to right its course, and even Jobs' panache for distorting reality did nothing to calm their nerves. 'The problem was that after a couple of years of being close to turning a profit, we were burning up the original $100 million from Canon,' says Rich Page. 'Month by month we would be spending more than we brought in ... Once we introduced the pizza box, the second-generation NeXT, and it didn't see market success, all of us could see the writing on the wall. There was no way the company could economically sustain itself or that we could go find another investor to put money in.' On 15 January 1993, Richard Page left the company. He was the seventh of nine vice-presidents to go. No replacement was immediately announced. What *was* announced was that NeXT was getting out of the hardware game, and would instead be focusing its efforts on selling software. With no hardware to work on, George Crow left. So did Matt Medeiros, with his last act being to close down NeXT's automated factory, ensuring first that everyone in the supply chain was paid off in full.

The following month the company officially changed its name to NeXT Software, carefully phrasing the decision not as the failure of a hardware manufacturer but as the democratisation of a software company. 'Today, we're letting out' the software that 'had been locked in a black box' gushed an advertisement which appeared in the *Wall Street Journal* on 18 February. There was another announcement as well. NeXT was making 330 of its 500 employees redundant, a decision that would affect almost every area of the business. More than two decades after Apple's original mass-firing of Apple II employees by Mike Scott, the NeXT lot – many of whom learned of their fate over the radio – took to referring to it as 'Black Tuesday'. Those still left in the marketing department described it as a change in strategy.

◻ ◻ ◻

The NeXT machines might have been an abysmal failure as mass-market saleable products, but they proved influential in kicking off a slew of new revolutions. Far from Redwood City, California, a then-unknown British computer scientist, Tim Berners-Lee, used two NeXT Computers to bring to life an off-the-wall concept he nicknamed the 'World Wide Web'. Not every NeXT user would create something quite *that* transformative, but there were certainly other less worthy, but equally fun applications. In the winter of 1991, two geekoid friends with the same first name, John Carmack and John Romero, had bought a NeXT Computer to put into practice an idea they had had for an entirely new type of computer game – a 3-D first-person shooter – in which the player would make their way around a castle, gunning down Nazis. They toyed with different names (*The Fourth Reich? Castle Hasselhoff? Luger Me Now?*) before eventually settling on *Wolfenstein 3-D*, as a tribute to a 1981 action game, *Castle Wolfenstein*, which had appeared on the Apple II. As two twenty-somethings in the early nineties, Carmack and Romero were part of a new generation of techies who had grown up with Apple as an important backdrop to their lives.* Perhaps too important, as it turned out. At the age of fourteen, Carmack had been caught trying to break into a local school to steal an Apple computer, resulting in a year-long sentence in a juvenile detention centre. 'Most of the kids were in for drugs,' noted author David Kushner in his excellent book *Masters of Doom*. 'Carmack was in for an Apple II.' Meanwhile, a teenaged Romero had had the code for

---

* Until IBM and Microsoft would make a concerted effort to catch up – and overtake – in the 1990s, Apple remained the province of true gamers. So what if the IBM PC seemed a safer bet for big businesses? It couldn't handle sophisticated scrolling *nearly* as well as an Apple II.

his first ever game published in the somewhat wittily titled Apple II magazine, *InCider*.

As much as both men loved Apple, however, the NeXT Computer was the only machine powerful (and, at a stretch, affordable) enough to do the kind of ground-breaking work they had in mind. A wintery trudge through the snow and an $11,000 cashier's cheque later, they had their computer and got to work.

If there is one thing that the Universal Law of Gravitation teaches it is that, sooner or later, the apple always falls. For Apple Computer the case can be made that the company stopped genuinely innovating the moment Jobs left, but it took until the 1990s for whatever problems there were to finally reach critical mass. *Time* magazine summed up the viewpoint of most outsiders (and a great many insiders) when it wrote, 'One day Apple was a major technology company with assets to make any self-respecting techno-conglomerate salivate. The next day Apple was a chaotic mess without a strategic vision and certainly no future.'

A recurrent theme during the 1990s was projects being developed and then unceremoniously dumped before ever making it out of the door. These ranged from a pen-based Macintosh, to the costly Taligent operating system, co-developed with IBM, neither of which saw the light of day. 'One colleague of mine at Apple had been there seven years and worked on seven cancelled projects,' one former employee says. 'That was part of Apple's dysfunction. They would start projects and put a lot of energy and money into them, get them to a certain point, and then realise that there wasn't the market for them that they had expected, or that it would be too expensive to finish ... It led to a lot of engineering disillusionment as people who came there to work on insanely great projects, realised that they could *work* on these things, but that we're never going to ship them.' The constant 'reorgs' (as reorganisations were referred to) meant that quite often the left hand didn't know what the right hand was doing. Often it seemed as if the passion of the engineers to make great products existed in the face of management's attempts to quash that passion.

Case in point was Ron Avitzur and the story of the graphing calculator. Avitzur had been brought into Apple to work on a portable, pen-based Macintosh project, which ultimately wound up being scrapped when managers realised that – while it was technically possible to shrink the Mac interface down into a miniature handheld device – it was next to impossible to read anything on the tiny windows. 'The project was so plagued by politics and ego that when the engineers requested technical oversight, our manager hired a psychologist instead,' Avitzur recalls. The aspect of the pen-based Macintosh that Avitzur was responsible for was the development of a graphing calculator, which allowed two-dimensional, non-linear mathematical equations to be entered, so that parts could be dragged around with far greater ease than on a traditional calculator. These, in turn, would manipulate the parameters of a real-time graph. To Avitzur, the cancellation of the project in August 1993 represented more than just six months or a year of wasted time. He had been working on some version of the calculator ever since he was at college. As the team went their separate ways, the jobless Avitzur took a bold decision: he would continue working on the graphing calculator, and he would do so at Apple. This proved oddly easy. 'They hadn't deactivated my security card, and everyone knew me because I had been working twelve-hour days for the last year anyway, so just having me continue to turn up, no one gave it a second thought,' he says. Avitzur didn't flaunt his outsider status, but he didn't exactly hide it either. If someone asked him what he was doing, he would sheepishly explain himself, in awkward conversations which progressed along the lines of:

'Do you work here?'

'No.'

'You mean you're a contractor?'

'Actually, no.'

'But then who's paying you?'

'No one.'

'How do you live?'

'I live simply.'

This form of covert activity went on for five months, with the 27-year-old Avitzur working for no pay. Since Apple's finances appeared to be in the doldrums, he joked that he was volunteering for a non-profit. 'I was living like a graduate student,' he says. 'I had no mortgage or family, and my rent was low.' At around the same time, Greg Robbins, a friend of Avitzur's who had been working on another cancelled project, lost his job with Apple. Rather than leaving, he simply claimed that he had been assigned to report to Avitzur. Nobody knew who Avitzur was, but assumed that he must be a manager working in a different part of the company. The concept worked both ways, as Avitzur was able to get people off his back by claiming to report to Robbins. 'I was twenty-eight and didn't have any debts so I wasn't concerned about the income,' Robbins says. 'It didn't seem dramatic at the time; it was just like continuing to go to work. It was more fun to be working on a project where we had very concrete goals for ourselves and no management.' It proved to be one of the most exciting times of both men's careers. Alongside the work at Apple, Avitzur was able to get the pair audiences with some of the most exciting start-ups in Silicon Valley, among them General Magic, the company started by former Apple engineers Bill Atkinson, Andy Hertzfeld and Marc Porat. 'Ron had amazing schmoozing skills,' says Robbins. 'People knew him. We would go out to dinner and invariably someone would come up to him and say hello – typically from Stanford or one of his previous contracting gigs in Silicon Valley. I've never been a social animal or the kind of raconteur that Ron is, so it was unique for me to be with someone who was able to open doors.'

After several months, Avitzur became bolder in his endeavours, asking other full-time Apple employees to lend them pieces of hardware for the graphing calculator or to come and help out with the programming for a couple of hours. 'For the most part when I explained to people that my project had been cancelled, Apple had stopped paying me, and that I'd been here for one, two, three or four months, and didn't actually report to anyone, but wanted to finish anyway, people listened with a very sympathetic ear,' Avitzur says. This wasn't always the case. On one occasion Apple's facility managers tried to move a new team into the apparently empty offices where Avitzur and Robbins were working. 'She asked what group I worked in, since it would be that group's responsibility to find me space. When I told her the truth, she was not amused. She called security, had them cancel our badges, and told us in no uncertain terms to leave the premises,' he recalls. They were saved – ironically enough – by the mass layoffs which began that month, when falling profits saw 20 per cent of Apple's 15,000 workers out of a job. 'Greg and I were safe because we weren't on the books in the first place and didn't officially exist,' he recalls. 'Afterwards, there were plenty of empty offices.'

When their entry badges eventually expired, Avitzur and Robbins took to waiting outside Apple headquarters and tailgating in through the door when genuine employees entered. Both would wear their old badges as decoys in case someone realised what they were up to. Towards the end of 1993, work on the graphing calculator began to come to an end as the job neared completion. Without a team of beta-testers to put the software through its paces, Robbins became increasingly edgy about introducing any further features which could negatively impact on the program's performance. 'I became paranoid about not doing anything that would hurt the software,' he says. 'If there were any serious bugs in our software we figured we would just be kicked off the machine, and our whole reason for working on this for most of the year was to ship. We became increasingly risk-averse and double-checked each other's work and left out things that we thought increased the chance of bugs being introduced.'

Their diligence paid off, however, as the graphing calculator proved to be one of the most bug-free pieces of software ever created at Apple. The final challenge was somehow finding a way to get their software on to the Macintosh. One of Avitzur's friends said that he could sneak the program on to the operating system just before it was due to be published. The engineer's job was making the PowerPC system disk master, which was then sent off to the factories where Apple's computers were made. Since the loading of the OS was the final stage of the PowerPC manufacturing process, this man essentially controlled what did and did not make it on to the final machines as they shipped. It was a tempting offer, but one which Avitzur and Robbins turned down, figuring that it was better to confess what they had been up to. 'Our real fear wasn't legal ramifications, it was that we didn't want to destroy our own future in the industry,' Robbins says. 'We had a lot of ideas; some of which we couldn't do because we didn't want to burn bridges. We didn't want to betray anyone who had trusted us.'

It came down to the moment of truth. Several of the engineers who had been helping Avitzur and Robbins were able to arrange for them to show the graphing calculator to some senior managers. Avitzur gave a twenty-minute demo and the reaction was positive. 'What group are you in?' asked one of the managers afterwards. 'Why haven't we seen this earlier?' Avitzur explained how he and Robbins had been sneaking into Apple and that the project didn't officially exist. The managers laughed, until they realised that he was being serious.

'Don't repeat this story,' they said, and left.

Eventually word reached the manager of the PowerPC division, who was slightly more understanding. The graphing calculator shipped. 'We wanted to release a Windows version ... but sadly Microsoft has effective building security,' Avitzur says.

☐ ☐ ☐

Microsoft's latest operating system was a joke. At least that's what Apple's employees told one another as they distributed car bumper stickers, reading 'Windows 95 equals System 7', suggesting that Bill Gates' 'Evil Empire' was still a good four years behind the Macintosh. Certainly Apple had been vindicated by the early reports coming out. The OS had already been delayed two years, and was apparently still full of glitchy applications and software bugs which quickly exhausted a computer's system resources. Apple's public response echoed the company's original reaction to the arrival of the IBM PC, with a double-page spread appearing in the *Wall Street Journal*, reading 'C:\ONGRTLNS.W95'.* But Bill Gates seemed strangely unfazed. 'The advertising that Apple's running is a very positive contributor because it serves to raise the awareness of Windows 95 with Win 3.1 users,' he told *MacWEEK* magazine, of all places. 'I'm not saying there's anything wrong with them. But they're good for us as well. Apple says, "Hey, Win 3.1 is crummy and it

---

* Apple had greeted the arrival of the IBM PC in 1981 with a sarcastic advertise-ment, placed in the *Wall Street Journal*, reading: 'Welcome IBM. Seriously. Welcome to the most exciting and important marketplace since the computer revolution began thirty-five years ago ...' Apple was not laughing for long.

always was crummy." And we say, "No, it *just* turned crummy. It was really great 'til Thursday".'

The trade response to Windows 95 might have been that it was good, not great, but the mainstream attention heaped on it far outstripped anything the industry had seen before. Buoyed by a $200 million marketing budget, the operating system became front-page news around the world as excited customers lined up to buy their shrink-wrapped copies. Microsoft even borrowed Apple's cool countercultural vibe for the event: paying the Rolling Stones a reported eight-figure sum to use their 'Start Me Up' single for the Windows 95 advertising campaign.* Windows 95 wound up selling more than three million copies in its first five weeks on the market. (By comparison, Apple sold 4.5 million Macintoshes during the entirety of 1995.) Back in Cupertino, Apple's employees filed out into the quad of research and development and had a barbecue, during which members of the team delivered morale-boosting speeches and gave demos of new products. The tone was defiant, but oddly muted. 'Dude, it was the only time I ever remember us having a rally,' says John Alfano, a product marketer at Apple. 'It was the first time I had ever experienced that – and probably the last time. It was a big event in the Apple psyche, at least from my perspective. Of course people were concerned, but the way that concern manifested itself was that we were going to be aggressive in our position against Windows 95 in the marketplace. We were going to do more ads, and we were going to do more marketing, to show how much better our stuff was than theirs. I don't recall there being any wailing and moaning and gnashing of teeth.'

---

* According to reports, Mick Jagger was against licensing the Stones' music for fear of impugning the group's artistic purity. Keith Richards, however, had a higher burn rate than Jagger and wanted the money.

Technophiles might have easily seen the advantages of the Mac OS, but Apple's lack of forward momentum was evidenced by the fact that, for the general public, differentiation between the two companies evaporated. Says Mitch Stein, then Apple's director of human-interface technologies: 'Microsoft was delivering a "look and feel" which, while we threw rocks and said, "it's not as good as ours", there was a real fear that it would be good *enough* that everybody who used Macs because they were easier to use would then be able to say that Windows was just as easy.' In some ways Apple had been too smart – and too arrogant – for its own good. The company had carried out extensive market research which stated that Apple users wanted muted tones onscreen, since too many bright colours distracted them from what they were trying to carry out and made the desktop look overly cluttered. When Apple engineers therefore first saw Windows 95 in action they snickered at its garish use of primary colours, and wondered whether Microsoft couldn't have put a bit of its R&D money aside to find out what people actually *wanted* from a computer. It quickly transpired that Apple was wrong. Users loved the friendly approach to computing – and if, long term, their eyes did get tired, then their eyes got tired. They turned off the computer, went and did something else, and then came back and loaded up Windows 95 again. Suddenly Apple began to realise that a lot of its too-cool-for-school market research had been wrong. 'We had this big crash course to make everything look sexier, to make it more colourful,' says David Curbow, one of Apple's human-interface engineers. 'I think that was the first realisation that we were not just building computers for people. We were building things that would look good enough that you wanted to buy them.'

Interestingly, the man responsible for Windows 95 – a 41-year-old Microsoft senior vice-president and product manager named Brad Silverberg – was a former Apple employee himself, with his first job out of graduate school having been as a programmer on the Lisa. Several weeks before the launch of Windows 95, Silverberg considered the impact that his new operating system was likely to have on his employer of twenty-something years earlier. 'I thought it was essentially game over for Apple,' he remembers. 'I didn't see how they could compete.' Silverberg, like many employees inside Apple, felt that the Cupertino-based company had simply rested on its laurels for too long and regarded themselves as invincible, even though all evidence indicated otherwise. 'It's not like Windows 95 was a huge secret in the industry,' Silverberg says. 'We were very open about what was coming. They should have realised, "this one's over and we've got to come up with the next thing. We had a good run for ten years. We were first with the GUI, but we can't compete. We can't fight a land war with Microsoft."'

Silverberg's attitude to computing was a populist, rather than elitist one. 'I didn't want to build Ferraris,' he says. 'I wanted to build the world's best Honda Accord; something that everyone would be able to use and afford.' He shared Gates's and Microsoft co-founder Steve Ballmer's dream of creating high market share. 'I wanted everybody to use this,' he says, 'not just a small number of rich people.'

□ □ □

'Apple had this incredibly dysfunctional culture,' says Michael Mace, a former marketing manager. 'It was a hard place to manage in those days. It was like ten thousand wild horses, none of which had been broken to the saddle.' Certainly it was a challenging situation for any company president to deal with. Some long-time employees claimed that it was as if Apple was haunted by the malevolent ghost of Steve Jobs, infusing the company with all the bad traits of its co-founder and none of the good ones.

After John Sculley left in 1993, he was replaced by Michael Spindler. Spindler had first joined Apple in September 1980 as the marketing manager for its European operations. For six months he went without pay because Apple couldn't work out how to transfer funds internationally from Cupertino to Belgium. An excellent strategist, Spindler was able to increase international business for the company at a time when its US sales were falling dramatically. Over a two-year period revenue from Apple Europe rose from $400 million to $1.2 billion: a quarter of Apple's total earnings. In 1990 he was moved to Cupertino and given the role of chief operating officer, before ascending to the position of president. Under Spindler, Apple slashed the R&D budget by more than $100 million per year and sacked a large percentage of its workforce to pretty the company up for a potential sale or merger. In protest against the layoffs, several employees ran screensavers with the names of those dearly departed, which they referred to as Spindler's List. Unlike Sculley, Spindler was no public speaker. Many fondly recalled the time he had been interviewed at length for a magazine, only for the article never to appear because the journalist couldn't understand a single word he'd said. Spindler would present long, meandering whiteboard talks during which he would ramble on interminably about the future of computers and engineering. One of his favourite sayings was about Gateway, a company based out of Nebraska, which made cheap Windows PCs which came in boxes with a cow-skin pattern.

'These guys out there in the middle of nowhere,' Spindler said in his mangled English, 'they're putting their pants on with screwdrivers.'

One of Spindler's concepts for turning around Apple relied on flooding the market with cheap, knockoff Macintoshes produced by companies like Power Computing. This was supposed to increase Apple's market penetration, but instead did nothing of the sort. 'They created this clone market that cannibalised their sales without expanding their base,' says Tim Dierks, an engineer with Apple at the time. 'It was just cheaper Macs, it wasn't more Macs.'

Meanwhile, a man named Gil Amelio joined the company. Amelio came from National Semiconductor, where he had made a name for himself as something of a turnaround artist. In doing so he had transformed the company from an unreliable third-tier chip manufacturer into an industry front-runner. When Amelio was asked to take a seat on Apple's board of directors, he was shocked to discover just how little control the board exercised. Everyone seemed happy to go along with the rosy picture of Apple's future as painted by the company's senior executives, which as far as Amelio could see had little to no basis in reality. He was also perturbed by Spindler's role of hysterical ship's captain, alternating as he did between total denial and outright panic as Apple slowly sank. At Amelio's second board meeting Spindler stood up and announced, 'We're not making it, the company has to be sold.' At this point there was cause for alarm. Spindler had been telling the board emphatically that Apple would make $150 million in profits for 1995. When the number finally came in it was less promising: Apple had lost $69 million. Then it turned out that the company had more than $1 billion worth of unsold merchandise, which was now outdated and would have to be written off. On 8 January 1996, Spindler checked into a hospital with heart palpitations. The Apple board of directors decided that there should be a change. Gil Amelio was offered the job.

Amelio proved no more successful. To many of the Apple employees, especially those whose hopes were raised initially by the idea of a techie CEO in their midst, it just appeared that Amelio was another highly paid stuffed shirt there to drain the lifeblood out of the company. 'He was doing his best impression of Nero while Apple was burning,' says Larry Yaeger, a former Apple Distinguished Scientist.

□ □ □

In 1996 Apple rescued humanity from certain destruction. In the climactic scene of that year's blockbuster film, *Independence Day*, Jeff Goldblum's plucky techno-geek uses a PowerBook 5300 laptop to upload a computer virus into the central mainframe of the evil alien mothership, thus lowering its defences for a nuclear attack. As far as product placements goes, there were few more worthy of celebrating. To mark the occasion Apple briefly switched its corporate advertising slogan from 'The Power to Be Your Best' to 'The Power to Save the World'. Unfortunately things were never that easy.

Apple had just shipped 1,000 PowerBook 5300s to eager dealers around the country when the shocking news came down the wire: two early production units had caught fire. One had done so at the home of an Apple programmer, while another ignited at Apple's factory in Singapore. 'The main hallmark for Apple is ease of use,' noted Pieter Hartsook, editor of *The Hartsook Letter*, at the time. 'If your machine doesn't work, it's certainly not easy to use.' Apple issued a recall for the one hundred machines that had already been sold. The cause of the fires was found to be the fault of an overheating battery. The offending item was replaced with a more reliable model, which turned out to have just two-thirds the capacity of its predecessor. Apple apologised again and lowered the price of the brand-new PowerBook by $100.

Meanwhile, the company scrabbled for inspiration. One idea, created under Spindler but taken forward by Amelio, was the opening of a string of Apple-themed restaurants, which would serve as rivals to the then popular Planet Hollywood franchise. The Apple Café concept was presented as a joint venture between Apple, the Landmark Entertainment Group and a UK-based real-estate company called Mega Bytes. Unlike the almost monastic setup of the later Apple Stores, the Apple Café would resemble a fifties diner, combined with an almost Jetsons-like projection of the future. 'The time is right,' claimed Apple's senior vice-president of marketing, Satjiv Chahil. 'Cybercafés are in. The technology finally is reaching out to "the rest of us". This will be a place to showcase our products in the real world.' At the Apple Café, patrons would be able to shop for Apple products, snack on a variety of dishes from a health-conscious menu, order up food using a graphical user interface and videoconference with neighbouring diners by way of a miniature kiosk screen built into each table. Potential locations were scouted in London, Paris, New York, Tokyo and Sydney, before the entire venture shuddered to a halt. This was not the doing of Apple, who remained keen on the idea until the bitter end, but, rather, a wavering confidence on the part of the partner companies, who sensed potential disaster by tying themselves in with the floundering computer company. *Would a Bill Gates Steakhouse carry more clout*, they wondered.*

□ □ □

If an Apple Café seemed like an odd bid to turn around an ailing computer company then at least it didn't cost Apple too much time or money. The same cannot be said for Copland, Apple's abortive attempt to create a new operating system. Named after the American composer Aaron Copland, perhaps best known for his 'Fanfare for the Common Man', Copland would be Apple's counterpunch to the hulking menace of Windows 95; *a fanfare for the common Mac*. Work had started in 1994, but a couple of years down the line was revealing itself to be a disaster. As with every project at Apple at the time, there was a team of incredibly

---

* As ridiculous as the idea of a computer company owning a restaurant chain would appear, it is worth noting that the popular US fast-food franchise Chuck E. Cheese's, which originally built its name on being the first family restaurant to integrate dining, animated entertainment and an indoor video arcade, was founded in 1977 by Atari, Inc. entrepreneur Nolan Bushnell: the same man who gave Steve Jobs his first break in the high-tech industry.

smart people, but the cumulative intelligence leading the project seemed somehow less than the sum of its parts. Chaos reigned at every stage. When Tim Dierks, a 25-year-old software engineer whose last project had been cancelled, started sniffing around looking for something to do, he was instantly promoted to the job of technical lead. 'In retrospect it was kind of an insane selection,' he notes. 'The fact that they gave me that job is representative, from my perspective, of a lot of management failures in Apple at the time.' Dierks was very bright, but was also just a few years out of college. 'I simply did not have enough understanding to lead a project with three hundred engineers on it,' he says. Finally the much delayed operating system was put out of its misery. 'They printed disks for the developer release but never sent them,' says Greg Robbins. 'The project was cancelled between the time the CDs were pressed and a few days later when they were to be mailed out.'

Copland's failure led to Gil Amelio looking around for another operating system that Apple could buy. One of the potential candidates was the product of a company called Be Inc., the creation of former Apple employees Jean-Louis Gassée and Steve Sakoman.

Gassée had put together a fifty-person team and was producing hardware in the form of a computer called the BeBox, which came packaged with a nimble operating system, named BeOS. For better or for worse it was not an entirely different setup from the outfit Steve Jobs had been running, although Gassée would hear none of it. 'For God's sake, don't compare us to NeXT,' he reportedly pleaded. 'We want to be a better tool for developers, not to be tasteful. We don't cost $10,000. We have a floppy drive. We do not defecate on developers.' Gil Amelio was impressed by what Be had on offer, while Gassée, who himself had been banking on the failure of Copland, saw a chance to broker the deal of the century. As far as he could see, Apple had no choice but to give him whatever he wanted. And what he wanted was $200 million, 15 per cent of Apple and a seat on the company's board of directors. 'A man in the desert doesn't bargain on the price of water,' he quipped.

Not everyone inside Apple was enamoured with Be. 'I went nuts because I'd taken a look at it,' says Mitch Stein, then Apple's director of human-interface technologies. 'It had some really nice concepts in the user-interface area and I was a fan of a lot of what they had done, but fundamentally there was nothing there. It wasn't a real operating system; there were just little bits and pieces. I thought that to turn over that kind of money and to risk the future of the company on what looked to be a fool's errand seemed insane.'

Finally, Apple turned Gassée down. Not only was the Frenchman asking too much, but he had also been outmanoeuvred. One day Steve Jobs phoned Amelio and casually noted, as he had done with IBM chairman John Akers about Microsoft, that BeOS was all wrong for Apple. What would be right was buying NeXT and bringing Jobs back to the company he had helped start. Amelio was agreeable to the suggestion. On 6 December 1996, Jobs met with him and two other high-ranking Apple executives in an eighth-floor conference room on the Apple campus. It was the first time Jobs a newly-minted billionaire thanks to Pixar going public, had set foot in an Apple-owned building since his ousting. As far as Amelio was concerned, buying NeXT was a no-lose situation. Even if the NeXTSTEP software (now an open, rather than proprietary, operating system called OPENSTEP) didn't pan out, he would be happy enough to have paid out for Steve Jobs alone.

After eleven years in the wilderness, Jobs was back at Apple in an informal advisory capacity. He was a man whom Amelio could trust, whom he could count on.

Wasn't he?

PART III

HOW STEVE JOBS REiNVENTED THE COMPUTER, THE CORPORATiON & GLOBAL CAPiTALiSM

DiFFERENT THiNKiNG

'One is Hip or one is Square ... one is a rebel
or one conforms, one is a frontiersman in the
Wild West of American night life, or else
a Square cell, trapped in the totalitarian
tissues of American society, doomed willy-
nilly to conform if one is to succeed.'

Norman Mailer,
'The White Negro' (1957)

It was a day in early July 1997, when Lee Clow, the chief creative officer of TBWA/Chiat/Day, bounded into his company's Los Angeles office and made a beeline for the desk of Rob Siltanen. With long, stringy hair and sandals, the 54-year-old Clow closer resembled an ageing surfer than he did the head of the world's hippest advertising agency. At the time, TBWA/Chiat/Day was in the middle of a hot streak. It had just been named 'agency of the year' by several top trade magazines and was picking up major accounts right and left without even having to pitch for them. The car manufacturer Nissan was spending some $400 million in media at the agency alone.

'Pack up your things,' Clow told Siltanen, who at thirty-three had already risen to the rank of creative director. 'You and I are flying out to Cupertino to meet Steve Jobs. We're going to sign Apple.'

Siltanen, of course, knew Apple. Even at a time when the company's stock was in the doldrums, it was a brand close to the heart of anyone working in a creative industry. There wasn't an ad agency worth its salt which didn't use Macintosh computers, and the seemingly imminent demise of the brand cast a curiously melancholic shadow. But Clow was confident. He believed in Jobs. Furthermore, as he explained to Siltanen on the one-hour flight from LAX to San Jose International, he believed that Jobs was going to do what was right for business. TBWA/Chiat/Day's business. Even though Chiat/Day had been the agency responsible for the ground-breaking '1984' Macintosh commercial, it had also followed up a year later with the disastrous 'lemmings' spot for Macintosh Office. After that, Apple's board of directors had opted to cut its losses and go instead with BBDO, who produced a series of more conventional, if largely uninspired advertisements focused on performance and price. Now that was going to change. Looking across at Siltanen, Clow relayed the story. Two days earlier, he had been driving through LA when his phone rang.

'Hi, Lee, this is Steve Jobs,' the caller said.

That alone was enough to grab anyone's attention.

'Guess what? Gil Amelio just resigned. I'm looking to change some things up. Can you come in for a meeting?' Forty-eight hours later and here they were. 'This isn't a pitch meeting,' Clow told Siltanen. 'We never deserved to lose Apple's business to begin with. I'm telling you, this is Steve's way of making up for us getting screwed over in the past.'

When they landed, Siltanen and Clow drove the nine miles from San Jose International to Apple's Infinite Loop campus. The two men picked up visitors' badges from the reception desk and were shown into a glass-walled conference room. Siltanen didn't know what to expect. Those who had worked with Jobs described him as a brilliant but difficult man. Once all was said and done, Siltanen's own appraisal would be that the Apple co-founder was a mixture of 'Michelangelo, Mies van der Rohe and Henry Ford – with some John McEnroe and Machiavelli thrown in'. Siltanen and Clow waited a few minutes before Jobs entered the room, wearing his soon-to-be iconic black turtleneck, a pair of shorts and flips-flops. He had been back at Apple since January that year, and his appearance at Macworld the same month had received a rapturous reaction, particularly coming after an almost three-hour address by Gil Amelio, which left pundits with no better idea of the company's future direction than they had had going in. (Wozniak also appeared briefly on stage at the event, suggesting a return of sorts to Apple's roots.) Following Macworld, however, Jobs' next few months back had been oddly muted, as if he was happy to join Wozniak as a mascot on the sidelines, to be wheeled out on high days and holidays. In fact, he was busy manoeuvring to oust Amelio, a situation that Amelio seemed completely oblivious to. 'Aren't you concerned that Steve will want power?' asked a journalist for the *Financial Times*. 'Steve is here to help,' Amelio responded. 'I had these same conversations with John Sculley,' the journalist said. The fatal blow came when it was announced that Apple had lost $708 million in the first financial quarter of 1997. The board, many of whom had been with Apple since the very beginning, voted to get rid of Amelio. Jobs stepped in to fill the void.

With no immediate plan in place, his first task was to finish the house-cleaning that Amelio had started. 'Steve was very good at killing things,' one former employee says. Amelio had taken an axe to Apple's list of research and development projects, cutting them down from 350 to 50. Jobs chopped it to just twenty per cent of that: ten projects were left. Among the casualties was the Newton product line, which had cost Apple around half a billion dollars in total. Jobs continued to believe what he had believed during the bad old days at NeXT: that there was never a better time to give the impression of success than when things were in the doldrums. This was where a new advertising campaign would come in.

'I figured that Steve would come and give Lee a hug and say, "hey, it's so good to see you again",' Siltanen recalls. 'But it wasn't like that at all.' Instead Jobs laid out a typically blunt diagnosis of Apple's chances for survival. The company, he noted, was haemorrhaging money. Unless a change was made, it would soon enough bleed to death.

'We have some decent product, but we need to get things figured out,' he said. 'I'm putting the advertising up for review and I'm meeting with a handful of agencies to see who gets it. I've already been talking with a couple of agencies that seem pretty good, but you're invited to pitch the account if you're interested.'

Siltanen was taken aback. It wasn't his place to say anything, but this wasn't quite the 'return to the fold' scenario that Lee Clow had described on the plane ride over. Jobs continued with a series of specifications for the campaign he envisaged. It sounded far from inspiring. For one thing, Apple didn't have any new products to announce. Nor did it have any real money to spend, and, besides, Jobs claimed that he wasn't looking for anything bigger than a few print ads in pro-Mac industry publications.

'A couple of adverts in computer magazines isn't going to change anything,' Siltanen said, speaking up. 'Half the world thinks that Apple is going to die. You need to show them the company is as strong as ever. Nobody stands around the water cooler and talks about print ads. You need to do TV and other things that are really going to give you some momentum.'

Jobs shot him a withering look.

'Fine, show me the ideas and executions that you guys think are best,' he said dismissively.

'Well, that's up to Lee,' Siltanen replied. He still expected that at any moment his boss was going to step in and put the kibosh on negotiations, cutting Jobs off with a curt 'thanks, but no thanks'. He and Clow did, after all, work for an agency built on its 'my way or the highway' approach to advertising. Jay Chiat had set the tone back in the 1960s with his outrageous behaviour in meetings, which once included cutting a client's tie in half because he thought it was ugly. Nothing even close to that happened here. In fact, Siltanen was surprised by the politeness of Lee Clow's response.

'If you like some of the other agencies you're talking with, why don't you just go with one of them?' Clow asked diplomatically.

Jobs responded that he was thinking about it.

□ □ □

In the taxi on the way back to the airport Siltanen confronted Clow.

'I thought you said we weren't going to pitch?'

'I've changed my mind,' Clow answered. 'If we can win this thing, we'll have a great story to tell. I want to get it back.'

When they returned to Los Angeles later in the day Siltanen gathered together the creative teams. 'Drop everything you're working on,' he said. 'For the time being I want all of you to focus your undivided attention on Apple. Lee wants to win this account badly. So do I.' He relayed what Jobs had told him regarding Apple's moribund state, but left out the part about thinking small. As far as TBWA/Chiat/Day was concerned, the more ambitious the idea, the better. Siltanen said that this was the agency's opportunity to put Apple back on the map, and to make some careers in the process. In terms of direction ... well, that was up to the creatives. Apple, of course, had no shortage of celebrity backers. Steven Spielberg, Sting and Bono were all well known to be fans of the brand. Perhaps they could play on that? Around the room brows began furrowing as people racked their brains for inspiration.

'We'll review your ideas in a week,' Siltanen finished.

□ □ □

Craig Tanimoto was a junior art director at TBWA/Chiat/ Day, one of the agency's hot young talents. When he heard the brief from Siltanen, he was determined that he was going to come up with the winning idea. Tanimoto had his own personal history with Apple. In 1984 he had been a student at University of California, Los Angeles, when the Ridley Scott-directed Macintosh commercial first played at that year's Super Bowl. Tanimoto had been blown away and decided then and there that he wanted to work for the agency that had created it. Several years later he was a fully fledged creative, the only one of the interns who had actually been successful getting a full-time job with the agency. He sat down at his desk, stared at the A3 sketch pad in front of him and started thinking. After a few abortive attempts, the first solid concept Tanimoto came up with was a variation on René Magritte's 'Ceci n'est pas une pipe' ('This is not a pipe'). 'I basically had the computer on a canvas and the tagline, "this is not a box",' he recalls. 'I thought that was really interesting because computers were boxy at the time, and this was something that wasn't going to *box* you in.'

He set the sketch aside and started coming up with alternatives. 'Another idea I had was a picture of Thomas Alva Edison, but instead of a light bulb illuminated above his head, there was a little Apple logo.' Tanimoto tore off the top sheet of paper and put it to one side. On the clean page he started writing a surreal, Dr Seuss-style poem about Apple computers. It wasn't as good, certainly not as visual, as the other two ideas and he started to lose focus. Then he noticed two words – 'think different' – that he had written as part of one of the lines. He circled them with his pen. 'My heart started racing because no one had really voiced that idea for Apple,' he says. 'I looked over at the picture of Thomas Edison and thought, "think different". I immediately drew a little sketch of Einstein and wrote "think different" next to him as well and drew a miniature Apple logo.' Tanimoto knew that he was on to something big, but he decided to sleep on it before telling anyone. The next day, he showed the idea to his writing partner, Eric Grunbaum. Grunbaum got it immediately and the two men spent the rest of the day blowing through the campaign, sketching out full-page pictures of Mahatma Gandhi, Albert Einstein and Thomas Edison, all with the new tagline and a miniature Apple logo. As a variation on the concept they drew several iconic scenes from history, including the famous snapshot of eighteen-year-old George Harris sticking carnations in gun barrels during an antiwar demonstration at the Pentagon in 1967. *Think different.*

When it came time for the review meeting, Tanimoto and Grunbaum took their stack of papers and pinned them up, alongside the work of the other teams at the agency, to be assessed. Just from scanning the competition, Tanimoto was quietly confident. 'Most people just had pictures of computers,' he remembers. 'Some had testimonials from athletes and artists; long copy ads about why they loved using Apple. Some people were doing graphic concepts with computers and arrows. But no one had anything like ours. I was just hoping that Lee would like it. If he hated it, who knows what would happen?' Rob Siltanen and Lee Clow walked up and down the room, spending a couple of seconds looking at each concept before moving on to the next. When they reached the end of the line, they paused, exchanged a few words and headed back to the 'think different' drawings.

'Whose are these?' Clow asked.

Tanimoto raised his hand. Clow turned back to look at the idea again.

'Ditch the other concepts,' he told everyone in the room. 'From now on, just work on this one.'

TBWA/Chiat/Day had its killer line for the campaign. Now it just needed to build on it and, beyond that, sell Jobs on the idea. With the billboards and print ads in place, the decision was taken to put together a draft television commercial, to show how the concept might transfer to the screen. Jennifer Golub, one of the agency's best broadcast producers, began searching around for footage of the various geniuses whose likenesses were being used. When she had compiled enough material, a talented editor by the name of Dan Bootzin cut all of it into a rough sequence, which those inhouse referred to as a 'rip-o-matic'. For the soundtrack Lee Clow suggested that they use the song 'Crazy' by the British soul artist Seal, which talked about how one needed to 'get a little crazy' in order to survive. That's exactly what Apple needed to do, Clow reasoned. At the end of the video, a series of title cards appeared:

There are people who see the world differently.

They see things in new ways.

They invent, create, imagine.

We make tools for these kinds of people.

Because while some might see them as the crazy ones,

We see genius.

The lines were written by Rob Siltanen. Normally when Chiat/ Day pitched clients, it was conducted by a team of three or four people, each taking turns to talk through one part of the campaign. In this instance, the decision was made that Lee Clow would give the entire presentation, partly because of his history with Jobs and partly due to the fact that he was still the most exuberant pitchman in the agency. A small team flew to Cupertino to present. This time Jobs seemed more enthusiastic than he had previously. Clow likened Apple to Harley-Davidson motorcycles, another countercultural brand whose target audience of baby boomers had long since grown up and sold out, only for a recent advertising campaign to help turn it around. He walked Jobs through every element of the 'Think Different' concept, explaining the reasoning behind each decision that had been made, and demonstrating how effectively the campaign could be adapted for different mediums. He ended by showing the mood video. The whole time, Jobs stayed silent. Finally he spoke.

'This is great, this is really great,' he said.

The team sensed that there was a 'but' coming.

'— but I can't do this. People already think I'm egotistical. If I put the Apple logo up there with all these geniuses I'm going to get absolutely skewered in the press.'

There was an awkward silence.

'What am I doing?' Jobs continued, looking around the room as if he were soliciting opinions from invisible parties. 'Screw it. It's the right thing. It's great. Let's talk tomorrow.'

He walked out. Chiat/Day had the account.

The 'Think Different' campaign wasn't close to finished by the time the Macworld Expo took place on 6 August 1997, but Jobs wanted to try out the message in front of an audience. Taking centre stage in a hotel ballroom in Boston, he had his chance. Aside from telling the world why Jobs was going to be a change of pace from the parade of disastrous CEOs that Apple had seen through the first half of the 1990s, there were added reasons why, on this day more than any other, it was important to drive home how Apple continued to think different. For starters, there was the now-forgotten issue of the clone computers. Although Jobs didn't mention them during his speech, and a large percentage of Apple's user base considered them an unnecessary blight on the company, there were certainly those 'pro-choice' advocates who worried that stopping third-party developers from creating licensed Macintosh clones would stifle innovation and price competitiveness. Fanning through the crowd at the Macworld Expo were people in black flak suits, representatives of Power Computing, handing out fliers which read 'WE DEMAND CHOICE' in a bid to start a grassroots campaign preserving their business.

More memorably, however, was Jobs' announcement that Bill Gates had agreed to invest $150 million in Apple. The cash injection had been under discussion during the Amelio days, but progress had faltered. The $150 million was a trade, in return for Apple stopping its ongoing patent lawsuit against Microsoft, as well as endorsing the company's Internet Explorer web browser. But that kind of straightforward explanation would have meant nothing to the ardent Apple advocates who had stuck with the company through its darkest years, in the face of Gates's advancing 'Evil Empire'. As far as they were concerned, Jobs had made a deal with the devil. They vociferously booed the announcement.

Those at Apple were clearly aware of what the news represented. Even though both Apple and Steve Jobs had, at separate times, teamed with IBM, it was now clear that Microsoft – not IBM – was the *real* computing monopoly. When Gates's grinning face appeared on the giant screen behind Jobs at Macworld, it recalled a potent image from Apple's history: that of the Orwellian despot of the original Macintosh commercial. With that symbolism exposed as a sham, it was the perfect time to introduce some new Apple imagery; something that would once again ally the company with the rebels and the dissidents. Closing his speech, Jobs gave as concise and poetic a summary of the Cupertino ideology as he would ever present:

> *I want to just talk a little about Apple and the brand and what it means, I think, to a lot of us. You know, I think you always had to be a little different to buy an Apple computer. When we shipped the Apple II, you had to think different about computers. Computers were these things you saw in movies [that] occupied giant rooms. They weren't these things you had on your desktop. You had to think differently because there wasn't any software at the beginning. You had to think differently when a first computer arrived at a school where there had never been one before, and it was an Apple II. I think you had to think really differently when you bought a Mac. It was a totally different computer, worked in a totally different way, used a totally different part of your brain. And it opened up a computer world for a lot of people who thought differently. You were buying a computer with an installed base of one. You had to think differently to do that.*
>
> *And I think you still have to think differently to buy an Apple computer. And I think the people that do buy them do think differently and they are the creative spirits in this world. They are the people that are not just out to get a job done; they are out to change the*

world. And they're out to change the world using whatever great tools they can get. And we make tools for those kinds of people. So hopefully what you've seen here today are some beginning steps that give you some confidence that we, too, are going to think differently, and serve the people that have been buying our products since the beginning. Because a lot of times people think that they're crazy. But in that craziness, we see genius, and those are the people we're making tools for.

□ □ □

Back in the Los Angeles office of TBWA/Chiat/Day, Rob Siltanen assembled a list of the various geniuses who had been agreed upon for the 'Think Different' campaign. Some, like Albert Einstein and Thomas Edison, had been part of Craig Tanimoto's original proposal. Others, like futurist Buckminster Fuller and photographer Ansel Adams (whose black and white prints had hung in the NeXT offices) were Jobs' suggestions. To get each person on board, Siltanen wrote a letter to their representatives or estates, offering to donate computers to a charity of that person's choice if they would allow Chiat/Day to use their likeness. Meanwhile, an advertisement was cut together, featuring footage of Albert Einstein, Martin Luther King, Jr, Richard Branson, John Lennon, Buckminster Fuller, Thomas Edison, Muhammad Ali, Ted Turner, Maria Callas, Mahatma Gandhi, Amelia Earhart, Alfred Hitchcock, Martha Graham, Jim Henson, Frank Lloyd Wright, Pablo Picasso and Jobs' beloved Bob Dylan. This last inclusion represented a significant change of heart on the part of the then 56-year-old singer-songwriter. Three years earlier Dylan had sued Apple for using his name for a new programming language the company had devised. (Apple, for its part, insisted that 'Dylan' was shorthand for 'dynamic language'.) Seeking unspecified damages, the lawsuit insisted that Apple was 'intentionally using ... the names of famous individuals, including [Sir Isaac] Newton, Carl Sagan and now Dylan, in conjunction with Apple's products in a deliberate attempt to capitalize on the goodwill associated with these famous individuals'.* According to popular legend, Dylan decided to sue when he realised that his counterculture brother Jerry Garcia of the Grateful Dead, was earning $250,000 per year in royalties from Ben & Jerry's, on account of their marketing a Cherry Garcia flavour of ice cream.

Jobs stayed heavily involved throughout the process. One day he lambasted Lee Clow over one of his choices of image. 'This is not the right picture of Gandhi,' he yelled. He proved particularly handy in getting clearances. With a not inconsiderable Rolodex at his disposal, he was more than happy to pick up the phone himself and call in personal favours if it meant the difference between securing a clearance or not. Joan Baez, for example, was an old girlfriend, while Yoko Ono had once been a neighbour. He phoned Yoko and asked for a picture of John Lennon. She sent him one, but it didn't come up to Jobs' standards. He arranged to meet her in a small Japanese restaurant in New York, where she handed over an envelope containing the photograph Jobs was looking for. In the end, the only people who turned both Chiat/Day and Steve Jobs down altogether was the family of Jacques Cousteau, who had died one month earlier.

---

* During development, the Power Macintosh 7100 was internally known as 'Carl Sagan'. Finding out about this – and perhaps not realising that it was a code name and not for public consumption – Sagan sued the company for libel, claiming that he was 'profoundly distressed to see ... Apple's announcement of a new Mac bearing my name'. In response, the engineers changed the code name to BHA, standing for 'Butt-Head Astronomer'. This time Apple's lawyers stepped in and the project was again renamed, finally settling on LAW, which stood for 'Lawyers Are Wimps'. Apple and Sagan came to an amicable settlement.

Jobs wasn't happy. Trying to convince a man used to distorting reality that the Seal song 'Crazy' wouldn't cut down to commercial length was no easy task. This was always the danger of using temporary tracks. No matter how much you reminded a client that they were there solely to establish mood, when they worked as well as this one did it was difficult to move forward. But Siltanen had an idea. 'Why don't you let me write a voiceover for the piece,' he said, telling Jobs that he could put together something that would act as a manifesto for Apple. What Siltanen was picturing as he spoke was a scene from the film *Dead Poets Society*, in which Robin Williams's inspirational teacher encourages his students to stand on a desk to gain a new perspective on the world. 'We must constantly look at things in a different way,' the character says. 'Just when you think you know something, you must look at it in a different way.' It was one of the most affecting moments of the movie and Siltanen felt that that same message could be made applicable to Apple.

'Have you seen *Dead Poets Society*?' he asked Jobs.

'Of course I have,' Jobs responded. 'Robin Williams is a personal friend of mine.'

Siltanen requested a week in order to come up with a voiceover monologue that Jobs would be happy with. As soon as he left the meeting, he began mentally building on the title cards he had written for the earlier 'rip-o-matic'. He especially liked the opening, 'Here's to the crazy ones ...', which he knew would have a certain shock factor when contrasted on screen with images of the most brilliant thinkers of the twentieth century. The ending was good, too, but what proved challenging was the middle. Siltanen did endless rewrites until he came up with a version that both he and Lee Clow were happy with. He recorded a rough version of himself reading the lines and overlaid this on Dan Bootzin's concept video. Then he showed it to several people he trusted, many of whom admitted to getting a lump in their throat when they saw how the commercial was coming together. Finally, Siltanen and Clow took the commercial to show Jobs.

□ □ □

Five meetings in, Siltanen felt that he was starting to get a sense of how Jobs worked. He had already seen in action Jobs' habit of shooting down an idea (in this case doing a television commercial along with print ads) and then parroting it back as if it were his own. 'I've decided that we're going to go into television,' he told Siltanen on one occasion.

When Siltanen started to say something, Jobs interrupted.

'It's not for the reasons you said,' he snapped.

He also knew that Jobs was direct in his dealings, and had already seen glimpses of the infamously abrasive behaviour. At an earlier meeting Jobs had suddenly decided that there were too many people in the room.

'You, you and you,' he said, pointing at several Apple PR officers. 'Get out!'

I'm glad that's not me, Siltanen thought at the time. For this meeting, the setting was more intimate. It was just Siltanen, Clow and Jobs in a conference room. The two Chiat/Day men introduced the spot and then played it for Jobs. As before, he remained absolutely quiet until the piece was finished, giving nothing away. Then he exploded.

'This is advertising-agency bullshit!' Jobs screamed. 'This is absolute crap! You said you were going to write something like *Dead Poets Society*. I don't know what this is, but I hate it!'

'I'm sorry you don't like it,' was all the cowed Siltanen could say.

'We should get the writer of *Dead Poets Society*,' Jobs said, ignoring him completely. 'Not this bozo!'

'I'll see what I can do,' said Lee Clow meekly.

The meeting was over.

□ □ □

Rob Siltanen returned to working on the Nissan account and never went back to Apple. 'I didn't know what to say,' he recalls of Jobs' outburst. 'He listened to it once and just went crazy ... I'd never seen anything like that in my life.' Chiat/Day also never put a call through to the agent of Tom Schulman, the screenwriter for *Dead Poets Society*. Instead, work on the commercial was handed over to Ken Segall, another creative director at the agency. Segall had first met Jobs during the NeXT days, when he had been working for the Ammirati & Puris ad agency. He had a rapport with the Apple co-founder and the hope was that he would be able to repair some of the damage that had been done. The first task was waiting for Jobs to cool off. 'When I came in the spot was in trouble,' Segall recalls. 'Steve didn't like and didn't want to run in, so I stepped in and started rewriting.'

Jobs soon calmed down, although he never apologised to Siltanen. Not long afterwards Segall stopped by Siltanen's desk.

'He's gone full circle,' he said.

'Who?' asked Siltanen.

'Jobs. I've made a few changes with your script, which I hope you're okay with. But we're going ahead with it.'

Segall had worked directly with Jobs, who allegedly penned several of the lines used in the finished commercial. Biographer Walter Isaacson credits Jobs with the line 'They push the human race forwards'. Siltanen disputes this. 'I had that in my typed version,' he says, remarking that he has kept copious notes throughout his career. 'I don't why that was – if [Steve] feels he came up with that, I don't know.' Segall doesn't comment either way. 'I think it was the perfect moment of a group of people coming together and doing something magical,' he says. Jobs also worked closely with director Jennifer Golub, who was piecing together the commercial. They spoke daily in order for Jobs to review the footage which was being used. Ever the perfectionist, he arranged for a satellite uplink between Apple and the Venice Beach offices of Chiat/Day so that he wouldn't have to rely on viewing drafts in a compressed video format.

One subject that provoked much discussion regarded who would narrate the finished advertisement. Fitting in with the *Dead Poets Society* concept, Jobs and Clow wanted to use Robin Williams. Williams's and Apple's connection went back years. After the original Macintosh had been completed in 1984, the comedian had been brought in to perform his stand-up routine for the Mac team. But Williams's agent claimed that the actor didn't do commercials. Taking the matter into his own hands, Jobs phoned Williams directly and got through to Marsha Garces, the actor's second wife, who refused to allow her husband to come on the line. Ditching that idea, Chiat/Day briefly considered the 69-year-old American author and poet Maya Angelou, and then actor Tom Hanks. Jobs even pulled President Bill Clinton aside at a charity fundraiser and asked him to phone Hanks directly, but nothing came of that either. Finally they settled on Richard Dreyfuss, a confessed Apple user, whose gritty and measured delivery perfectly suited the advert's visuals. For Craig Tanimoto, Dreyfuss was the most exciting choice of all. To a comic-book and sci-fi geek, who had grown up with *Close Encounters of the Third Kind* posters up on his wall, there was absolutely no one with more gravitas.

At one point there was talk of Jobs doing the narration himself. 'The person who can really say these words like he means them is Steve,' Clow enthused. Jobs didn't like the suggestion but reluctantly agreed to have a go. Segall personally recorded the reading.

'I don't have much time,' Jobs said, walking into the room. 'I'll do this once.'

In the end he read it three times. 'We tried to sell it to him but he didn't want to be a distraction,' Segall recalls. 'He believed so much in the message that he didn't want people debating whether or not he should have done the voiceover.'

□ □ □

Once the 'Think Different' campaign had been successfully launched, Jobs threw a party for the people at Chiat/Day to celebrate. He attended, bringing along his wife, Laurene Powell-Jobs. Various attendees were perturbed to find that, at Jobs' request, all the food being served was vegan. Jobs stood up and gave a speech. He started by talking about the positive reaction that the commercial had received. 'Steve said that the feedback on the ad was about 75 per cent favourable,' recalled one Apple employee who was there. 'The other 25 per cent of negative reactions to the ad had that, "come on, let's show 'em why we kick Microsoft's butt!" flavour.' Neither reaction would have been totally alien to Steve Jobs.

'Apple spends a hundred million dollars a year on advertising and it hasn't done us much good,' Jobs said. 'We'll continue to spend the same amount. Not much more or less. Only we'll spend it better. Because our brand is the most, or at least one of the most, valuable things we have going for us right now.'

He finished by reading out a letter which had been sent to him by the mother of a child who was, in her words, 'different', describing the impact that the advert had had. It was an emotional moment and some people present blinked away tears as the Apple co-founder spoke.

Later in the evening Jobs cornered Craig Tanimoto at the bar.

'So *you're* the guy that can't use grammar correctly?' he growled.

Tanimoto had been steeling himself for this moment. 'Think different' is, of course, grammatically incorrect. If 'different' is supposed to modify the word 'think,' then it should be an adverb: 'think differently'. Tanimoto had had this argument numerous times, both with people inside and outside the agency. The way he figured it, 'think different' was an order, along the same lines as 'stand straight'. Arguing with a fired-up Steve Jobs, however, was a different story.

'It's ... err ... supposed to be, like, a command,' he stammered. Jobs cracked a smile.

'I know,' he said. 'I've been defending it myself for the past couple of months.'

□ □ □

Little changed outwardly at Apple following the 'Think Different' campaign. There were still no new products, and wouldn't be until the arrival of the Power Macintosh G3 in November 1997. At Apple's Infinite Loop campus, whose north side runs parallel to Interstate 280, large-scale photographs of Albert Eistein and aviation pioneer Amelia Earhart were hung from the back of the building known as IL-3. However, these offered only the slightest of hints to passing motorists that the drab series of buildings they were whizzing by housed one of the world's most exciting computer companies, which would soon reclaim its position atop the high-tech heap. What was most important about 'Think Different' was that it again cast Apple as the countercultural rebel, locked in a perpetual struggle with the technocratic elite, now firmly epitomised by the respective hardware and software of Intel and Microsoft: the so-called Wintel alliance. (The campaign attacked some old enemies, too: lost on few people was the fact that the slogan also functioned as a neat subversion of IBM's 'Think' tagline.)

'It only took fifteen, thirty, maybe sixty seconds to re-establish Apple's counterculture image that it had lost during the nineties,' Jobs said. At the following year's Emmy Awards, the television advert won the prize for Best Commercial. 'It was the first time in my life that the work I was doing could impact the culture,' Jennifer Golub said. 'With every frame I knew I had the opportunity to be provocative.'

'Provocative' certainly proved an apt word. Like the original Macintosh spot, there was an underlying arrogance to the commercial, which led to some accusing Apple of co-opting the identities of people who, in many cases, were no longer around to defend themselves. 'In today's society you see these giant billboards in cities, which celebrate athletes,' says Siltanen, defending his intentions. 'Nike will go and have a wallscape with a picture of Michael Jordan or Kobe Bryant. You see the same thing with rock stars and actors, these larger-than-life images plastered all over the place. What I thought was interesting is that we could take these figures from history – far more important people – and celebrate them in the same way that we do with these athletes and rock stars. In some weird way the athletes and rock stars have been elevated in the mindset of society above some of these figures that literally changed the world. So I thought if anyone came after us, we would have that to talk about.'

In his book *The Power of Persuasion: How We're Bought and Sold*, Robert Levine writes that the 'Think Different' campaign:

> *... was an acrobatic display of spin control: 'We're not going broke because of an inadequate product,' they implied, 'but because we won't sell out to convention.' [It] was also an interesting solution to another Herculean advertising contradiction: blatantly telling people what to do ('Buy Apple') while at the same time urging them to think for themselves.*

Levine ends his analysis by likening this scenario to a scene in the Monty Python film *The Life of Brian*, in which Graham Chapman's mistaken messiah pleads with his crowd of misguided disciples, 'Don't follow me! Don't follow anyone! Think for yourselves! ... You are all individuals', to which the crowd replies in unison, 'We are all individuals.' Certainly there was an unintentional irony in the fact that Apple encouraged its customers to think differently while, at the exact same time, Jobs was going through the rank-and-file troops at Apple and ruthlessly sacking anyone whose vision didn't exactly mesh with his own. *At Apple you could think different from the herd, so long as you thought the same as Steve Jobs.* In other words, Apple might have been heading back to its hippie roots, but there was no doubting just who was in charge of the commune.

Ironically, one of the few people who came forward and registered their dislike of 'Think Different' was an associate of Jobs from way back. Bill Kelley, the original copywriter who had worked on Apple's first print ads for Regis McKenna in 1977, sent Jobs an email stating that in his view the campaign was a mistaken direction to head in. 'I still feel that,' he says. 'It was a huge masturbatory ego boost for Chiat/Day and Steve Jobs, but I don't think it did the company any good at all. I think it delayed their eventual success. The whole attitude of Apple's advertising has to be this whole insider "we-know-you-know-we're-better" wink-wink, and it had a counterproductive effect in the marketplace, I believe.'

Jobs never replied to the email, and he never spoke to Bill Kelley again.

14

THE CUPERTINO iDEOLOGY

'At such moments of profound social change, anyone who can offer a simple explanation of what is happening will be listened to with great interest. At this crucial juncture, a loose alliance of writers, hackers, capitalists and artists from the West Coast of the USA have succeeded in defining a heterogeneous orthodoxy for the coming information age: the Californian Ideology.'

Richard Barbrook and Andy Cameron,
'The Californian Ideology' (1996)

It was an evening in late 2001, and 31-year-old software engineer Michael Lopp was drinking in a Mexican food joint called Chevys Fresh Mex, in Sunnyvale. It was a turbulent time in America's history. The dot-com bubble that had started several years earlier, and which had looked set, in the words of *Wired* writer Kevin Kelly, to confirm the existence of a new economy – combining a sort of cybernetic communism with a mode of high-tech neoliberalism – had well and truly burst. The NASDAQ had fallen 50 per cent from its peak. At that year's Super Bowl, seventeen years after the hammer-hurling launch of the original Macintosh, just three high-tech companies advertised, compared to the seventeen that had done so one year earlier. If the 1990s had been viewed as the start of a new Age of Affluence, then the early years of the new millennium set that record straight once and for all. On 23 October, just six weeks on from the terrorist attacks on the twin towers of the World Trade Center, Apple had released its portable music player, the iPod. Like Jobs' previous new product line, the iMac, a colourful upgrade of the original Macintosh, the iPod was definitely interesting, but although it seemed to be selling well it was far from being viewed as the panacea that Apple needed.

Drinking with Lopp was the CEO at Icarian, the start-up at which Lopp was employed as director of engineering. Icarian was named after the 1800s utopian movement that had established egalitarian communes in the United States. Lopp had been working there since 1998, having joined from Netscape where he had previously been an engineering manager. At first things had been great at Icarian, although now the dot-com implosion suggested that this was going to be a rough patch for the company. The business was burning through VC cash like there was no tomorrow, which there very well might not have been. Now the two men were out drowning their sorrows. Lopp was just starting to feel the effects of his drink when his mobile phone rang.

'Just let me get this,' he told his boss.

He picked up the call and found himself speaking to a recruiter from Apple.

'How would you like to come and work for us?' they asked him.

Lopp glanced over at his boss, considered the extreme awkwardness of the situation, and then said, 'Work for Apple? I've only wanted to do that since I was a kid.'

It was absolutely true. The son of a Hewlett-Packard engineer, Lopp had grown up in a house in which the Apple II was kept ruthlessly up to date, with the latest models and peripherals picked up as soon as they were available. Almost immediately he joined Apple as the manager of the team responsible for the Mac OS X server. It was the beginning of a new era, both for Lopp and for Apple. And a lot of it was thanks to Steve Jobs.

□ □ □

Steve Jobs' return as CEO in 1997 might have come at the culmination of a bad period for Apple but, at least in retrospect, it came at the perfect time historically. Since John Sculley had left the company in 1993 (and arguably before), Apple had been without a truly visionary CEO, able to bring the company into line with the zeitgeist. And if there was one idea circulating in the zeitgeist at the time it was revolution. Only this time, literally echoing the suggestion that the nineties were the sixties turned upside down, the people spouting tear-it-up-and-start-again rebel yells were not unwashed flower children traipsing to Woodstock, but, rather, heads of state and billionaire CEOs. For a few years it seemed that everyone was thinking different. Across the pond in the United Kingdom, Tony Blair was able to downplay his comfortably middle-class, Oxford-educated background long enough to ally himself with the rising tide of rebellion, born under the stifling conditions of Margaret Thatcher and John Major's Britain. In the United States, Bill Clinton did much the same, albeit with more pelvis and obvious sex appeal. At Harvard Business School a professor instructed, 'You must ruthlessly trash outmoded obstructions to creativity: standard operating procedures, protocols, norms of behaviour'. Meanwhile, management guru Tom Peters embraced his inner anarchist by informing a room full of elite business professionals that 'destruction is cool', while Burger King adopted 'Sometimes You Gotta Break the Rules' as its corporate slogan. If there was a reason for the party it could be found at the tail end of the previous decade. When the Berlin Wall came down in 1989, the culture of late capitalism became, for a whole generation of consumers, the only App Store in town. Socialism had seemingly suffered a fatal blow and was quickly retconned into a state of never having happened. Francis Fukuyama cheerfully referred to this as the 'end of history' and noted how liberal capitalism had triumphed, with no real alternatives. It was against this backdrop that the conflation of New Left and New Right, of countercultural ideals and free-market privatisation, occurred.

As a direct product of the sixties counterculture – and a freshly minted billionaire to boot – no one better summarised this new ruling creative class of boomer bobos (that's 'bourgeois bohemians') than Steve Jobs. In his book *Bobos in Paradise: The New Upper Class and How They Got There,* David Brooks sets out the tenets for bobodom in a similar way to the manner in which Steven Levy had set out the hacker ethic years earlier. According to Brooks:

> *They are prosperous without seeming greedy; they have pleased their elders without seeming conformist; they have risen towards the top without too obviously looking down on those below; they have achieved success without committing certain socially sanctioned affronts to the ideal of social equality; they have constructed a prosperous lifestyle while avoiding the old clichés of conspicuous consumption.*

Jobs followed many of these rules to the tee. As the countercultural rebel who modelled himself on traditional leaders of industry, and the $1 a year CEO who, nonetheless, would later hold out for lucrative stock options and a private jet, he encapsulated many of the paradoxes of the new bobo class. He certainly looked the part. During the 1990s Silicon Valley had witnessed the unassailable rise of youthful CEOs like Amazon's Jeff Bezos, Yahoo co-founders Jerry Yang and David Filo, and Sun Microsystem's Scott McNealy. Meanwhile, Apple's aged board of directors had continued to hire one company president after another who looked – and thought – like they did. Now Steve Jobs was in, and the board was out – to be replaced by different-thinking members of the hip Establishment, like Al Gore.

If there was another reason to be utopian in the nineties, and especially in Silicon Valley, it was because of the new technological paradigm that was the Internet, which seemed for many to represent a whole new world order; a digital frontier country in which self-sufficient individuals could carve out a rugged life for themselves as pioneers in an untouched digital utopia. Suddenly long-abandoned sentiments last heard by the mainstream in the 1960s came flooding back into circulation. In Nicholas Negroponte's book *Being Digital*, the founding director of MIT's Media Lab wrote how the triumphal ascendance of the Net would welcome a new generation 'free of many of the old prejudices' in which the power of 'digital technology can be a natural force drawing people into greater world harmony'. As Jobs himself had said in February 1996, one year before returning to Apple, 'The Web is an incredible democratiser. A small company can look as large as a big company and be as accessible as a big company on the Web. Big companies spend hundreds of millions of dollars building their distribution channels. And the Web is going to completely neutralise that advantage.' If ever there was a time for a company freshly realigned with 'the misfits, the rebels [and] the troublemakers' to rise to dominance, this was it. Jobs already had his award-winning ad campaign to reignite interest among Apple's long-suffering customers.

Now all he needed was a new computer to sell to them.

□ □ □

Many employees within Apple had long felt that the company needed to get back to its roots with a computer that would once again be centred around low-cost ease of use. In the mid-1990s Mitch Stein, Apple's director of human-interface technologies, had bought his son a toy called My First Sony, which was a miniature cassette recorder designed for children, constructed out of tough ABS plastic, with large buttons and a colourful exterior. He thought it was the neatest thing and took it into work to show some of his colleagues. 'It was a brilliant leveraging and repurposing of all this technology, which opened up a whole new marketplace,' Stein says. He started work on a project called My First Apple, which he soon retitled the Macintosh JFW – standing for 'Just Fucking Works'. Stein talked with Jony Ive and was assigned a member of the industrial-design department, Tim Parsons, to create a prototype JFW. Chief among its features was the aspect of personalisation. Where previously users would be able to customise their machines by naming the hard disk or coming up with a bold new configuration for the onscreen icons that defined them as an individual, here was a computer that physically reflected its user in terms of aesthetics. Users would be able to snap on different coloured case parts to get the colour computer they wanted. 'A cool twist was the idea that when you snapped the case parts on, there would be moulded into the case parts themselves a barcode or RFID (radio-frequency identification) tag that would communicate to the machine itself what it was wearing, and the user interface that was on the screen would adapt to show that,' Stein says. 'So you'd snap on the Barney purple case parts and the interface itself would change so that it looked as if it was all carved in one piece.'

Mitch Stein had left Apple before Jobs' return, but the idea hadn't. Gil Amelio has claimed that a low-end consumer Macintosh was 'within three or four months of being finished when Steve took over'. Jobs initially wanted to create a stripped-down network computer (or NC), but Apple's chief financial officer, Fred Anderson, instead insisted that the company focus on the Mac. To try and align Apple with the idealism of the new online digerati, Jobs made the decision to market it as an Internet computer. He decided that it would be called iMac.

This in itself was a brilliant marketing decision. There was nothing innately superior about the iMac over any computer created by Dell or Compaq in terms of its ability to connect to the Internet, but Jobs saw that, as had been the case with personal computers twenty years earlier, there were plenty of cautious but interested parties who would happily embrace a painless 'plug and play' entry point to the medium, complete with plenty of handholding. There was also an intensely personal aspect to his wanting to align himself with the utopianism of the Net. When Jobs was asked by an interviewer about what it represented, his answer betrayed a new degree of contemplation. He was older, married and a father. 'The rewarding thing isn't merely to start a company or to take it public,' Jobs said. 'It's like when you're a parent. Although the birth experience is a miracle, what's truly rewarding is living with your child and helping him grow up. The problem with the Internet start-up craze isn't that too many people are starting companies; it's that too many people aren't sticking with it. That's somewhat understandable, because there are many moments filled with agony and despair, when you have to fire people and cancel things and deal with very difficult situations. That's when you find out who you are and what your values are.'

Jobs went on to discuss how people who had got 'fabulously rich' from their Net start-ups were too often 'gypping themselves out of the potentially most rewarding experiences'. It was as if he was drawing a parallel with himself as a young man; stumbling on to a world-changing experience with Apple, but taking a self-imposed exile as the result of his own misplaced arrogance and pride. He wouldn't make the same mistake again. Apple's new Macintosh would return Apple to its roots, while righting the wrongs of the past. The advertisements for the computer stressed Apple's new quality of thoughtfulness; updating René Descartes' famous statement by suggesting, 'I think, therefore iMac'.

□ □ □

Like the original Macintosh, the iMac's name radiated friendliness, with a toyetic quality that stood in stark contrast to the complex strings of letters and numbers which had made up Macintosh product names for the last decade. The moniker was the creation of none other than Ken Segall, the man who had previously worked with Jobs on the 'Think Different' campaign. Inside Apple the iMac had been known as C1 – standing for *Consumer 1* – during development, although this name was for internal consumption only and never intended to make it out of the door.

'We already have a name we like a lot,' Jobs told Segall and several other creatives one day in the spring of 1998. 'But I want you guys to see if you can beat it.'

He then dropped a name so horrible that, in Segall's words, it would 'curdle your blood' just to hear it.

'It's called the MacMan,' Jobs said triumphantly. A shudder went through the group. Almost turbocharged, Segall raced back to his office to come up with a more suitable title, which would banish the MacMan to the same parallel universe in which the original Macintosh was launched as the Bicycle. It wasn't going to be easy. MacMan was the creation of Phil Schiller, Apple's worldwide marketing manager, and one of Jobs' best lieutenants. Also Jobs' guidelines for naming the project seemed to be at odds with the actual name he had picked out. MacMan sounded like a cross between the video game Pac-Man and Sony's handheld music player, the Walkman. 'Don't make it sound like a toy, and don't make it sound portable,' Jobs had instructed. No one had argued. Instead Segall launched himself into the task, plastering his office wall with myriad scraps of paper on which were scrawled various naming suggestions, including MiniMac (years before the Mac Mini) and iMac. Not realising at the time that the 'i' prefix would come to be a mainstay on Apple's product line, Segall saw that the 'i' could stand for Internet, along with terms such as *imagination*, *individual*, *inspire* and *instruct*. A week later he pitched Jobs, who was far from enamoured with any of the ideas.

'I hate them all,' he said, curtly. 'You've only got one week left or it's going to be "MacMan".'

Back to the office. Seven days later Segall pitched Jobs again, casually dropping in iMac for a second time.

'Well, I don't hate it this week,' Jobs said. 'But I still don't love it.'

The very next day Segall heard that Jobs was having the name silk-screened on to prototypes. Jobs never explained his change of heart. 'You don't get a lot of personal credit working with Apple,' Segall says, 'but I was really happy that my naming suggestion had been accepted.'

☐ ☐ ☐

In terms of aesthetics the iMac may as well have been a coming-out party for Jony Ive as a designer. Ive had been frustrated before Jobs returned to Apple. Design was not viewed as particularly important under Gil Amelio and Michael Spindler, and Ive had spent much of his time since arriving from the UK four years earlier working in a studio on Valley Green Road on less than inspiring projects. 'It was a difficult time for him,' says Peter Phillips, his former colleague at Tangerine. 'He was designing beige boxes for the current CEOs ... the honeymoon period was very firmly over.' After the NeXT buyout, Ive decided that enough was enough. He spoke with his new boss, Jon Rubinstein, the new head of hardware engineering, about finishing up his obligations and leaving. Rubinstein convinced him to stay, offering him both a pay rise and – more importantly – the promise that things were going to change. The iMac certainly delivered on that promise. In form it was as perfect an articulation of the utopian spirit of high-tech in the nineties as existed anywhere. Not everyone initially saw it that way, of course. Many were shocked by the machine's physical appearance which, in Jobs' words, 'looks like it's from another planet. A good planet.'

Distinguished scientist Larry Yaeger was rehired by Apple after a five-month break prior to the iMac's release. An authority on graphics and computational ecology, Yaeger had asked that his salary be 'halfway decent' for his return to the company. Jobs asked him to come for a private meeting in his office. 'I'm not sure how much was him wanting to entice me back, or just wanting to kick the tyres of this person they were about to give a fair bit of money to,' Yaeger recalls. Although he had joined Apple in 1987 – after the Apple co-founder's first departure – Yaeger had always held Jobs in the highest regard. He knew that this would make him weak when it came to negotiations, however. 'I had a great deal of respect for him and a love for the history of the company, so I was probably all primed to have the reality-distortion field work,' he says. He found Jobs highly enthusiastic about Apple and its future. Jobs must have been happy with the direction the meeting was heading in also, since before Yaeger left he unzipped a bowling-ball bag that was sitting in his office and pulled out a pre-release iMac prototype.

'What do you think?'

Yaeger was taken aback. 'It hadn't hit the market yet, so this was the first time I'd seen anything remotely like it,' he says. 'I wish I could say that the scales fell from my eyes and I knew immediately that the company was saved, but actually I looked at it and thought, "Whoa – that's weird." To him, of course, I said, "Ah, very pretty, very nice, very interesting."'

In a sense Yaeger wasn't the iMac's target audience. Like the original Macintosh, the iMac was a tool for those unempowered individuals who had yet to shell out for a personal computer. If its name conveyed reassuring friendliness, then so too did its design.

For a start it was distinctly unthreatening in appearance. From the bulbously colourful, Jony Ive-designed transparent casing (surely an in-joke at the expense of those still complaining at the lack of transparency in Apple computers) through to the simplified aesthetic of its one-button, hockey-puck mouse, the iMac resembled nothing so much as a protective bubble that would shield the user taking his or her first steps into cyberspace. Perhaps the term which summed it up best was 'blobject' a combination of the words 'blobby' and 'object', coined by the designer-author Steven Skov Holt. Bearing a curvaceously amorphic, somewhat gelatinous look that science-fiction writer Bruce Sterling once described, in an essay entitled 'Blobjects and Biodesign', as 'blown goo', the blobject briefly became the face of hip industrial design in the final years of the twentieth century. The iMac was, of course, one prominent example – as was Volkswagen's New Beetle, the Gillette MACH3 razor with its curvilinear grip, an Oral-B toothbrush and Marc Newson's Orgone Chair. All fitted together as lifestyle products that looked back fondly on the multicoloured plasticity of 1960s design, while also celebrating the arrival of a new technological age in which such seemingly unique, amoebic designs could be cheaply and easily mass-produced. In a sense it could equally be taken as symptomatic of the formlessness of a world in which market values ruled over all else; in which a crate of nappies might be equal in value to a single washing machine, or a curvaceous new computer to two weeks' holiday in Spain. The blobject's looseness of form paradoxically existed as a form in and of itself. It was representative of the decoding and recoding of our attitudes to individuality in a competitive world of capital.

To signify this momentous moment, Apple made the decision to move beyond the world of earthy tones that had previously typified its computers, into a veritable kaleidoscope of colours; starting with tropical Bondi Blue and expanding outwards to include Grape, Strawberry, Tangerine, Lime, Blueberry, Graphite, Indigo, Ruby, Sage, Snow, Blue Dalmatian and Flower Power. 'It doesn't have to be beige,' read Apple's pre-release materials. It proved a phenomenally successful proclamation. Prior to going on sale on 15 June 1998, 150,000 pre-orders had been placed for the iMac. Within its first six weeks it had sold 278,000 units. The candy-coloured iMacs served a dual purpose. For one thing, colours sell. When Heinz issued its EZ Squirt Blastin' ketchup in luminous green several years later, the brand saw its highest sales spike in history. But Apple's campaign relied on a neat piece of revisionist history. Far from being an authoritarian edict enforced from on high and ruthlessly held in place by an unfeeling technocracy, beige had originally been selected as a base colour for personal computers because it was unthreatening and would blend in when placed into an office or home environment. By reappropriating the reasoning behind the original creative decision, Jobs tapped into a sentiment surrounding the old Henry Ford maxim that 'the customer can have any colour so long as it's black' and turned the choice to buy a computer in Blue Dalmatian into a statement of individualisation.

With the Internet so ubiquitous today (and having long since lost much of its 'brave new world' lustre) it is almost difficult to remember where people's anxieties lay about going online. Certainly one of the fears related to the possibility of picking up a virus; a fear that no doubt tapped into the well of public concern about the global spread of HIV. It is ironic that the iMac – coming closer to resembling the boxiness of Volkswagen's 'Love Bug' than any previous model – should also be presented as a safeguard against the dark side of post-Kinsey sexual promiscuity. In the same way that the graphical user interface centred around inviting metaphors of desktops and offices, so malicious self-replicating computer programs disseminated online all too easily carried over to the terrifying image of sexually transmitted disease. Terms like 'epidemic' and 'infection' were thrown around in the popular press, while even the anti-virus programs that sprang up adopted similar cyber-medical rhetoric with names like Flu Shot+, Vaccinate and Antidote. On *Saturday Night Live*, comedian Dennis Miller quipped, 'Remember, when you connect with another computer, you're connecting to every computer that computer has ever connected to.'

Although they are not immune to viruses, Macintoshes have always been less susceptible to them than Windows PCs. (A combination of relatively few such viruses being written for the platform and, today, Apple's usage of the more secure, UNIX-based system.) Since the iMac represented for many their first experience with the Apple product line, the surprisingly conservative metaphor couldn't have been any better: Macs do not suffer from viruses. By staying with one partner (Apple) you too could stay disease-free.

This idea gained momentum as the year 2000 crept closer, and the now forgotten menace of the Millennium Bug threatened to throw all of us into the kind of techno-apocalypse that the hippies had warned us about. All this, somewhat bafflingly, was supposed to come about because programmers in computing's early days had chosen to save space coding by presenting years in their binary format, i.e. '85' instead of '1985'. At the end of the twentieth century, we were now warned that the repercussions of this simple lapse in judgement could range from the systemically destructive (the freezing of bank accounts) to the absolutely destructive (the accidental launch of nuclear warheads). While not even the combined brainpower of those engineers at Infinite Loop, Cupertino, could save us from the latter, Apple was quick to dispense the blame, as well as to highlight that it was among the safest systems out there, once again adopting popular sci-fi imagery to make its point. In this case the imagery came from Stanley Kubrick's space-age epic *2001: A Space Odyssey* (incidentally a popular acid-trip movie in the sixties), with Apple putting together a spot for the 1999 Super Bowl featuring HAL-9000, the sentient computer who turns against humanity.

'Hello, Dave,' the familiar voice intoned during the commercial. 'You're looking well today.' The machine then goes on to ask whether Dave recalls the year 2000 in which computers began to misbehave, as the result of their programmers failing to make them recognise the year 2000. As a result, with the coming of the new millennium, they had no choice but to wreak global economic disruption. All, naturally, except for the Macintosh, which had been adequately future-proofed for just such an occasion. Coming fifteen years after the '1984' Mac commercial, it was a neat reprisal of the anti-IBM feeling from Apple's first Super Bowl ad. Although both writer Arthur C. Clarke and director Stanley Kubrick strenuously denied it – claiming that HAL stood for 'Heuristically programmed ALgorithmic computer' – it is interesting to note that HAL's initials also happen to be one letter removed from that of IBM. Jobs, certainly, would have been happy for that to be less than coincidental, although he also would probably have widened the field to include Microsoft. 'HAL is the perfect spokesperson to address the Y2K issues because he lives in the year 2001 and speaks from experience,' he said at the time. 'Plus HAL is the foremost expert on things that can go wrong with computers.' On the other hand, Macintosh computers, Apple claimed, were more than capable of internally handling generated dates through to the year 29,940, just so long as the software running on them was also capable of doing so.* It was another example of how the Establishment had let people down – and how Apple was happy to come to the rescue.

For everything that it did well, the iMac was not the drastic systemic shake-up that Jobs was looking for. It built on what computers were already doing at the time, but gave the medium a new lick of paint and attached it to an excitingly utopian movement. As such it represented the next link in computing's evolutionary chain. What Jobs wanted, meanwhile, in the words of a senior vice-president at Home Depot at the time, was to 'think revolution not evolution'. He began looking around at the developments in consumer electronics at the time to find something that was truly revolutionary. Both digital cameras and camcorders were briefly considered, but were deemed too competitive a market. One possible option was the personal data assistant, also known as the palmtop computer. In 1998 Jobs made a concerted effort to buy the Palm Pilot product line from 3Com; a move that would have allowed him to release a handheld PDA device without first having to sit through the development process. But he was rebuffed. One of the people negotiating with him referred to their dealings with the Apple CEO as 'the worst experience of my life'. In the end nothing came of it. When Jobs was interviewed by *Fortune*

---

* As a neat Easter egg, paying homage to the classic *2001*, engineers working on Siri, the voice-control personal-assistant feature introduced in the iPhone 4S, featured three references to the film: a modem resembling HAL's faceplate; if asked to sing it might reply 'Daisy, Daisy, give me your answer do'; and it responds 'I'm sorry, I can't do that' when asked to 'open the pod-bay doors'.

magazine several years later in 2001, he downplayed the PDA's revolutionary status as being, really, something of a fad. 'I won't lie,' he said. 'We thought about [building a PDA] a lot. But I started asking myself, "How useful are they, really? How many people at a given meeting show up with one?"'

Aside from being turned down by Palm, the reason Jobs had changed his mind about the PDA was because he had come across another idea, one which tapped far more directly into the Dionysian qualities of rebellion. 'I don't think early cultures had [pocket] organisers, but I do know they had music,' he said. 'It's in our DNA. Everybody loves it. This isn't a speculative market.' Apple was about to enter the music business.

If this shift in product-line approach was about trying consciously to tap into some spirit of revolutionary zeal, however, then the thought process he had undergone to get there was equally countercultural. After fighting a land war against Microsoft for the longest time, Jobs was willing for Apple to try something new. Publicly he had denied this, suggesting that it made no sense whatsoever for Apple to deviate from what it had already proven to be good at. 'If Mercedes made a bicycle or a hamburger or a computer, I don't think there'd be much advantage in having its logo on it,' he told an interviewer for *Fortune* in 1998. 'I don't think Apple would get much equity putting its name on an automobile, either. And just because the whole world is going digital – TV, audio, and all that – doesn't mean there's anything wrong with just being in the computer business. The computer business is huge.' Certainly Apple had no plans to release a car. Jobs was more than happy driving his Mercedes for which – through a legal loophole – he was able to get away without having number plates. (He also got around Apple's democratic lack of fixed parking spaces by parking in the disabled bays – often diagonally across two spaces – leading to one brave employee one day sticking a note under his windscreen wiper reading, 'Park Different'.) At the same time he was happy to embrace the role of smaller, faster underdog, ducking and weaving against a much larger foe. This notion brought Apple into line with a host of other high-tech companies which were creating a new kind of DIY culture in their fight against the homogenising forces of Microsoft. Netscape, for example, was giving away its web browser for free in order to compete with Microsoft's Internet Explorer. Linux, meanwhile, had created a viable alternative to Windows by constructing an operating system pieced together by various users from around the world employing collaborative, open-source code. Nowhere was the rebel yell better summed up than in an article by co-author of the *New Hacker's Dictionary*, Andrew Leonard, entitled 'Let my software go!':

> *Hi there, Mr. CEO — tell me, do you have any strategic problem right now that is bigger than whether Microsoft is going to either crush you or own your soul in a few years? No? You don't? OK, well, listen carefully then. You cannot survive against Bill Gates playing Bill Gates' game. To thrive, or even survive, you're going to have to change the rules. I'm here to show you how.*

It was a strategy that, in some cases, paid off handsomely. In *How the Weak Win Wars: A Theory of Asymmetric Conflict*, the author, Ivan Arreguín-Toft, describes how strong actors are increasingly losing to weak ones over time.[*] In fact, Arreguín-Toft suggests that in the last fifty years the weak have won a majority of engagements against opposition ten times more powerful. The

---

[*] 'Actors' in this sense generally means states or coalitions of states, although Arreguín-Toft refers also to business competition throughout the book.

reason for this is that, while strong will almost always defeat weak within the confines of a conventional, direct confrontation, they will also fare less well when it comes to indirect, guerrilla fighting, where conditions strongly favour insurgents. David over Goliath; the Vietcong over the United States; the Afghans over the Soviets – *why not Steve Jobs over Bill Gates?*

There was another battle that Apple could win also, and this one Jobs was more confident of. Although Apple was far from the first company to take a crack at the MP3 market with its iPod, it was the first to do it well. Early players in the field had included Saehan, Pontis and Diamond Rio. By 1999, around twenty-three other companies had joined in with their own versions. With the exception of a few – namely Sony, Samsung and Thompson/RCA – these were mostly small and medium-sized entrepreneurs and new arrivals to the industry, the majority of whom relied on sales through the Internet. It was clear to Jobs that digital music players were the future. By 2000, the market in the US alone was valued at $80 million. The following year it would rise by 25 per cent again to $100 million. In terms of units sold this represented an increase from 510,000 to 700,000 MP3 players in the hands of paying customers. No one company had yet managed to crack the magic formula, and nor had any one player gained a substantial lead. It was equally apparent to Jobs that, while Apple was a small fish in a big pond compared to Microsoft, it was a whale in a teardrop next to some of these start-ups. In other words, Apple could afford to throw its weight around.

Jobs' first move was to survey the scenery. His industry scouts pointed him in the direction of a music-software product called SoundJam MP, being marketed by a small company called Casady and Greene. C&G, while not a part of the Apple blood family, was related by marriage. The company had had some success over the years developing games for the Macintosh, including 1987's Crystal Quest, the first full-colour game for the platform. SoundJam itself was the creation of Jeff Robbin and Bill Kincaid, two former Apple software engineers who had gone on to join C&G to make MP3 software that would allow Mac users to manage the songs on their computer. Before long, SoundJam had 90 per cent of the Macintosh market and was bringing in the bulk of C&G's $5.5 million annual revenues. Jobs could spot its potential a mile off and approached C&G with an offer that would be difficult to refuse. In the words of C&G president Terry Kunysz this amounted to little more than, 'Sell the rights, or we'll develop a competitive product and put you out of business'. Kunysz sold the rights. Jobs renamed it iTunes. 'Join the music revolution,' he told those attending the Macworld Expo in January 2001.

If iTunes was immediately recognised as a neat accompaniment to the Macintosh's existing iMovie software, from which the development borrowed the brushed-metal interface aesthetic, then the iPod, by contrast, was initially viewed as a disappointment. Upon its announcement in the middle of 2001 (teasing invitations to the unveiling noted, 'Hint: it's not a Mac') Apple's portable music player was met with barely concealed dismay by many analysts. Looking back, it is easy to see why it raised eyebrows. After all, every impression that Jobs had given was that he was going to streamline Apple's product lines and strip away any and all unnecessary drains on resources. 'No wireless. Less space than a NOMAC. Lame,' wrote the website *Slashdot* with a slapdash reaction that would have its authors cringing years later.* 'All that hype for an MP3 player?' scoffed one forum visitor on a normally Apple-friendly website. 'Break-thru digital device? The Reality Distortion Field™ is starting to warp Steve's mind if he thinks for

---

* The Creative NOMAD was a range of digital audio players designed and sold by Creative Technology. They failed to catch on and were discontinued in 2004.

one second that this thing is gonna take off.' Some were put off by the initial price of $399, which certainly seemed excessively steep for an MP3 player.

The iPod may have been expensive, but it also proved immensely successful. Indeed, it wasn't long before the degree to which Apple's latest gadget had been upgraded from major disappointment, to must-have, iconic technology became apparent to everyone. In Britain, the Queen revealed that she had bought one, although she failed to report what was on it. George W. Bush also had an iPod, and his 250-song playlist included 'My Sharona' by the Knack, along with former Creedence Clearwater Revival singer John Fogerty's baseball anthem 'Centerfield'. (CCR's anti-Vietnam War song 'Fortunate Son' did not make the cut, although Bush's mountain-biking partner sagely noted, 'If any President limited his music selection to pro-establishment musicians, it would be a pretty slim collection.') Jobs had finally got the wish he had made years earlier when he told the people at Regis McKenna that he hoped his company's products would become so popular that *Apple-ing* would take over from 'computing' in the mainstream parlance. That obviously had not happened, but a version of it certainly did as the term 'iPod' became shorthand for any technology that was in equal parts innovative, stylish and ubiquitous. (Today, iPod creator Tony Fadell is reportedly working on what he hopes will become the 'iPod of thermostats'.)

As with every major new product line that Steve Jobs embarked on during his stint as Apple CEO, the iPod was a personal project for the company co-founder at every level, one that would allow Jobs to tap into the world of his youth. As he noted, 'We love music, and it's always good to do something you love.' On a literal level, of course, Jobs' formative years hadn't been marked by a plethora of portable music players. It was 1979 – three years after the founding of Apple – before

Sony marketed its first generation of Walkmans. But music *had* served as the backing track to the counterculture, whether it be the folk protest ballads of Bob Dylan or Joan Baez, or the acid rock, which reflected the surrealistic turn America had taken with the Vietnam War. The iPod was a chance for everyone to have their own personal soundtrack. In fact, linking Apple's portable music player to its previous innovation, the iMac, the iPod took up the reins of providing a device to grant users the ability to tap into some form of comforting hyper-reality, in the days before the Internet became transportable. In the essay 'iPod ... iCon', which appeared in the Italian journal *Studi Culturali* in 2008, Tim Dant quotes one iPod user as noting that, 'Once I put the headphones on, it's like there's an invisible wall all around me. I don't really feel like I exist in the same way any more, like I'll touch something and almost be surprised I can feel it.' George Prochnik expanded on the idea in his book *In Pursuit of Silence: Listening for Meaning in a World of Noise*, in which he noted how:

> *The iPod is touted not for its ability to mask over old infrastructure sound but for the way it blocks out the discretionary din that got plastered on top of that layer and defines the new noisiness of the digital age. Based on how people speak about their iPods, it would seem that today the nexus of pain is not so much any single loud noise as it is our diffraction between multiple stimuli engaged in a continual 'boom-off.' Rather than ranting about the 'alienation of the iGeneration,' we have to ask ourselves what we did to make our larger soundscape so disposable that it became the sonic equivalent of fast food.*

Whilst wearing the iPod's trademark white headphones, users were able to retreat into a bubble of individual meaning, often within public spaces. If the shared experience of owning an iPod immediately made one a member of an increasingly popular subset, suggesting that the user was hip, youthful and in-the-know, then the listening itself was a fundamentally lonely endeavour; more about blocking others out than letting them in. Interestingly the iPod's obvious predecessor, the Sony Walkman, was originally envisaged as a sociable device, able to remove the antisocial noise pollution of their less portable rivals, the ghetto blasters, while retaining the element of shared experience which made that technology so memorable. In the first wave of 1980 Walkmans (originally known as Sony Stowaways in the UK) one of the features, along with two sets of headphone sockets, was an internal microphone. When a button was depressed, users were able to talk with one another, with their voices overlaid on to whichever music they were listening to at the time. As it became apparent that Walkmans were far more likely to be listened to by one person at a time, however, this rather unique function was eliminated for future models. The iPod continued this trend and thus became perhaps the ultimate embodiment of Jobs' individualistic mantra of 'one person, one [machine]'.*

Unlike the iMac's attaching of itself to the utopian spirit of the Internet, the links between the iPod and the rekindled countercultural ethos were not always immediately apparent, but they were there if one looked for them. One television commercial, entitled 'Pods Unite', showed the iPod being listened to in a VW Beetle, the iconic car of the sixties generation. Similarly, third-party vendors, ever keen to spot

---

* This changed slightly with the release of iPod docks that would allow playlists stored on the device to be enjoyed by more than one person. On the whole, however, I would still argue that the iPod is primarily intended as a solitary listening experience.

a gap in the market, began manufacturing straps that would allow the iPod to be affixed to the upper arm; a visual reference, perhaps, to actor James Dean's famous placing of a pack of cigarettes exactly there, rolled into the sleeve of his t-shirt. Dean in *Rebel Without a Cause*, like Marlon Brando in *The Wild One*, epitomised the heady joys of (white) youthful rebellion. When one character asks Brando's Johnny Strabler in the latter film, 'What are you rebelling against, Johnny?' he throws out a stony glance and says, 'What have you got?' Naturally, they had no answer to his ice-cool, anti-Establishment chat, but Apple did. If you had already bought a first-generation iPod, then there were other iterations of the product line around the corner ...

At least initially, the iPod was designed to be available only to Apple users. 'I'm never taking this to the PC,' Jobs claimed to those inside the company, seeing it as a gateway drug into the Apple ecosystem rather than a device that might be a business unto itself. Although he later changed his mind, what allowed Jobs to do this was the device's reliance on the iTunes software in order to manage the tracks which were stored on it. iTunes, in turn, would allow Jobs once again to link his device to the revolution that was taking place online.

As a tool substantially developed equally by hippies, academics and computer geeks, the Internet's first iteration took the form of a high-tech gift economy, not dissimilar to the archaic gift economies that pre-dated capitalism, in which exchange was reciprocal, but payment was not demanded in a predetermined form or on a specified timescale. Since it was funded by the US military, there was no demand on Net users to turn their intellectual work into any form of marketable commodities. But as the Internet gained momentum with users in the nineties, this notion – once seen as a great method of collaboration across physical boundaries – took on a worrying turn for the heads of major companies as they realised that this hyper-connected model of free-market capitalism had one major problem: there was no way to capitalise on it. It was one thing when all that was being traded were stodgy academic papers and arguments on Usenet groups about who was the better *Star Trek* captain; another altogether when increasing numbers of people decided that that same model of gift economy might as well also apply to traditional copyright laws. After all, since information wants to be free, and music – at least on a binary level – is simply information, why should anyone be forced to pay for it? If every user uploaded all of their music to the Net then everyone would have the equivalent of the world's best record collection. What could be more egalitarian than that? Suddenly the utopian idea of the Internet as a world 'not [of] haves and have-nots [but have-nows and] have-laters' took on frightening new possibilities for big business. Enter peer-to-peer file-sharing services like Napster, the notorious scourge of the Recording Industry Association of America.

The irony that a technology like the Internet so heavily subsidised by the American defense department would come to be viewed as the tool of decentralised liberation did not go unnoticed. At least initially, however, the Net appeared to be the dream of the techno-hippies writ large. Like one giant global commune, on the Internet everyone would have an equal voice. The rise of e-commerce, of course, was always going to necessitate the eventual need for an effective legal regulation of the Net. After all, even the frontier towns of the Old West could only fully prosper once a sheriff was installed. But, it didn't do anything to hurt the Net's early utopian atmosphere that it appeared pleasingly difficult for governments or large corporations to police it in any meaningful capacity. (An early, failed attempt to do so, in the form of the 1996 Communications Decency Act, which aimed to ban pornography, and to punish the use of the word 'shit' online with a fine of $250,000, prompted *Wired* editor Louis Rossetto to liken it to 'the illiterate [trying to] tell you what to read'.)

As noted, both Apple and Steve Jobs had missed out on the initial party that had been the Internet during the first half of the 1990s. Jobs had promised to rectify that error. In a February 1996 interview with *Wired*, while he was still at NeXT, Jobs noted that the main long-term beneficiaries of the Net would not be those willing to enter into a high-tech gift economy, but, rather, those with something to sell. ('To share?' asked *Wired*. 'To sell,' Jobs repeated.) 'It's commerce,' he explained. 'People are going to stop going to a lot of stores. And they're going to buy stuff over the Web.' Jobs, however, had himself failed to understand exactly what was going on upon his return to Apple. In a rare instance of not having his finger on the technological pulse, he missed out on the burgeoning popularity of users downloading music and then burning it to CD. The early iMacs shipped without a CD-RW ('read-write') drive. While this was a lesser-known feature in Windows PCs during the late 1990s, by the winter of 2000, 40 per cent of new personal computers included the function. No longer the moocher he had been in his youth – when he slept on dorm-room floors at Reed College and walked seven miles across town to get a meal at a Hare Krishna temple – Jobs didn't pick up on the trend. Billionaires don't, after all, have to steal inexpensive things. It didn't take long for him to realise what had gone wrong. 'I felt like a dope,' he admitted.

iTunes had initially been solely a media player, allowing users to save, organise, play and download music files on to their Apple computers. Controversially, Jobs had chosen to ally Apple with the online utopians by launching the software with the slogan 'Rip. Mix. Burn'. Then Disney CEO Michael Eisner objected to the campaign in the strongest possible terms, singling out Apple's slogan while testifying in front of the United States Senate Commerce Committee as an example of how piracy had become the computer industry's latest 'killer app'. Eisner claimed that 'Rip. Mix. Burn' was an implicit suggestion on the part of Apple that its customers 'can create a theft if they buy [a Macintosh] computer'.

Eisner's reaction smacks of justifiable frustration. Despite the fact that it meant going directly after their own customers, the recording and music industries saw the only possible way around the problem of piracy being by tyrannical means; to punish digital theft with hefty lawsuits that would be taken as far as the legal system would possibly allow. Jobs saw another answer. His solution – like Apple's best – would please both the world of big business, as well as making Apple look more humanistic than the other corporations out there. 'These kids are using the best product,' he told *Esquire* regarding the illicit downloading of music. 'Until [the launch of iTunes] Kazaa was the best product.* Why is that? Because the minute you get your music over the Internet and experience that instant gratification, there's no going back. ... I don't blame these 15-year-old kids. I blame us – for not coming up with a better product that was legal.' Years later Jobs delivered a stirring sermon to biographer Walter Isaacson on the notion that information, in reality, does not want to be free at all. 'From the earliest days at Apple, I realised that we thrived when we created intellectual property,' he said. 'If people copied or stole our software, we'd be out of business. If it weren't protected, there'd be no incentive for us to make new software

---

* Kazaa, like Napster, started life as a peer-to-peer file-sharing application, commonly used in the exchange of MP3 music files and other documents.

or product designs. If protection of intellectual property begins to disappear, creative companies will disappear or never get started. But there's a simpler reason: it's wrong to steal. It hurts other people. And it hurts your character.'

Whether or not a breakdown in traditional copyright laws does, in fact, lead to a similar decline in creativity and innovation remains a hotly contested debate.* Regardless of this, Jobs' comment did require a bit of moral reversioning on the part of the Apple CEO. This was, after all, the same Steve Jobs who had flown a skull and crossbones above the Macintosh building while he was working on that project, and told anyone who would listen how 'It's better to be a pirate than to join the navy'. Similarly, a large part of the 'Microsoft as Evil Empire' myth relied on Bill Gates's insistence that the ethereality of software be something that users should pay for. When Jobs, again in Isaacson's biography, referred to Gates as 'basically unimaginative' and commented that he 'never invented anything', he was wrong. What Gates had invented was the notion that software (be it entire operating systems or simple files) could be sold. Jobs merely reframed the idea as a necessary protective measure for creativity.

The iTunes Store itself came together over the course of 2002 and the early part of the next year. In January 2002, a team of executives representing Warner Music, AOL Time Warner and Sony flew to Cupertino to see Jobs. Jobs, who never liked sitting through presentations, was bored and annoyed by their meeting. He began rocking backwards and forwards in his chair in frustration as the Warner executives talked through their digital strategy. Finally he had had enough.

'I don't want to talk about what you guys are doing,' he said, bluntly. 'You guys have always had your head up your—'

---

* Both Jaron Lanier's *You Are Not a Gadget* and Lawrence Lessig's *The Future of Ideas* make interesting reading on this topic.

He trailed off, diplomatically leaving the word unsaid. There was a long pause.

'You're right,' said Paul Vidich, the executive vice-president of the Warner Music Group, finally. Vidich was suffering from a bad cold and had brought with him a deputy to speak on his behalf, but now he decided to interface with Jobs directly.

'We don't know what to do,' he reiterated. 'You need to help us figure it out.'

□ □ □

The result was the iTunes Store, an online shop in which songs, at least initially, would retail for 99¢. The reasoning behind pricing individual tracks at just under a dollar was that this would be cheap enough that they would qualify as straightforward impulse purchases. Since the iTunes interface would be infinitely more straightforward than anything offered by a peer-to-peer network, Jobs rationalised that the majority of users would think it a better use of their time simply to download legally. According to the deal that Apple hammered out, the company would receive around 30 per cent of each sale, while the balance would go to the music company that owned the rights.

The iTunes Store opened 28 April 2003. To commemorate it, Apple put on an event in a San Francisco convention hall. After Bob Dylan's anthem 'The Times They Are A-Changin'' had played, Jobs appeared on the stage to deliver a speech. He was unshaven, clad in a black knit shirt, jeans without a belt and a pair of New Balance trainers. 'Consumers don't like to be treated like criminals and artists don't want their valuable work stolen,' he said. 'The iTunes Music Store offers a ground-breaking solution for both.'

One of the true revolutions that accompanied the launch of the iTunes Store was certainty its democratisation of distribution. While it has hardly turned popular taste on its head it has produced a viable distribution method for music and other media which previously would not have had the opportunity to be exposed to such a mass audience.

At least in theory, users today have more say than ever on what it is that they listen to and purchase. One illustration of iTunes being used as a medium to send a message to traditional media came in December 2009, when users in the United Kingdom, alerted to the cause courtesy of a Facebook group, began purchasing *en masse* the 1992 song 'Killing in the Name' by rap metal band Rage Against the Machine, in order to prevent the winner of the hugely popular TV programme *The X Factor* from achieving the Christmas number-one slot in the music charts for the fifth year running*. Similarly, by unbundling the tracks on an album (confirming what Napster had already hinted at: that the natural measuring unit for music is not the album, but the song) consumers have the ability to stick it to greedy record executives who had loaded albums with one or two hit tracks and a lot of filler.

At the same time this unbundling raises certain questions. Some of it is down to preference. Isn't an album like *Sgt. Pepper's Lonely Hearts Club Band*, for example, designed to be listened to as one cohesive experience, rather than thirteen bite-sized chunks? This is fast becoming an outmoded argument. After all, if you can restructure society and your own consciousness, why shouldn't you reorder your own music?† But some of it hints

---

* Being a major musical act themselves, of course, Rage Against the Machine was not necessarily the best choice of band to pick in order to protest music-as-microcosm-of-capitalism. Making the move more ironic was the fact that the song picked was a Sony BMG catalogue track, just like 'Hallelujah', 'Don't Stop Believin' and the various songs regularly featured on *The X Factor*.

† This idea was seemingly cemented by the arrival of the iPod Shuffle in January 2005, which, as its name implied, took advantage of the iPod's popular ability to allow the user to move quickly between songs, artists and even entire genres, so that the previously linear experience of consuming music turned into the auditory version of hypertext.

at bigger problems, yet to be solved by the techno-utopians. iTunes' unbundling is, essentially, symptomatic of a much larger cultural shift. Over the past decades changing economies of production and consumption have led to an unbundling of various entities across a variety of media. Consider the decline of newspapers, whose readership has fallen as the influence of personal computers has grown. After peaking at sixty-three million copies per day in the Macintosh's launch year of 1984, the daily circulation of American newspapers has declined with increasing rapidity ever since. The rise of the Net has created more news outlets than ever, but these no longer function as a cohesive whole. Because of the way online advertising is structured, the most successful articles are the ones that will provoke a high click-through rate and thus secure lucrative banner advertising. In the days before unbundling occurred, an overseas correspondent or investigative journalist's salary might be paid by the revenue generated by the pages of the classifieds. Today, in an era in which stories are individualised, every webpage must stand on its own. Advertisers must no longer pay for bundled ads, but, rather, pick and choose their placements.

This, of course, smacks more of Google than Apple, but iTunes is symptomatic of the same trend for unbundled information. Ironically, as Daniel Akst, pointed out in a *New York Times* article entitled 'Unbundles of Joy', the unbundling of information is one of the most significant results of bundling the world's computers into one single network. By making it possible for users to download just one song from an album, the iTunes economy has done for music what Google has done for advertising, or Wikipedia for the encyclopaedia, or YouTube for video content.

□ □ □

The day after the iTunes Store was launched, Jobs was interviewed for *Esquire* magazine. Walking into the main lobby of Apple's

Infinite Loop campus, journalist Andy Langer noticed – *it was hard not to* – a billboard 'the size of a basketball court' hanging from the ceiling, upon which was emblazoned the words: 'YOU SAY YOU WANT A REVOLUTION?' He was quickly met by an Apple publicist and taken to a conference room. Jobs entered a few minutes later, having picked up some soup and a salad for lunch. Langer had been warned that Jobs was an intensely private individual and that, while he would be more than happy to talk about Apple and the music industry, Langer was not to enquire about Jobs' vegetarian diet or anything of a personal nature.

Jobs began by discussing how the iTunes Store had come about. The conversation then moved on to the accusations that the original incarnation of iTunes seemed to advocate piracy. 'People equated burning CDs with theft,' Jobs said, claiming that 'Rip. Mix. Burn' had simply been a matter of mistaken intent. He explained that many people burn CDs for their own use, to create custom compilations. Theft only enters the equation when people visit sites like Kazaa.

Jobs, who had often admitted to listening to Bob Dylan bootlegs in his youth, was clearly uncomfortable about discussing Apple's relationship with music companies, or what it suggested about Apple as a media publishing company in its own right.

'We don't own media,' he said. 'We don't own music. We don't own films or television. We're not a media company.'

The interview had only been going on for a short time but Jobs was getting fed up.

'We're just Apple,' he finished, and reached over and turned off Langer's tape recorder.

'Thanks for coming out to California,' he said, getting to his feet and reaching out a hand to the somewhat bewildered journo. Then he turned around and strode out of the room.

15

APPLETOPiA

'With the coming of a wired, global society, the concept of openness has never been more important. It's the linchpin that will make the new world work. In a nutshell, the key formula for the coming age is this: Open, good. Closed, bad. Tattoo it on your forehead. Apply it to technology standards, to business strategies, to philosophies of life. It's the winning concept for individuals, for nations, for the global community in the years ahead.'

Peter Schwartz and Peter Leyden,
Wired magazine (July 1997)

The ninth of January 2007 was a day Steve Jobs claimed to have been looking forward to for two and a half years. Standing in front of a crowd of Apple fanatics at the Macworld expo in the Moscone Center in San Francisco, he announced that he had three new devices to unveil. The first was a revolutionary mobile phone; the second, a new widescreen iPod with touch controls; the third, a breakthrough Internet device. Jobs repeated the trio, this time speaking more slowly. 'An iPod, a phone, and an internet communicator,' he reiterated, patiently allowing for those watching to join the dots. 'Are you getting it?' Jobs said, his gee-whiz-boy inventor voice rising in volume. 'These are not three separate devices; this is one device. And we are calling it *iPhone*.'

Like the deceptive simplicity of the device itself, nothing about the iPhone's development had been straightforward. Jobs had wangled a terrific deal with AT&T (then Cingular), which would give him absolute control over the way the iPhone would look, how it would be manufactured and the means by which it would be marketed, along with $10 a month from customers' bills as the result of a revenue-sharing agreement. In return, AT&T would receive five years' exclusivity, around 10 per cent of iPhone sales in AT&T stores, and a small chunk of revenue from iTunes. Next had come the tricky part: *building the damn thing.*

Inside Apple the slog towards the iPhone's product launch was termed by those working on it as the 'death march': that final, trudging push before the technology – in First World War parlance – went 'over the top' to face its eventual fate in the eyes of the buying public. Codenamed P2 (for Purple 2) the project had taken the form of a deadly superbug; stealing away the most talented engineers from each division and resulting in whole sections of the Infinite Loop campus being sealed off. In 2006, the year in which Daniel Craig took up the role of James Bond in *Casino Royale*, Apple employees had the chance to play a secret agent themselves. Staying in hotels, they would check in as employees of the semiconductor company based in Neubiberg, Germany, which happened to be building the transmitter for P2. Even inside the Apple campus, secrecy reigned supreme. Hardware engineers slaved away with circuitry loaded up with fake software. Software engineers worked with circuit boards set into wooden boxes. As they laboured, they were advised to keep what they were working on under a black cloak, as if they were developing a roll of light-sensitive film. If they ducked out of the room for just a few seconds to visit the bathroom or to ask a question of their colleagues, they pressed a button, triggering a red light that warned everyone in the vicinity to be extra vigilant. Fuses ran short, but deadlines ran shorter. Those working on P2 became absentee parents, husbands or wives.

'What was so important that you had to miss [a particular event]?' concerned spouses would ask.

'I'm sorry,' came the standard reply. 'That's classified.'

❑ ❑ ❑

Apple had long been rumoured to have been working on a smartphone. The suggestion that it could enter the mobile-phone market with a device called the iPhone dated back at least as far as 2002, the year after the iPod was released, when reports circulated that Jobs had reunited with Steve Wozniak to explore the options presented by a next-generation cell phone. 'We know that [Apple is] going to build a smartphone,' said then Motorola CEO Edward Zander in 2005. 'It's only a matter of time.' That year Apple had dipped its toe half-heartedly into the water with the ill-fated ROKR, which licensed iTunes technology to Motorola. It disappointed. What, if anything, proved a surprise about Apple's eventual arrival in the mobile-communications industry, however, was just how ill prepared the industry was for the ensuing systemic shake-up. It didn't hurt (or help, depending upon whose side you were on) that Jobs' rivals had none of his messianic presence. Where the Apple co-founder talked about changing the world, his opposite numbers seemed content to play comedic buffoons. When Motorola's profit margins had collapsed in 2006, Ed Zander appeared at the Las Vegas Consumer Electronics Show, riding a bright yellow bicycle on to the stage for his keynote speech to indicate how 'everything is mobile'. (A month after Apple unveiled the iPhone, Motorola struck back by dressing its phones up in red lingerie as a Valentine's Day promotion.)

With the iPhone Jobs once again painted himself in the role of leader for the forthcoming technological rebellion. With the words 'Revolutionary UI' emblazoned in giant letters on the screen behind him, he made his case for shaking up the Establishment. In this instance the Establishment was portrayed by four phones: the Moto Q, the BlackBerry, the Palm Treo and the Nokia E62. Jobs referred to these as 'the usual suspects'. The problem with all of them, he explained, lay in the bottom 40 per cent of each device: the omnipresent built-in keyboard which was there 'whether you need them or not'. Calling up images of conformity and the technological priesthood, he noted how these existing smartphones had 'control buttons ... fixed in plastic, which are the same for every application'. *Didn't every user, as with every application, deserve their own unique user interface?* 'What happens if you think of a great idea six months from now?' Jobs asked, rhetorically. 'You can't run around and add a button to these things, they're already shipped.' The BlackBerry and the Palm Treo were stuck as a product of the time in which they had been created, while the iPhone – with its bitmapped screen and Multi-Touch technology – could effectively evolve for any purpose.

Of course, the iPhone was absolutely a product of its time, although Jobs failed to point this out. Technology, and most assuredly technology which gains the kind of marketplace momentum that the iPhone has, does not exist in a world free of social context. Successful technologies are such because they function within a social world, and the lessons from that world impact upon the form of that particular device. Perhaps the best way to think about it is in terms of the famous M. C. Escher image *Drawing Hands*, in which a three-dimensional left hand, rising out of a two-dimensional sheet of paper, draws a right hand, which itself draws the left hand. If we imagine one hand to represent technological development and the other society, then it is the entangled hierarchy of these two images (which itself links back to the programming in-joke – Infinite Loop – that gives the Apple campus its name) which summarises the symbiotic relationship between both entities. They are a co-construction, running hand in hand with one another. This is why the hippie's dream of liberating one from the other (that is to say, to take the technology and leave the technocratic society in which it was created) appeared such a striking bit of cultural jamming when it was first proposed.

Among Jobs' most brilliant traits was an ability to select the perfect moment to seize upon a particular concept and reinvent it. Try something too early, before the technology is ready, and it will disappoint. Leave something too late and it is difficult to establish a foothold, regardless of quality. As had been the case with the iPod, whose tiny 1.8-inch hard drive, capable of storing five gigabytes' worth of music, was fresh off the conveyor belt when Jon Rubinstein convinced Jobs to buy it for Apple's use, the technology behind the iPhone was also brand new. Even the material from which the screen was composed was original. At first the plan had been simply to use plastic for the iPhone screen, as had been the case with the iPod, but Jobs changed his mind at the last minute. What was needed, he decided, was glass – but more than that, *unbreakable* glass. The search led to a company called Corning Incorporated, the inventor of Pyrex dishes. In the sixties, Corning had manufactured a material called 'muscled' glass, which was nigh unbreakable and used for the windows in cars and aeroplanes. However, muscled glass had long since been discontinued. Jobs told Corning's chief executive, Wendell Weeks, to make everything that they could in six months.

'We don't have the capacity', he was told. 'None of our plants make the glass now.'

But Jobs cajoled. 'Get your mind around it,' he said. 'You can do it.'

Six weeks later Corning was back in the new, improved muscled-glass business. Apple nicknamed it Gorilla Glass for its super-strength. When it hit the shelves, the iPhone thus looked like no smartphone the world had seen before. Aesthetically it could scarcely have been more different from the blobjectine iMac of less than a decade earlier. Gone were the soft edges, the artful formlessness and the neon colours. In their place was a striking industrial minimalism, which appeared to have been infuenced by the work of Dieter Rams, the legendary designer at German consumer-products company Braun. Apple's newer products still referenced the sixties, when Braun was at the height of its powers prior to being absorbed by the Gillette Group, but it was a sixties image of functional austerity more than of light-hearted utopianism.* Simultaneously the iPhone managed to look strikingly modern; certainly more suited to its time than its chief rival, the relatively formless BlackBerry, whose combination of straight edges and unnecessary curves spoke of a device with no real identity of its own. Apple, by contrast, had an identity. Dropping the term 'Computer' from its name in January 2007, in an effort to reflect the company's continued move towards more generalised consumer electronics, it was absolutely sure of its place in the world. It was also getting older.

In 2007 David Kelley was sick. The 56-year-old head of IDEO had been diagnosed with cancer, a stage-four squamous-cell carcinoma, which had gone misdiagnosed as 'inflamed fish gills' for an entire year and a half. Now it had migrated to his lymph nodes and was serious enough that his oncologist sat down with

---

* For an example of the influence of the work of Braun's Dieter Rams on Jony Ive look no further than the iPhone's calculator app, whose design strongly echoes that of Braun's ET44 and ET66 calculators, originally created in the year of Apple's incorporation.

him and explained that he had a 40 per cent chance of surviving. What had followed was worse than any illness Kelley had previously known. He had surgery and underwent chemotherapy. He lost fifty pounds and suffered from mouth sores, nausea and a throat so raw that he could hardly swallow. Steve Jobs stayed with him every step of the way. Having worked together for years, the two men had become good friends. Jobs had even introduced Kelley to his wife. 'Steve had had his own experience with cancer a couple of years earlier, so the minute that I was diagnosed he was over there telling me what to do, calling doctors, and threatening the hospital – telling them that they had to take care of me, or else,' Kelley recalls. 'I don't know what "or else" meant! ... He also warned me about alternative medicine, telling me that I should go right to the Western medicine.'

Kelley was convalescing at home when Jobs introduced the iPhone. The next day he brought one over for his friend. 'It was one of the first iPhones anyone had seen and he gave it to me,' Kelley says. 'It was very touching.' Jobs stayed with Kelley. Looking for something to busy himself with, he decided that he would call AT&T and get the iPhone up and running. Jobs stepped outside to make the call. Kelley could hear him speaking on the phone for what seemed a long time. 'I would have loved to hear the person on the other end of the phone,' Kelley remembers. 'Steve would say "look, I'm Steve Jobs" and the other person probably said, "look, I'm Napoleon".' Eventually Jobs re-entered the room, looking flustered.

'I couldn't do it,' he admitted. 'They wouldn't set it up for me.'

The next day Kelley's wife called AT&T and got the iPhone activated.

□ □ □

Not that customers knew it at the time, but the iPhone's natural companion piece was another similar-looking device, albeit just over three times the size, so that it appeared, in the words of *Wired* writer Steven Levy, like an Apple smartphone designed for the 'supersize inhabitants of Pandora'. In fact, this was the wrong way to look at it. The iPad was not so much a larger iPhone, as the iPhone was a smaller iPad, with the ability to make telephone calls. Jobs had come up with the idea of being able to type directly on to a glass display in the early 2000s and taken this to the people on his team who worked on concepts such as inertial scrolling. By this point, Apple had already been in the process of developing its smartphone, so Jobs' idea for a tablet was instead applied to that platform, while the tablet concept was temporarily shelved. The idea went further back than that, however. Despite publicly claiming that tablet computing was not something he was interested in, Jobs had a long-held fascination with an idea dreamed up in 1972 by Alan Kay, while Kay was at Xerox PARC. Calling it the Dynabook, Kay wrote about the concept in a paper entitled 'A Personal Computer for Children of All Ages'. When he found out about it in the eighties, Jobs was hooked. Elements of the Dynabook concept had appeared in Apple's laptops over the years, but the iPad represented its most fully realised execution yet.

Alan Kay had been among those at the iPhone's unveiling in January 2007. By this point the 66-year-old Kay had been out of Apple for a decade. Sensing the writing on the wall for his Advanced Technologies Group upon Jobs' return, he had left the company for a job with Walt Disney Imagineering in Burbank, California, where he became a vice-president of the R&D division. Still, he retained enough interest in what was going on at Apple to turn up for this particular event. Sometime during the course of the day, Jobs turned to him and casually asked, 'What do you think, Alan? Is it good enough to criticise?'

The question was a reference to a comment made by Kay almost a quarter of a century earlier, when he had deemed the original Macintosh 'the first computer worth criticising'. Kay considered Jobs' question for a moment. In answer he held up the Moleskine notebook he was carrying for the Apple CEO to see.

'Make the screen at least 5" x 8" and you will rule the world,' he said.

When the iPad was first launched on 3 April 2010, it was hailed as a new technological paradigm. It received the requisite revolutionary treatment. In its accompanying television advertisment, Jobs drafted in actor Peter Coyote to deliver a rousing narration. Coyote was a fascinating choice. He had previously been a member of the anarchic sixties counterculture theatre group the Diggers, named after a seventeenth-century English sect of religious radicals. Although subtle, the actor's inclusion was another link to the countercultural backdrop against which Apple had been forged; a brief nod to customers about where the company had come from.*

Like the iPhone, the iPad was an exercise in studied simplification. It stripped away everything expendable in order to find out what the purest vision of a computer could be. It was the *essence* of technology. There was no mouse, no keyboard; nothing other than that which could be found in the equally minimalist cyberspace world of the iOS interface. The decision had inevitable casualties. The USB connection was now a thing of the past, as was the stylus, which had previously been viewed as a crucial element of the tablet paradigm. Jobs, the man who had taken his love of calligraphy into the mainstream digital realm by incorporating fonts into the original Macintosh operating system, was now responsible for also killing the pen. Taken as twin devices, the iPad and iPhone represent perhaps

---

* When I asked Coyote about the links between the counterculture and the high-tech movement he admitted that he did not have a great deal to say on the subject, 'except that some of my old counterculture friends sold a lot [of] pot down in Silicon Valley to these computer guys'.

the two most singular encapsulations of Jobs' vision for the perfect computing form factor achieved during his career. There are few images more iconic, nor more potent, than a simple screen, augmented by one singular button. For the French philosopher Gilles Deleuze, the notion of the screen suggests a new means by which to conceive the human brain as a surface upon which the materiality of images and sensations enters into productive relations. For Slavoj Žižek, it suggests the fantasmatic window, requiring a cultural criticism emboldened by the insights of psychoanalysis. As the gravitational pull of cyberspace has grown stronger, so life itself has disappeared into the screen.

Another of Jobs' most brilliant traits (something that he shared with Jony Ive) was an absolute appreciation of the tactile element of computing. Both the iPad and the iPhone took this one step further than had ever been the case previously. By removing the stylus and instead asking users to touch the screen with their fingers, an astonishing intimacy is built up with the devices. Despite the presence of an interface (by definition, a mediating level between the user and the machine) it does not feel as if the user's experience is being mediated. As a purely emotional reaction it feels like we are melding with the machine. Prior to this, and no matter how much we liked to think otherwise, the relationship between the user and his or her computer went in one direction only. We could move the icons on screen, or rename the hard disk drive, but essentially the computer stayed as it was. Even the iMac's multitude of available colours was merely the tiniest of nods towards the computer reflecting us, the user. The iPhone and iPad changed all of that. Thanks to both devices' inbuilt accelerometer and gyroscope, they understand where they are in a spatial dimension. We tilt the screen to one side and the image adjusts accordingly. It is truly personal computing. Jobs understood this better than anyone. When he introduced the iPad onstage for the first time he did so reclining in a comfortable leather Le Corbusier chair, kicking back as if he were sitting in his living room in Palo Alto.

Without a doubt the iPhone and iPad represented Apple technology at its friendliest. These were devices so straightforward and intuitive that even a child could use one.* Indeed, the accompanying videos for the iPad made liberal use of footage showing children interacting with the device. Jobs himself was personally touched when he read a *Forbes* article by the writer Michael Noer. Entitled 'The Stable Boy and the iPad', the piece was exactly the kind of story that wouldn't have looked out of place appearing in an issue of *The Whole Earth Catalog*, or perhaps as an additional verse in Richard Brautigan's poem 'All Watched Over by Machines of Loving Grace'. Noer described how he had been staying on a dairy farm forty miles north of Bogotá, Colombia. He'd been playing around with his iPad when he was approached by one of the young boys who helped out in the stables. Noer estimated the boy's age as being around six-years-old. For whatever reason, he handed over his iPad and watched, amazed, as the boy began using it without a single word of instruction. 'Steve Jobs has designed a powerful computer that an illiterate 6-year-old can use without instruction. If that isn't magical, I don't know what is.'

---

* A study carried out by branding consultant Martin Lindstrom gave one BlackBerry phone to each of a group of twenty babies between the ages of fourteen and twenty months. According to Lindstrom, 'No sooner had the babies grasped the phones than they swiped their little fingers across the screens as if they were iPhones, seemingly expecting the screens to come to life. It appears that a whole new generation is being primed to navigate the world of electronics in a ritualized, Apple-approved way.' This would seem to be backed up by a somewhat alarmist headline which appeared in the *Daily Telegraph* in March 2012, claiming, 'Schoolchildren can use an iPhone but cannot tie their shoelaces, poll finds'.

Many writers have tried to put their finger on exactly why Apple has had such success over the past decade. The majority of answers centre on the fact that the products the company produces 'just work'. 'People like to say that about Apple products and for good reason,' says Larry Yaeger, former Apple Distinguished Scientist and currently a professor of informatics at Indiana University Bloomington. 'It was always Steve Jobs' vision that technology should be as easy as a toaster; that it be something that would improve your life, but not something that you had to spend your life working on.' Especially given today's world there is something comforting about this notion. If the 1990s had been a ten-year party in which technology across the board looked set to cure whichever political, spiritual, economic, cultural and ecological problems were faced by mankind, then the 2000s was the bracing comedown from that party, as talk of cyber-utopianism was fast replaced by jihadist terrorism and imperial adventures. From the hijacked aeroplanes that caused the collapse of the twin towers of the World Trade Center, to the failure of the technologically unparalleled American military to immediately right that wrong, the decade seemed nothing more than a non-stop testimony to the fatal flaws inherent in modern technology, along with its unnerving ability to turn against us at a moment's notice. A series of devices, like those created by Apple, which promise to do exactly as we tell them, by contrast seems almost impossibly utopian.

In her excellent book *Alone Together: Why We Expect More from Technology and Less from Each Other*, MIT's Sherry Turkle relates conversations she has had with American teenagers on the subject of cell phones. What emerges is an overriding fear of disconnection, often centred on the aftermath of the 9/11 terrorist attacks. Turkle notes how many of today's teenagers, primary-school-level students in 2001, were taken out of their classrooms and hidden away in basements, the iconic hiding place of the Cold War.

Meanwhile, the parental gift of the cell phone often carries with it the stipulation, either tacit or overtly stated, that 'if I ring, answer the call'. The resultant message is, essentially, that this is a device that will keep someone safe. Little wonder that teachers have such a hard time confiscating phones. The idea of a piece of consumer high-tech possessing such qualities is something that even the most enthusiastic of the seventies hacker-hobbyists might have found hard to believe, but anecdotal evidence suggests that this may, in fact, be the case. Certainly there can be few more striking examples of the kind of technological DIY culture that Stewart Brand dreamed about while mimeographing copies of *The Whole Earth Catalog* than the story of Dan Woolley. A US filmmaker working for a mission organisation called Compassion International, Woolley was shooting a video about poverty in Haiti in January 2010 when the earthquake struck. 'I just saw the walls rippling and just explosive sounds all around me,' Woolley later recalled. 'It all happened incredibly fast. [My colleague] yelled out, "It's an earthquake," and we both lunged and everything turned dark.' When he awoke he found that the back of his head was gushing blood, the hotel lobby he was in was pitch-black, and his right leg was badly broken. Although he had no phone signal to call for help, Woolley was able to utilise an iPhone first-aid app that he had downloaded in advance to help him. The app taught him how to fashion a bandage and tourniquet for his leg, as well as how to stop the bleeding from his head wound. Because it warned him not to fall asleep if he felt that he might be going into shock, Woolley set his iPhone's alarm clock to sound every twenty minutes. Sixty-five hours later, he was saved by a French rescue team.

Woolley's story is incredible because of what it suggests about a world in which the user is 'always on'. When he was interviewed about his ordeal, Woolley drew attention to one almost inarguable fact: 'For people who pointed out I should've had a pocket first-aid kit, the reason they're wrong is I wouldn't have it in my pocket. How many people have gone out of their way to add one more thing to their pocket? What was valuable about the iPhone is it was already in my pocket. And I thought, it would probably be a good way to have some first-aid tips in here, so I downloaded that [particular] app. That's the value of this utility.'

In the MIT Media Lab in the mid-1990s, Sherry Turkle encountered young researchers who carried with them computers and radio transmitters wherever they went. Some went so far as to clip digital displays into the frames of their glasses, which enabled them to be permanently connected to the Internet, via wireless connection, wherever they went. This group proudly referred to themselves as 'cyborgs' and saw their augmentation not as a defacement of the human form, but as an enhancement.[*] Twenty years on – courtesy of devices such as the iPhone – some 316 million of us live what the MIT Media Lab of the nineties would have considered a cyborg lifestyle.[†] In fact, more iOS devices sold in 2011 than all the Macs that Apple had sold in twenty-eight years. We have become a cyborg planet.

What is interesting is the manner in which public perception of this phenomenon has shifted in recent decades. Just as the hippies recontextualised computers in the sixties and seventies, so today we are far more welcoming of a world in which we

---

[*] Turkle's work built on that of Donna Haraway, whose 'Cyborg Manifesto: Science, Technology, and Socialist-Feminism in the Late Twentieth Century' was laid out in the 1991 book *Simians, Cyborgs and Women: The Reinvention of Nature*.

[†] This number is based on the number of iOS devices sold over the past several years, according to the website Asymco.

are always on than would ever have been the case previously. Look no further than the character of Darth Vader, whose first appearance in *Star Wars* took place in 1977. As a half-man, half-machine Vader is the quintessential cyborg nightmare of the decade that marked Apple's creation. Seven years later, when the Macintosh hit, this idea was examined in Apple's '1984' commercial. A key tenet of the ad was the techno-hippie adage of humans using technology as peripherals, and not the other way around. The Big Brother figure of the commercial appears to be a creature of the television screen, as opposed to a real flesh-and-blood persona. The viewer is left unsure if he, in fact, lives anywhere other than in an interface fantasy. By comparison the sledgehammer-tossing female figure emblematic of the Macintosh represents the way in which technology can add to, rather than detract from, humanity. She is what Ted Friedman refers to in his book *Electric Dreams* as the 'Reebok feminist':

> *Compared to [*the other figures in the advertisement*], the running woman, unencumbered in shorts, sneakers, and tank top, might seem to represent the body freed from technology. But that's not completely right. Rather, she's an example of the new kind of athletic ideal which emerged in the 1980s: the athlete who employs Nautilus, Stairmaster, and the other technologies of exercise to hone the body to perfection. This 'robo-cized' athlete is the flip side of the cyborg nightmare. Rather than the technology wearing her, she wears the technology.*

Steve Jobs' second stint with Apple coincided with the next iteration of this idea. Following the original iMac – a computer so beautiful and striking that it could not be ignored – Apple's products, despite their increasing popularity, started to vanish. 'Where did the computer go?" asked the advertising for 2005's iMac G5, whose internal components, ranging from its hard disk drive to its microprocessor, were integrated into a monitor not more than two inches thick. Despite the superficial suggestion, it was, paradoxically, not indicative of a world in which the computer became less omnipresent in our lives, but, rather, one in which the computer's physical presence diminishes as its ubiquity increases: a perverse spin on Moore's Law. Apple's products anticipated a world long dreamed about by science-fiction writers, in which technology – perfectly melded with the environment so as to be almost invisible – can cater to our every whim, from communication to entertainment. Not for nothing were Apple's three major new product lines of the decade: the iPod, the iPhone and the iPad all portable devices that the user could carry with them wherever they went.

There is, of course, another side to the 'always on' debate. The hippies dreamed of a world in which the traditional barriers between work and play would come down. Work would be so enthralling, so can't-wait-to-wake-up-in-the-morning that it would become play. Play, for its part, would become work. That same sentiment is expressed in the euphoric cyborgism of the iPhone and iPad. Thanks to our constant degree of connectivity we are always within reach. Just like the iPod, there is no on-off button.

Nothing sums up this notion better than the concept of apps. Apps, by their name alone, are presented as endless tools of productivity, perhaps the purest articulation of the Apple ideology in existence. They are tools *by* knowledge workers, *for* knowledge workers.

And, surprisingly, at first Steve Jobs didn't even want them.

To say that Jobs didn't want user-created apps on iOS would be an understatement. He *hated* the idea. Others tried to talk him down from the ledge, most notably marketing chief Phil Schiller and Apple board member Art Levinson. Levinson phoned Jobs repeatedly, lobbying for the Apple CEO to rethink his position. As Levinson saw it, if Apple didn't get in on the act, another smartphone company certainly would. Still Jobs resisted. The way he viewed it, loosening his grip on the perfect Zen aura that surrounded his devices would simply mean allowing them to be compromised by someone with less exacting standards than his own. It would be like letting fly-by-night decorators help Michelangelo finish painting the ceiling of the Sistine Chapel. No: better to stay in total control than run the risk of viruses or – worse – subpar applications ruin the integrity of what he had created.

In fact, this attitude spoke volumes about Apple's relationship with hackers. While Steve Wozniak has always happily classed himself as a hacker, Jobs never did; happy to embrace the freedom-fighting hacker-chic of computing's rebel class, but little else. This attitude put Jobs in line with mainstream thought on the subject. It is interesting to note that as companies such as Apple (in addition to the rise of video games and Hollywood movies like the Pixar films) have aided public acceptance of computing, so hackers have been increasingly demonised. In court we witnessed the criminalisation of hacking during Apple's lifespan. Meanwhile, films like *WarGames* (in which a high-school hacker comes close to accidentally starting the Third World War from the bedroom of his suburban home), *Die Hard 2* (cyber-terrorists hack into an air-traffic-control system to crash a plane filled with passengers) and *The Net* (Sandra Bullock has her identity erased online) represented the hacker ethic as filtered through the lens of popular media. This is nothing, however, compared to the real-life computer hackers, who were vilified almost beyond belief. One illustration is Kevin Mitnick – at one point the most wanted computer criminal in the United States for hacking into various government agencies – who was likened to a serial killer when one publication referred to him as 'a real electronic Hannibal Lecter'.*

As Jobs viewed the iPhone, it existed not as a generative platform as, for example, the Apple II had been, but, rather, a device whose functionally was locked down. Only Apple would be able to alter it through the use of remote software updates. One such update noted, in bold capital letters, 'IF YOU HAVE MODIFIED YOUR iPHONE'S SOFTWARE, APPLYING THIS SOFTWARE UPDATE MAY RESULT IN YOUR iPHONE BECOMING PERMENANTLY INOPERABLE'. This wasn't just about ego; Jobs was steadfast in his belief that taking unnecessary choice (and therefore risk) from the user would lead to a better overall experience. 'We define everything that is on the phone,' he said. '... You don't want your phone to be like a PC. The last thing you want is to have loaded three apps on your phone and then you go to make a call and it doesn't work any more. These are more like iPods than they are like computers.' More than thirty years removed from his phone-phreaking days as a seller of Blue Boxes, Jobs was no longer the freeloader looking to subvert the system, but, rather, a figure firmly entrenched in the Ma Bell camp. 'People will try to break in, and it's our job to stop them breaking in,' he said as a warning to would-be iPhone hackers.

The irony of Jobs' desire to turn the cell phone into a computer was that by doing so he was almost implicitly inviting someone to hack it. He didn't have to wait long. Within several months of

---

* The case of Kevin Mitnick perfectly illustrates the dichotomy between the Two Steves, Jobs and Wozniak. While Jobs wanted nothing whatsoever to do with hackers, Wozniak agreed to write the foreword to Mitnick's 2011 autobiography, *Ghost in the Wires*, in which he referred to Mitnick as 'one of my best friends'.

the iPhone's 2007 launch, a 23-year-old from Norway, Jon Lech Johansen, created one of the first hacks for unlocking the device; allowing users to activate it using Windows, without being forced to subscribe (or '[give] any of your money or personal information') to Apple's then mandatory service provider AT&T. Using revolutionary language, Johansen published the solution free on his blog, under the legend 'iPhone Independence Day'. Johansen's hack emphasised the degree to which the iPhone was a computer. Using his hack, users would not be able to make phone calls, but would be able to use the phone's other functions. As subversions went, this one was brilliant. Where Jobs thought he was making a high-end smartphone, the hackers showed that, in fact, he had manufactured a reasonably priced computer.

Eventually Jobs saw the degree to which he was missing out and agreed that Apple should release a Software Developer Kit (SDK) to allow those outside Apple to create their own apps. One early winner was a man named Steve Demeter, the 28-year-old independent programmer behind Trism. A sliding-blocks puzzle game, Trism made innovative use of the iPhone's accelerometer. 'As a non-corporate entity you have to get your product out there and get it seen,' Demeter says. 'Not having a huge marketing budget, I knew I had to have an idea that was instantly sellable, that utilised technology that had never been mass-marketed before.' The result was Trism, which was coded by Demeter with help from a friend and a designer, who were paid $500 for their efforts. In true hacker fashion, Demeter initially released the app as a free native application in the Jailbreak community, meaning that users could play Trism only if they first hacked their iPhones. When Apple finally released its SDK, however, he was among the first to get to the platform and began charging $5 per copy of Trism. Within two months he had made a profit of $250,000. The best part? Like Steve Wozniak more than thirty years earlier, Demeter wasn't even trying to strike it rich. 'I really didn't think about the money,' he has said. 'I got an e-mail from a lady who's, like, a 50-year-old woman who says, "I do not play games, but I love Trism." That's what I did it for.'

Six months later Demeter's story was trumped when another independent developer, Ethan Nicholas, earned $600,000 in one month with his tank-artillery app, iShoot. If Demeter's story had parallels with Wozniak's, then Nicholas's bore faint echoes of the romantic image Jobs painted of himself as a penniless youth. A Sun Microsystems engineer, Nicholas worked on the app for eight hours a day, holding his one-year-old son with one hand and coding with the other. When iShoot jumped to the No. 1 spot in the App Store, Nicholas quit his day job. 'I'm not going to be a millionaire in the next month, but I'd be shocked if it doesn't happen at the end of the year,' he said.

Ultimately both men were in the right place at the right time. Trism sales eventually slowed down, while Nicholas was unable to come up with another hit to rival iShoot. This didn't matter, though. What mattered – to Apple at least – was the myth-making that made both stories so tantalising to would-be entrepreneurs. Imbued with a homespun DIY ethos, these tales served as Web 2.0 examples of the self-made 'men in a garage' legend which continues to present Silicon Valley as the high-tech dream factory. As of the first months of 2012 Apple announced that it had paid out a total of $4 billion to developers for its iOS App Store. Since the company takes a 30 per cent cut of all revenue for apps sold, this means that Apple has thus far earned around $1.71 billion from user-generated material.

But in the same way that Silicon Valley has more millionaires per capita than anywhere else in the United States, while also experiencing some of the greatest decline in wages for working-class residents of any city in the nation, so, too, are the rags-to-riches fables disproportionate to the number of those who just break even – or even lose money – developing for the platform.

Trip Hawkins, the founder of Electronic Arts and a former Apple employee, calculated that each app in Apple's store earns, on average, around $4,000. 'That doesn't even pay for a really good foosball table,' he told a conference he was speaking at. Paul Lutus, the author of Apple Writer, describes the App Store as 'a classic marketer's dream, [with] too many programmers with too many programs, chasing too few buyer dollars, and the marketer in the middle the only one really cashing in. There are a handful of success stories. They are enough to keep the system working, for the same reason that stories of lottery winners keep that system working, but only because most of the players are too young or too inexperienced to understand the system.'

Meanwhile, if Apple appeared to have ceded control to outside developers, this was done only on the company's terms. '[The App Store] was an absolutely magical solution that hit the sweet spot,' said Apple board member, Art Levinson. 'It gave us the benefits of openness while retaining end-to-end control.' In the same way that prior to the development of the personal computer, enthusiasts had to work within a system (be it corporate, academic or otherwise) to gain access to computing equipment, so today do developers have to work within Apple's rule set which, on occasion, can be every bit as rigid and regulated as any military-industrial complex. Possessing the power to edit, publish, distribute and – if desired – pull anything it wishes that is for sale in its stores, Apple has *carte blanche* to behave as it pleases. Apps can be removed because they offer a service too similar to one that Apple provides (or might wish to provide in the future); because they're juvenile ('We have over 350,000 apps in the App store. We don't need any more fart apps'); because they're overly amateurish; or simply because they offend Apple's sensibilities. 'If you want to criticise a religion, write a book,' notes Apple's official guidelines for developers. 'If you want to describe sex, write a book or a song, or create a medical app. It can get complicated, but we have decided to not allow certain kinds of content in the App Store … We will reject Apps for any content or behaviour that we believe is over the line. What line, you ask? Well, as a Supreme Court Justice once said, "I'll know it when I see it".'

This last point is the one that has prompted the most outrage, with a vocal minority of customers claiming that it represents Apple turning its back on its founding libertarian ideals. Certainly some of the censorship appears patently ridiculous, and is the result of technical error more than it is human fault. For example, in 2010 Apple's 'naughty words filter' took issue with Herman Melville's *Moby-Dick*, on account of Captain Ahab's vendetta against a sperm whale – or, as Apple referred to it, a s***m whale. (This is not dissimilar to the classic Scunthorpe glitch, in which the English town is frequently blocked by various internet filters since it contains a string of letters spelling out the word 'cunt'. Penistone, South Yorkshire; Lightwater, Surrey; and Clitheroe in Lancashire all face similar problems.) Not all are machine-made missteps, however. A digital edition of *The Kama Sutra* was initially labelled as 'obscene', leading to at least one publication quipping that the second-century Hindu text might have put Apple in an awkward position. Similarly, a comic-book adaptation of the Oscar Wilde play *The Importance of Being Earnest*, which took the step of replacing Wilde's original cast of characters with an all-male ensemble, was heavily censored, with black bars being placed over images of men kissing. In another instance an illustrated adaptation of James Joyce's *Ulysses* was held back because it contained an image of a woman's bare breasts. When the artist responsible offered to pixelate the image, or else to obscure it with a well-placed fig leaf, Apple informed him that it was unacceptable to show any part of the woman's body at all and that he should instead crop the image to show only her face. (The irony here is that Joyce battled censorship for a decade over *Ulysses*. The eventual court verdict, reached in 1933, was a landmark ruling in favour of free speech.)

In each of these instances, Apple was heavily criticised in the press, leading to the company backtracking from its original position. When satirical cartoonist Mark Fiore's work was censored because his submitted app contained 'content that ridicules public figures', the fact that he went on to win a Pulitzer Prize for the work meant that the material could be resubmitted, at which point it passed without question.

Apple is not the only high-tech company guilty of censorship. Microsoft, Google and Facebook have all been accused of varying degrees of censorship, with the suggestion made that high-tech libertarianism apparently leans heavily towards the puritanical.

Making Apple's situation more challenging for the company to manoeuvre, however, is a business model for both iTunes and the App Store that, at its root, is relatively old-fashioned. Unlike YouTube, for example, which does not directly profit from any one individual piece of content (and thus is only obliged to remove genuinely illegal material after complaints have been made) Apple takes a cut of every item sold. 'To use an analogy, it's like the early days of television in the United States, where each programme was sponsored by a particular company,' says Bruce Tognazzini. 'Those sponsors often produced the television show, and there was very heavy censorship as a result, because the sponsors couldn't take the heat. If Chevrolet was sponsoring a show which contained R-rated material, the people that were against that material may stop buying Chevrolet. That's the situation Apple is in. With [Google's Android platform] anything can go on it, whether it works or doesn't. It might have a virus on it, but that doesn't matter. They're just the publisher.' Throwing fuel on the fire was Jobs' insistence that what Apple was doing was not framed simply within legal terms, but ethical ones as well. 'We believe we have a moral responsibility to keep porn off the iPhone,' he wrote in an email to one customer. 'Folks who want porn can buy an Android.'

□ □ □

The comparison to Google is an apt one. Of all the companies in Silicon Valley none appears to borrow more from the libertarian dream of Apple, the Homebrew Club and the techno-hippies than the multi-billion dollar Search leviathan.* Even the story of its founding, by two twenty-something visionaries working in a Silicon Valley garage appears strangely familiar. Although neither Sergey Brin nor Larry Page possesses the evangelical leadership qualities of Steve Jobs, both grew up idolising the Apple co-founder and the company he helped to create. In fact, Jobs was Brin's and Page's first – and, initially, only – choice for Google CEO when the pair was told that they needed to hire one. Seeing as Jobs was already the CEO of two publicly traded companies at the time (Apple and Pixar) there was no way that was ever going to happen, but he did go on to become something of a mentor to the two Search impresarios, seeing Google as a potentially powerful ally in his then-raging war against Microsoft. The cooperation continued in August 2006, when Eric Schmidt, Google's chief executive, was invited to join the board of directors at Apple. Schmidt gushingly referred to Jobs as 'the best CEO in the world today'. Even when the two

---

* If Google leads the way, then certainly other companies show hints of this same dream. Facebook (which has the youngest workforce among prominent Silicon Valley companies, with a median age of twenty-six) carries the chaotic feeling of a college dorm room writ large, reminiscent of the Apple of the late seventies or perhaps the early eighties. Its walls are covered in graffiti, abuzz with revolutionary proclamations about changing the face of technology. 'This is what it must have been like building the first Macintosh,' thought Dave Morin, a man who had worked for Apple before joining Facebook, as he drove home from the Facebook offices at four o'clock one morning.

companies later fell out, it was not over a fundamental difference in direction, but, rather, a similarity: the move into smartphones.*

Today many of Jobs' former colleagues roam the halls of Google, like the ghosts of Apple past. 'There's lots of similarities [between the two companies],' says one current Google engineer. 'I think the founders of Google and the founders of Apple are similar in terms of being incredible creative geniuses, and as a consequence of that wanting to do things better than anybody else has ever done them. Both Apple and Google throw out the conventional rulebook. Every problem, there's always a better solution. Both companies are idiosyncratic and deeply committed to doing the best possible thing.' The engineer speaks excitedly, with a 'this is the coolest thing ever' type of voice that brings to mind an excitable eleven-year-old at his first monster truck rally. It's none other than Andy Hertzfeld, the former Software Wizard for the original Macintosh, who jumped from Apple to Google, via the briefly red-hot nineties start-up General Magic. At Google, Hertzfeld became one of the main software engineers for Google+, the company's stab at a social-networking service to rival Facebook.

For all their apparent similarities, however, Google and Apple possess some striking differences. Unlike the stripped-down, functional look of the Apple offices – where minute touches like the use of Susan Kare-style iconic mosaics on doors are the only real nods towards Silicon Valley culture – Google wears its geekoid engineer credentials on its sleeve at every turn, as if it were the Apple of some alternate dimension where Steve Wozniak, instead of Jobs, had risen to the rank of CEO. Google prides itself on being as egalitarian as possible, in the manner of a Xerox PARC or Atari. Its employees work hard and play

---

* In fact, Android's co-founder, Andy Rubin, was an Apple veteran of the early nineties; yet another reminder of the cross-pollination between the two companies.

hard; lounging on bean bags as they program, or grabbing a healthy gourmet meal from the Google restaurant, whipped up by a former travelling chef for the Grateful Dead. Googlers are able to see a doctor, enjoy a Pilates class, get their hair cut, or receive a Swedish massage without ever having to leave the campus. Any errands which do require leaving can be carried out by someone else for the modest fee of $25 per hour. To all intents and purposes, it is a beacon of cyber-libertarianism in action. By stumbling upon its AdWords algorithm – its pay-per-click advertising system, which earned the company $28 billion in 2010 – the company has, quite literally, discovered a way to profit from pure information. Every time we click a link we are making Google and its shareholders a little bit richer.

The concept of cyber-libertarianism draws on the wave theory of technological development as proposed by Alvin Toffler. According to Toffler there have been three such waves. The first was agricultural in nature, centred upon human labour. The second was larger in scale, built around massive industrial machinery. The third would be dedicated to knowledge, primarily disseminated through networked computers. Apple's rise to prominence has taken place on the crest of this third wave. Two years after the release of the original Macintosh, in 1986, US Secretary of State George Schultz declared that the 'Industrial Age is now ending', to be replaced by the Information Age. From that point on, he argued, power would be determined not by who controlled gold or manufacturing, but, rather, by who had control of knowledge. Even better, Toffler theorised that the worst aspects of life, the conformist nature of mass society, was a by-product of the second wave of technological development, and would be swept away altogether by the third. After all, the Industrial Age had required, by its very nature, mass production, centralised corporate bureaucracies and large governments, in order to function. All would crumble with the arrival of the Information Age, under which a glorious 'demassification' would occur.

Apple, like Google, is a believer in demassification. Entire books have been written on Apple's constant attempts to simplify every aspect of its operation. Within the company the idea of simplification is almost holy, to the point where the company's marketing and communications team has the word written three times in bold silver-white letters on its office wall, a constant reminder of the company's founding premise. True to form, in an effort to uncomplicate even this, the first two repetitions of the word have been crossed out: ~~SIMPLIFY~~ ~~SIMPLIFY~~ SIMPLIFY. Jobs believed that the demassification process could be applied, with equal success, to every aspect of life. When he arrived back at Apple in 1997 he discovered a company overrun by internal fiefdoms. Far from competitive in the greater world of high-tech, the only competition going on within Apple was among the separate divisions. 'Apple had a lot of smart people, but it had an insane amount of politics,' recalls Jim Reakes, an employee who witnessed Apple both before and after Jobs' return. 'The politics were horrible and you had a bunch of really not so bright people in charge of things. The bureaucracy was just ridiculous. There was no real system to get anything done.'

Jobs restructured the entire internal organisation of the company around an idea he had had in the 1980s. When Jobs had originally led the Macintosh division he told then CEO John Sculley that he wanted the department to have only one hundred people, since this was the number of names he could remember, and he didn't want to head a division in which he did not know the name of everyone working for him. Today Apple hosts an annual offsite meeting, called the 'Top 100', during which the executive team (consisting of, surprisingly enough, one hundred people) plans Apple's strategies for the coming year. By limiting the decision-making process to only the people Jobs would take with him on a hypothetical lifeboat to his next business venture (as he described it to biographer Walter Isaacson) Apple was able to avoid unnecessary levels of bureaucracy. 'We're the biggest start-up on the planet,' Jobs proudly noted at an *All Things Digital* conference in 2010.

As with almost every aspect of its business, Apple only releases limited information about its organisational structure. What has got out, however, paints it as a company with an extraordinarily flat hierarchy for a corporation that employs more than 60,000 people. 'Apple is organised more like a cult than a company,' says cyber-libertarian John Perry Barlow. 'There was Steve and the people that worked immediately with Steve, and then there was everybody else. It wasn't a very vertical structure.' A simple corporate structure might be seen as solely an extension of the kind of maniacal thought process that would see value in painting the internal circuit boards of a computer black to match its exterior, but the reality is that, as Jobs points out, it allows Apple to operate in the kind of streamlined, integrated manner that a large number of other companies (many smaller than Apple) could never hope to emulate. Ken Segall experienced this firsthand when he later went to work for Intel. He recalls:

> *Intel used to have around thirty [advertising] agencies they dealt with. They had these big conferences where all of these agencies would get together to try and get in sync. The woman who was the head of Intel's graphics group stood up one day and gave a speech saying, 'Everyone should look at an Apple launch to see how it should be done.' First of all, there's all this anticipation. Steve Jobs gets up on stage in San Francisco, and when he leaves the stage and people leave the convention centre, suddenly these giant posters that weren't there*

when they walked in have been unfurled, with lines
and images showing what Steve's just been talking
about. Then you walk out into the street and all the bus
shelters and the billboards that were blank two hours
earlier have posters up. Then you walk into the Apple
Store and all of these images are on the wall. Then you
go to the newsstand and buy a magazine, and it's got
an ad or an insert for the product. Then you turn on the
TV and there's the commercial. None of this was there
two hours before, and it's like Apple just flip a switch
and this massively integrated campaign goes into effect.
It's really extraordinary just how well it's done. It's a
major miracle and companies like Intel can point to
it and say, 'that's what we've gotta do' but they just
can't. They make things so difficult. Steve would never
let it get so complicated that things like that couldn't be
pulled off.

Apple's best products are similarly the result of this same demassification approach. Following the 'company within a company' conditions under which the original Macintosh was created, Apple applies this unique pressure-cooker environment to all its major projects. 'Both the iPod and the iPhone were created as a skunkworks in a separate building to the rest of Apple,' recalls George Crow, who worked on the former. 'They were sequestered away; they were secretive, and most of the company had no idea what they were doing.' This applied equally to the iPad team, the OS X group, and to almost every other Apple department of any note. Apple even has its own wing of advertising agency TBWA/Chiat/Day – dubbed the Media Arts Lab – which is separated from the company's Los Angeles offices and dedicated purely to producing Apple's award-winning advertisements. 'What Steve [did was] to isolate a very small group of people and make them work fourteen, eighteen hour days,' says Bruce Tognazzini of Jobs' production process. 'Only the most innermost brains at Apple know about the upcoming technology and they're sworn to absolute secrecy. They know that they'll be fired if anything leaks out, and they value their jobs.' According to popular lore, only thirty people had seen the finished iPhone by the time it was publicly unveiled. 'Steve fostered a sense of boutiqueness in individual groups,' says Mac OS X technical lead Michael Lopp. 'It doesn't feel Big Company. I mean, [Apple] *is* a big company and there were a lot of visual cues telling you that, but they do a really good job of making teams feel small even among all of that bigness.'

However, if Apple is the product of the same libertarian, demassified dream as a company like Google, it is also considerably more strait-laced. 'I was used to a culture in the computer business where, to use an old saying, job descriptions are what you do until you figure out what you *should* be doing,' says Hank Shiffman, a Silicon Valley engineer who had previously worked at Sun Microsystems and Silicon Graphics before joining Apple in 2002. Shiffman was surprised by what he discovered. 'At Apple things were much more buttoned down,' he says. 'They allow very little individual freedom among their employees.' At Google, new recruits – called Nooglers – wear beanie hats with propellers on top for their first day, and receive a round of applause from existing employees as they stand up to introduce themselves. At Apple, the first experience new employees have when joining is the so-called Indoc (standing for 'indoctrination') during which they are briefed about the paramount importance of security, and warned that Apple has a zero-tolerance policy towards employees speaking out of turn. This is rule number one for anyone who works for Apple. 'You were ultra-paranoid [about speaking to outsiders],' says Michael Grothaus, a consultant who worked at Apple for five years in the mid-2000s. 'One colleague of mine showed the general manager of a Best Buy store one of our spreadsheets ... It didn't have that much significant information on it; just some sales number for that region for the quarter. She thought she was doing the right thing by getting the general manager involved, but when her boss found out it came down through the channel that she was to be let go the next day.'

The inside of the Apple campus reflects this desire for confidentiality at every turn. A world away from the open-plan dream of NeXT – where employees were able to find out everything they wanted, down to how much each other was earning – at Apple people are unable even to move around the campus freely. Almost no employees' card will open every door on Infinite Loop and sections of the different buildings become classified and declassified, seemingly at random, as different divisions move about. 'Apple security was always a pain in the ass,' says former employee Jim Reekes. 'You had to have a badge with you at all times, and the only way to get through any door was to flash this badge to the card-reader so the door would unlock. Then there would be a security person at each door watching you to make sure you'd waved your badge.' Steve Jobs, of course, was the exception to the rule: not only allowed wherever he wanted, but also refusing to wear any sort of identification badge whatsoever. The knock-on effect of this, naturally, was that he was unable to use his ID badge to open doors. Instead he would simply rap once on the door he wished to go through, at which point someone would run and open it for him. 'He was the one guy at Apple who could get away with that', one former employee says. 'He didn't care. He thought the rules didn't apply to him.'

Nor does Apple go in for the kind of employee-pampering that is *de rigueur* in Silicon Valley. 'Apple perks are crappy,' says former Mac OX engineer Michael Lopp. 'There's no free soda, nothing like that. The cafeteria is subsidised, but you're still paying there, right? The way it was always rationalised to me though, was would you like your perks to be stock, or do you want free soda? And that shut everyone up. You hear about Google and these other start-ups with all of their perks, but with Apple stock doing so well, why would you go anywhere else?' Apple had never been as magnanimous a company as Google, but the perks that were once on offer were gradually stripped away during the nineties as costs were cut. That process only speeded up upon Jobs' return. Today Apple's headquarters is a pleasant enough building, but certainly nothing to write home about by Silicon Valley standards. There is an outdoor basketball court and foosball, snooker and ping-pong tables in the employee rec room, but these are a far cry from the indoor rock-climbing walls at Google. At one point the Apple gym (on-site, although employees are charged for using it) was almost closed altogether, until Jon Rubinstein – then senior vice-president of the iPod division – lay down on the tracks to protest the decision. At Apple the motivation is equal parts belief in what people are working on and absolute fear.* This in itself is a drastic change from many Silicon Valley companies, which stress employee empowerment of the kind first preached by Maria Montessori and Anne E. George in the 1912 book *The Montessori Method*, in which the authors claim:

> *Discipline must come through liberty ... We do not*
> *consider an individual disciplined only when he has*
> *been rendered as artificially silent as a mute and*
> *as immovable as a paralytic. He is an individual*

---

* One Apple employee I spoke to mentioned that he would always take the stairs to reach his office, despite there being an elevator, for fear of finding himself trapped in a lift with Jobs and having to justify his position in the company. Although Jobs' former assistant has denied it, the rumour that Jobs was a man who could fire someone over the course of an elevator ride has never lost steam.

annihilated, not disciplined. We call an individual
disciplined when he is master of himself.

Although widely ignored in the pervasive attitude of Taylorism in the first part of the twentieth century, in which workers were treated as interchangeable cogs, this notion gained momentum when it was picked up and dusted off by those in the Beat Generation. Writing in the late sixties, Theodore Roszak, author of *The Making of a Counter Culture*, jumped on Montessori's suggestion that it was in a childlike state of willingness to accept new ideas that utopian societies of the future would be created. 'The expansion of the personality is nothing that is achieved by special training,' he notes, 'but by a naive openness to experience.' For Roszak it was the surrendering of control to embrace a state of higher cosmic consciousness that would ultimately cause the existing organised structures of American life to crumble. 'We must be prepared to trust that the expanded personality becomes more beautiful, more creative, more humane than the search for objective consciousness can make it,' he argued.

Roszak was far from alone in taking up the cause. During the fifties and sixties, as the disparate threads that would combine to form the backbone of the counterculture were beginning to come together, business-management thinkers went through their own version of the mass-society critique. First came the eulogies to entrepreneurship, whose different ways of thinking had been crushed in the deathly grip of the technocracy. Bestselling books such as William H. Whyte's *The Organization Man* bemoaned the failings of a corporate culture in which risk-averse executives were content simply to conform to the status quo, rather than to attempt any kind of shake-up. Alfred P. Sloan, Jr's *My Years with General Motors* (written in the 1950s, but withheld from publication until 1964) recalled in stultifying detail the endless, bureaucratic levels of management in a large American corporation as if they were particularly dull badges of honour.* Next came the proposed cures. In Douglas McGregor's 1960 tome, *The Human Side of Enterprise*, the author laid out the benefits of a friendlier approach to management in which workers were recognised as individuals and the traditional climate of fear that pervades most large organisations was erased. Thanks to increased automation and better education, many predicted that the traditional industrial working class was being replaced by a new producing class of mental, rather than manual, workers. Peter Drucker gave this new group a name, calling them the 'knowledge workers'. In his 1966 book *The Effective Executive*, Drucker noted how, 'The knowledge worker cannot be supervised closely or in detail ... he must direct himself.'

This shake-up of the traditional business hierarchy largely took place in the first fifteen years of Steve Jobs' life. Looking at these examples it is easy to see which tenets he chose to abide by and which he opted to ignore completely. The apparent contradiction is clear. 'Apple is a very Prussian kind of place with an obsession with secrecy and control that I think is very anti-countercultural, but the fact remains that they make products that I think humanise technology,' says cyber-libertarian John Perry Barlow. 'It's ironic that that's the case, given that it doesn't seem a very human institution. Nevertheless, there is a significant difference between their software and products and everybody else's.' In a 2008 article for *Wired*, entitled 'Evil/ Genius', journalist Leander Kahney came to a similar realisation:

---

* None other than Bill Gates commented that *My Years with General Motors* 'is probably the best book to read if you want to read only one book about business' – a comment that reveals more than a bit about the opposing Microsoft ideology.

> *[B]y deliberately flouting the Google mantra, Apple has thrived ... It's hard to see how any of [Apple's success] would have happened had Jobs hewed to the standard touchy-feely philosophies of Silicon Valley. Apple creates must-have products the old-fashioned way: by locking the doors and sweating and bleeding until something emerges perfectly formed ... [W]hile Apple's tactics may seem like Industrial Revolution relics, they've helped the company position itself ahead of its competitors and at the forefront of the tech industry. Sometimes, evil works.*

Kahney, of course, is not referring to Apple itself as 'evil' so much as he is subverting Google's trademark mantra 'Don't Be Evil'. Because, as Steve Jobs knew all too well, sometimes being nice only leads to compromise. With the failure of the Apple III, he had learned early on that a committee-driven creation without a singular vision behind it becomes not the sum of its parts, but, rather, lowest-common-denominator stuff. As one former Apple colleague told me, 'the creation process can be driven by committee; the product definition can't be'.

This notion, widely remarked upon in other creative mediums such as cinema, is somewhere all too easily overlooked by the cyber-ideologues, many of whom appear to believe that so long as work is done democratically – an extension of the old commune spirit – the widest possible audience will be reached by the end result. The idea that users should be able to decide on what they want in terms of aesthetics, interface and otherwise would appear to be the ultimate example of Silicon Valley libertarianism in action. The social-networking tool MySpace most famously tried this, granting users the ability to customise everything from their backdrop images to onscreen fonts. While some undoubtedly appreciated the freedom on offer, others quickly realised the downside. A very small percentage of MySpace users were professional designers and, in the majority of cases, pages wound up looking unnecessarily cluttered as users collectively learned that hot-pink writing on leopard print may not be the ideal paradigm. *En masse*, those in want of a social-networking site in which they could truly ascribe themselves individualistic online profiles jumped ship to Facebook, whose rigid design to pages allowed far less customisation.* Google's engineers picked up where MySpace left off, albeit with a more complex, algorithmic approach to user design. Since the entire basis for Google's search-engine business revolves around user-generated databases, a logical extension of this idea appeared to be: why not allow the interface to be determined in that same democratic manner? As a result we have situations like the one in which maths whiz Google employee Marissa Mayer instructed her team to user-test forty-one gradients of blue for one interface element. At least one designer left the company as a result.

This concept of hive-mind democracy is about more than simply what colours appear on the Google homepage. For many (not to Steve Jobs) it is a key to cyber freedom as integral to the digital constitution as anything. A practical illustration of this was the mass public-gaming experiment carried out at the 1980 Siggraph and the Arts Electronica Festival in 1994 by none other than Pixar co-founder Loren Carpenter and his partner Rachel. By distributing red and green handheld paddles whose colour changes were picked up by a camera, Carpenter arranged a means by which hundreds of participants would be able collectively to control the two onscreen paddles in a giant reproduction of

---

* Apple was an early supporter of Facebook, whose original mandate as a social-networking website aimed at those in higher education perfectly meshed with one of Apple's target demographics. In December 2004, when Facebook was still known as Thefacebook, Apple agreed to pay $1 per month to the site for every user who joined its group, with a monthly minimum of $50,000. It was, up until that point, the largest financial development in Facebook's short history.

the 1972 arcade game Pong. Their democratic decisions would determine how the game played out. Struck by the concept of mass participation, digital guru Howard Rheingold labelled these the first *smart mobs*: a veritable commune of collective hive intelligence, possessing the cumulative ability to 'cooperate in ways never before possible because they carry devices that possess both communication and computing abilities'. While it is possible to see where this idea would sit in a world populated by iPhones, iPads and even the tradeable playlists seen on iTunes, the reality is that Apple's approach is markedly different.

On projects that Jobs was interested in, he was involved at every stage of execution, from industrial design, to interface, to product marketing. 'When the iPhone came out, Steve insisted on writing the webpage for it, because he felt it was really important that the words be just right,' Ken Segall remembers. 'I was doing the web materials for iPhone and I was given the text and told, "This is what Steve wrote." To be honest I read it and thought, "Huh, some of this seems a little [over the top] to me" but that's what he wanted.'* Meanwhile, David Kelley discusses Jobs' uncompromising approach to design:

> *In almost every company that I know about, they're*
> *reasonable about design – and that may be the right*
> *way to be. But what happens is that you come up with a*
> *vision of the way that something should be, some dream*
> *about the future, and then as you go into production,*
> *as you make it real, you have to compromise – for cost*

---

* Nowhere is Jobs' dedication to perfection on a universal level highlighted more than in the case of Mac's OS 10.7 'Lion'. The operating system's standard backdrop image depicts the Andromeda Galaxy, truly elevating the user to the level of an Immortal. When the astrophotographer Robert Gendler lined up the Apple wallpaper with an image of the real galaxy, however, he discovered that Apple had opted to delete several stars and galaxies from the sky, purely for aesthetic reasons. Evidently Apple's vision of a perfect world extends far beyond earth's atmosphere.

reasons, or to make it easier to service, or whatever.
Each time the project is passed down the chain there's
a little compromise that's made in order to make it,
in companies' minds, better. Steve never allowed that
compromise to happen. Once the vision was set, once
the design was worked out, he wouldn't allow anyone
to compromise it. All of us in design have benefited
because of the examples that Steve has put out into the
ether over the years about how great design can be if
you don't compromise it. It's like a thirty-year science
experiment where you say, 'What if we don't have any
give-and-take regarding what the vision is? Now how
does it come out?' And the answer is – just look at the
Apple Stores, look at Apple's products. Nobody would
have said, for example, let's hack our laptop cases out
of big pieces of aluminium. Everybody would have
said that's a bad idea. First off, you'd have to buy
every milling machine in the world to get the job done.
Everyone would have agreed that it wasn't reasonable.
But Steve's distortion field allowed him to believe that
these things were possible, and as a result we got a lot
closer than we would have ever thought we could.

Despite holding 313 patents, Jobs was no inventor.* Unlike his
Apple co-founder, Steve Wozniak, he might instruct his team to
get the parts number down on a certain product, but he would
not offer his own solutions as to how this might be achieved.
Like Pixar director and chief creative officer John Lasseter – who

---

* In addition to the obvious patents for the likes of desktops, iPhones and iPads, Jobs
also owns those for the freestanding glass staircases used at Apple Stores, product
packaging and power adaptors. Bill Gates, by contrast, has only nine patents to his
name. Google's Larry Page and Sergey Brin have slightly more than a dozen.

has gone on record with his desire not to know too much about computing, so as not to have his decisions governed by what is and is not possible – Jobs didn't care about what engineers thought could be done; he cared about doing right by the user. Given his ability to reimagine previously unsuccessful (or less successful) technologies as fully articulated user experiences, some have claimed that Jobs' 'real genius' lay not in his ability to invent new technologies, but, rather, to tweak existing ones. In an article published in the *New Yorker* shortly after Jobs' death, Malcolm Gladwell likened Apple's current dominance of the high-tech industry to Britain's regional advantage during the Industrial Revolution over countries such as France and Germany. Gladwell argued that this owes not so much to the traditional explanations put forward by historians – Britain's plentiful coal supplies, its strong patent system, its high labour costs which speeded the search for labour-saving devices – but, rather, to the country's large number of skilled engineers who were able to take promising ideas and to tweak them until they were just right. Gladwell's suggestion (via economists Ralf Meisenzahl and Joel Mokyr) is that it is history's tweakers – more so even than its inventors – who truly define the age:

> *The visionary starts with a clean sheet of paper, and re-imagines the world. The tweaker inherits things as they are, and has to push and pull them toward some more nearly perfect solution. That is not a lesser task.*

But if Jobs was a tweaker, he was a masterful one. What came out of Apple's workshops after the reimaging process looked nothing like what went in. (This goes for the entire industry also. Compare, for example, the number of touch-screen smartphones – *sans* keyboard – which existed post-iPhone to that which existed before.)

Arguably the best way to describe Jobs was as an *auteur* of his chosen medium, a term normally more commonly found in film criticism than high-tech. The auteur theory was an invention of French critics of the fifties and sixties who maintained that directors are to films what a poet is to poems. While this might seem obvious today, previously film directors were viewed as simply another part of a collaborative process. They would have their ideas, while a producer might have his, and a lighting cameraman might have his. Post-auteur theory, however, it is taken as given that a film can be viewed as an individualistic piece of art, with its own quirks and lines of exploration unique to the director. This is the creative role in which Jobs cast himself at Apple. As a micromanager extraordinaire, no personality was more deeply entrenched in Apple's products than Jobs' own. When we look at the interface or aesthetics of one of the company's products from the last fifteen years, we are intimately acquainting ourselves with a (quasi)physical manifestation of the dreams, hopes, ambitions and desires of one Steve Jobs. 'It's almost like all the products are his own appearance,' Steve Wozniak told Tom Junod when the writer was researching his Steve Jobs profile for *Esquire*.* In the article, one of the most poignant print portraits of Jobs ever published, Junod went on to ruminate on this idea:

> *It's a Dorian Grayish fable, transposed to the twenty-first century: Steve Jobs has become Steve Jobs by doing what nobody else has done before – by treating computers not just as tools but as mirrors, by making*

---

* Jobs himself used the metaphor on occasion. In an interview with technology writer Steve Levy in July 2004, he commented, 'I'm going to be fifty next year, so I'm like a scratched-up iPod myself.' At the time, Levy took the quip as an offhanded joke. Later, when he discovered that the interview had taken place just weeks before the Apple CEO underwent cancer surgery, it took on a new poignancy.

technology not just the engine but the emblem of transcendence.

Junod was not the first person to posit such theories. Having spent most of his adult life among techies (many of whom are also science-fiction fans) Jobs would have been intimately acquainted with the sci-fi staple of the man who manages to transport his 'spirit' beyond the corporeal body into the metallic lattices of a computer. Writing several years before the original Macintosh was unveiled, in *The Enchanted Loom: Mind in the Universe* NASA's Robert Jastrow waxed lyrical about his hope for a day in which such transportations would be possible. In such a sense life might go on for ever.

> *Because mind is the essence of being, it can be said that this scientist has entered the computer and that he now dwells in it. At last the human brain, ensconced in a computer has been liberated from the weakness of the mortal flesh ... It is in control of its own destiny. The machine is its body; it is the machine's mind.*

This may be one of the reasons why the death of Steve Jobs hit so hard. How could the man responsible for devices that seem so alive in almost every sense no longer be living? It is also why, in the immediate aftermath of Jobs' passing, when more and more stories – many of them unflattering – began to circulate about Jobs, such a large number of people had a hard time believing that this could have been the same Steve Jobs responsible for not only such beautiful gadgets, but for Apple users' own means of personal liberation also.

There is perhaps nowhere on earth better designed for a discussion of utopianism than Silicon Valley. 'Utopia' as a term was largely derived from the 1516 book of that title by the humanist and philosopher Thomas More. The word itself is a pun, drawn equally from the Greek words *ou-topos* ('no place') and *eu-topos* ('good place'). The virtual world may, depending upon one's perspective, be a good place, but it is, by definition, a no place; being located nowhere more than on various scattered servers across the planet and, more aptly, behind the interface fantasy of the screen.

At the start of this chapter I quoted from an article, 'The Long Boom: A History of the Future, 1980–2020', which appeared in *Wired* magazine in July 1997, the month that former CEO Gil Amelio was ousted from Apple and Steve Jobs became *de facto* chief. Following on from the passage I reproduced, the article's authors, Peter Schwartz and Peter Leyden, continued to compare possible worlds of the future, which take the 'closed' or 'open' routes to their natural conclusions. In the 'closed' instance, nations would turn inwards, dividing into blocks. Rigidity of thought, stagnation of economy and an increase in both poverty and intolerance would naturally follow, resulting in the vicious cycle of a yet more closed and fragmented world. On the other hand, should society adopt a more 'open' model, Schwartz and Leyden argued for the creation of a utopia in which innovation and new ideas are the norm, a new Age of Affluence results in increased tolerance, smaller economic units, greater transparency and – ultimately – a more integrated world. A nice idea, certainly. Today that notion is most widely explored through the activities of the Open Source community. Like a latter-day Homebrew Computer Club, the Open Sourcers (whether it be in terms of Linux or the development of online editable 'wikis') develop their projects *en masse* with no (immediate) eye on commercialisation, through a dedicated community of users around the world, each correcting, building upon and documenting one another's work in order to create the tools that will deliver us from closed corporate slavery.

A question to be asked of any utopia is: whose utopia is it exactly? Jobs' vision might have differed from many of his peers, but that doesn't make it any less utopian. Anyone who talked about making computers that would 'put a ding in the universe' was, by very nature, a world builder. This was spotted early on in Jobs' career. The first book written about Apple, initially published in 1984, was originally entitled *The Little Kingdom*. Jobs' methodology never changed. Where a person like Bill Gates saw computing as a way to make a great deal of money, very quickly – at which point he could give back through philanthropy – Jobs cut out the middleman. Even here he simplified. Jobs felt that by building insanely great computers he could change the world every bit as much as Gates could, perhaps more so. Apple's success exists more as a paradox than it does a contradiction. If it were a contradiction, it would not work – and Apple works. In Jobs' own words, 'it just works'. Jobs went against popular Silicon Valley wisdom by creating a company that is, by any measure, extraordinarily secretive; producing devices that encourage the user to stay on the surface rather than delving into their inner workings. But by doing so it has become a high-tech dream factory, out of which a seemingly endless line of new products is unleashed upon the world, fully formed, as if by magic.

The overwhelming majority of popular retellings of the sixties counterculture end the same way: with its protagonists dropping back and selling out. As always, the Apple story is more nuanced than that. The countercultural vision that Jobs saw from Apple's earliest days was one fundamentally rooted in selling mass-produced items, but also paradoxically in individualism. Taking up Alan Kay's rebel yell that 'People who are really serious about software should make their own hardware', he created a company in which even the concept of design (which is viewed as 'how something works' rather than 'how it looks') bridges the gap between the Industrial and Information Ages. 'Apple creates beautiful, wonderful new products that capture everybody's imagination,' says Bruce Horn. 'It's all about passion for the actual product rather than making a buck. Making a buck comes automatically out of having beautiful, wonderful products that serve people's needs and wants and desires – and makes them happier.' Such sentiments are perfectly in line with the hippie fantasies of the Information Age; in which capitalism can occur without exploitation. But while this might be a possibility when you are dealing in the ethereal software dreams of pure Information, like a Google or Facebook, it is more problematic when at least one part of your business is a publicly-traded Fordist company involved with the manufacture of hardware, aimed at securing the highest possible profit margins. Indeed, in recent years Apple – like many other large high-tech companies – has come under increasing scrutiny for the means by which its products are made. Outsourcing to a Taiwanese company called the Foxconn Technology Group, whose subsidiary Hon Hai Precision Industry assembles Apple's devices in factories in China, reports have circulated in the press concerning subpar conditions for employees, possible underage workers, excessive overtime, harsh punishments and worker suicides. After a spate of deaths in 2010, special suicide nets were erected to catch jumpers.

Is this, like the surface-level sheen of an Apple interface, the inevitable hidden side of producing mass-market tools of empowerment? Certainly the workers themselves are far from empowered. As noted, Apple is not the only company to manufacture its products through Foxconn, but its countercultural posturing makes it an obvious target.* Through the wish fulfilment of its interface, Apple is a company which deals – quite literally – with desire. As every social critic from Marx onwards has pointed out, ideology critique attends to the notion that what fuels the economy are forces of production which remain obscured from view; the political economy being a shadow-play of the libidinal economy. Ultimately it raises the question of just who Apple's products are empowering. As relate to the devices themselves, these discussions continue to rage online, most notably in the flame wars between Apple-boosters ('fanboys') and detractors ('haters'). For those that praise Apple, Jobs succeeded in executing the hippie dream of demystifying computers. He made complex technologies simple without being reductive. For those who bash the company Jobs was merely a man whose cult of personality allowed him to sell overpriced and overly simplistic fetish objects to a gullible fanbase. 'Apple products,' writes a pseudonymous blogger on *Madame Pickwick Art Blog*. 'iWant, iWant, iWant, iDesire, iDesire ... the [epitome], the sweet spot of American middle class values. Hegemonic, expansive and absorbing pervasiveness of American culture. Beautiful design, beautiful advertising. All to sell a stupid computer. They did it. More cash in the bank than Bernie Madoff absconded with. The cult of cool around Apple new product releases seem almost mystical imbued with religious [fervour]. Messianic and utopian. It's the fascinating cult of Apple that defies the coherent criticisms of their products, the necessity – the iPad – boutique designer pricing, and functionality compared with

---

* Apple, to its credit, is taking this matter seriously and, as of this writing, was putting new codes of practice in place to address the issues. 'We want everyone to know what we are doing, and we hope that people copy,' current Apple CEO Tim Cook said in a 2012 interview with Kara Swisher and Walt Mossberg. "We've put a ton of effort into taking overtime down.'

alternatives. Apple and its corporate culture; one of the clearest examples of white patriarchal and colonialist culture, is the most visible example of commodity fetishism in America.' Both camps are so ardently entrenched in their respective thought processes that any attempt to change them does little good.

In an article called 'iFascism?' John Marshall, a writer for the website TPM, noted that 'the interplay of aesthetics (which Mac has in spades) and centralised control (which Mac also has in spades) is an interesting one'. This is an incredibly complex issue because of what it suggests about the thin divide between giving the user a straightforward, user-friendly experience and being overly proscriptive in terms of what one can and cannot do.* Essentially there is no difference between the fundamental view held by the Apple fanboys and the Apple haters. As Jean Baudrillard noted about technology, it is its enigmatic opacity that makes it seductive. There are always going to be people wondering about the hidden degree of technocratic control lurking one step below interface level – and the techno-hippies would not have had it any other way.

In his book *Insanely Simple: The Obsession that Drives Apple's Success*, former Apple employee Ken Segall relates the story of a disastrous meeting in which a new agency planner – the person hired to represent the potential consumer's point of view, rather

---

* This line between straightforwardness and a counterproductive degree of simplicity was cleverly parodied in 2009 by the satirical news website *The Onion*. In a deadpan video, *The Onion* supposedly unveiled Apple's latest innovation: the MacBook Wheel. According to an Apple 'Senior Product Innovator' (read: a paid actor) this would mark the next leap forward in computing paradigms, by replacing the classic keyboard with a sleek, touch-sensitive iPod-style click wheel. Cutaway footage shows users attempting to type emails, using the wheel to scroll to each individual letter and then centre-clicking to select it. In case this takes too long, the MacBook Wheel also boasts a predictive sentence technology, including such timeless examples as 'The aardvark asked for a dagger' and 'The amiable crocodile brushed his teeth with a toothbrush'. The video concludes with the announcement that Apple is already hard at work on the next generation of the MacBook Wheel, which will weigh four ounces less than its predecessor due to its lack of screen, hard drive or wheel.

than that of his boss or the client – made the fundamental error of turning a 'brand audit' for Apple into a long and decidedly East Coast presentation for Steve Jobs, complete with spreadsheets, charts and graphs highlighting the different demographics interested in Apple products. As Jobs sank lower into his chair, a bored and impatient look settling across his face, the planner – oblivious to the fate that was about to befall him – continued to discuss his findings.

'We even went to the ghetto in Harlem and asked the kids on the basketball courts what *they* thought of Apple?' he said.

'Why'd you ask them?' Jobs snapped. 'They don't have any money!'

Five minutes after the meeting ended, Segall was on his way back to San Jose Airport when Jobs called him on his cell phone. 'Look, I don't know who that guy was or why you brought him, but I'm not paying a cent for anything he just did and I never want to see him at Apple again,' he said. Jobs' comment was not about race; it was about money. You either had enough of it to be part of the Apple revolution or you didn't. If you weren't going to be buying the new iPad or MacBook Air then who cared what you thought about it? This idea was elaborated on in an essay by Gilles Deleuze, who foresaw with remarkable prescience the emergence of a 'society of control', in which our credit cards and social-security numbers are more important in establishing our standing within society than anything else. One could look at this bit of free-market societal restructuring either positively or negatively. Deleuze opted for the latter. The way he saw it, one 'apparatus of capture' has simply been succeeded by another. Today we are not segmented by our gender, race or class, so much as by our line of credit or debt. We are no longer 'a man confined but a man in debt'.

❏ ❏ ❏

On 12 June 2005, Steve Jobs delivered the commencement speech for graduating students at Stanford. It was the 114th such address in the university's history and among its best. In a sense it was typically Jobsian. Even bedecked in a purple and black graduation gown as opposed to his uniform ensemble of black turtleneck sweater and blue Levis, Jobs retained his role of human logo for his twin companies, Apple and Pixar. When he spoke idealistically of his childhood, the 1985 casting out of Apple, and his triumphant return to the company he had co-founded, he '[connected] the dots' in such a way that his story resembled that of Joseph Campbell's monomythic hero's journey, comprising the call to adventure, the road of trials and, eventually, the return from without. Even without its lyrical flourishes, however, the speech was extraordinarily revealing. While he clearly realised, and relished, the iconic value that he held, Jobs had never enjoyed speaking about his personal life. That same year he would attempt to block the publishing of a less-than-glowing unauthorised biography of him. When that didn't work he banned all future titles from the publisher from appearing in Apple's digital retail stores. Jobs' discomfort was evident. Although he had help writing the speech from a close friend, he appeared somewhat unsure speaking the words written, strangely seeming at his most comfortable when discussing death, which he did in the manner of a new Apple product.

Shortly after Jobs' commencement address, he received an email from *The Whole Earth Catalog* founder Stewart Brand. Brand had been touched by Jobs' mentioning of the *Catalog* in his speech and thought it was high time that the two caught up for a reminiscence. Jobs could be tricky to pin down, however, and it took until early 2006 before the Apple CEO was able to find a time in his schedule that suited him. When he did, he agreed to meet Brand at an upmarket sushi restaurant on the Sand Hill

Road in Menlo Park, an area notable for its high concentration of venture capital companies. Brand, who was sixty-seven and still in rude health, brought up Jobs' recent medical issues.

'I dodged a bullet,' Jobs admitted.

'God, pancreatic cancer,' Brand replied. 'It's more like you dodged a bazooka.'

The conversation moved on to other topics. They talked about Brand's philanthropic work, Bill Gates's failure to do right by his users, and the Iraq War. Brand found Jobs' views on the latter subject interesting. Jobs thought the whole thing rather trivial and noted that he couldn't see what all the fuss was about.

'America's not even breaking sweat over there and yet we're all getting worked up about it,' he said.

As ever, it was Jobs at his most paradoxical: the left-leaning billionaire who nonetheless refused to toe the accepted liberal line.

Almost forty years had passed since Brand had started *The Whole Earth Catalog* and the former Merry Prankster had long since moved on to other things. (The *Catalog*'s successor, the virtual community, the Whole Earth 'Lectronic Link – normally shortened to the WELL – was sold to Salon.com in 1999.) Certainly both he and Jobs were a world away from where they had begun, but Brand felt a definite kinship with the Apple CEO. 'We were cohort members in a sense,' he says. 'He was far younger than me, but the areas and people we'd been interested in, and been working with, had intersected a lot over the years. I had been keeping a pretty good eye on what he was up to and he had a corner of his eye on me, I think.'

The discussion about Iraq and the changing face of global politics returned the conversation to more Silicon Valley-centric matters. At the time, Apple seemed tipped to take over the high-tech world. On the back of its United States vs Microsoft antitrust suit, Microsoft's influence was receding, while Apple's reach was just as quickly advancing. Having seen one 'Evil Empire' go, Brand was keen to make sure that Apple would prove to be capable of running things in the egalitarian way that the techno-hippies had always wanted.

'Can you make sure that Apple becomes a benign hegemon,' he asked Jobs across the table.

Jobs looked at him, and then diverted the question.

'We're far from being a hegemon so far,' he said.

Five years later, on 26 May 2010, the NASDAQ markets closed for the day with Apple shares having risen while Microsoft's floundered. With a market value of around $222 billion – compared with Microsoft's $219 billion – Apple was now the world's biggest tech company.

There was still time.

Epilogue

¡VIVA LA REVOLUCIÓN!

'Make war on machines. And in particular the
sterile machines of corporate death and the
robots that guard them.'

Abbie Hoffman,
Steal Yourself Rich (1971)

One unseasonably sunny morning in early October 2011, an expectant crowd of journalists gathered in Cupertino. They had been brought there by a typically smart – and particularly tantalising – piece of guerrilla marketing issued by Apple's PR firm. Each attendee clutched an invitation upon which the image of four iOS app icons had been printed. One showed the familiar calendar graphic, marked with that day's date; the second, the clock, with the hands showing 10 a.m.; the third, the map with an arrow pointing towards the Apple campus; and the final one the phone icon showing a single call. Beneath it were the words: 'Let's talk iPhone'. As ten o'clock approached there was a palpable energy in the air. This was widely expected to be the unveiling of the iPhone 5, the latest iteration of Apple's industry-reshaping smartphone. The Town Hall auditorium in which the journalists were congregated was three miles (or eight minutes' drive) from the garage on 11161 Crist Drive, Los Altos, in which the Two Steves had brought the Apple I kicking and screaming into the world. Some things had changed; others were strikingly similar. What had changed was the scale. Far from the yawning indifference of the Homebrew Club members, that had marked the first showing of the Apple I in 1976, whatever Apple was springing that October morning would be transmitted all over the globe within minutes of its announcement, and reported as a serious news item on television alongside that day's top stories. What had stayed was Apple's identity. Thirty-five years after its founding, the company continued to pride itself on its countercultural roots. The four songs which played over the PA system as the army of journalists took their places were 'Under My Thumb', by the Rolling Stones; 'Whole Lotta Love', by Led Zeppelin; 'Can't Explain', by The Who, and 'Jumpin' Jack Flash', again by the Stones. All four had been released between 1964 and 1969. Something else stayed the same that day, too – disappointingly so.

Instead of a revolutionary new iPhone 5, what was announced was the iPhone 4S, a marginal upgrade of the existing iPhone. What did the 'S' stand for, pundits asked. Was it 'Shutter', a reference to the new eight-megapixel camera built into the phone? Was it 'Siri', referring to the smartphone's new voice-activated virtual personal assistant? Was it 'Speed', since the 4S's processor was twice as fast as the one used in its predecessor? Or could it be 'Steve', a tribute to the man that had stepped down from his position as CEO in late August, and a semi-return to the company's early days when products were named after individual people? Although the iPhone 4S was met with generally favourable reactions its timing was surprising in many ways. In contrast to previous iPhone launches which came at times of declining sales, sales of the iPhone 4 had not yet climaxed when the 4S was released. Meanwhile, some people were disconcerted by its lack of a sizeable leap forward in either technology or design. 'Although services such as the iCloud and

Siri voice control are enticing features, the hardware upgrade is simply not Apple worthy,' complained one online commentator. Tim Stevens of the technology blog *Engadget* summed up the reaction as well as anyone when he noted, 'the iPhone 4S does *everything* better than the iPhone 4, but it simply doesn't do anything substantially *different*'. Had Apple, a company known for painstakingly mulling over each one of its products, simply rushed one out the door? *Where was the Revolution?*

One possible answer emerged the following day when another change revealed itself. Eight years after being diagnosed with pancreatic cancer – the same malignancy that had killed original Macintosh founder Jef Raskin in 2005 – Steve Jobs passed away at his Palo Alto home, at the age of fifty-six. According to several sources, he had been watching the 'Let's talk iPhone' event from his sofa via a special video feed. Even with Jobs' prominence as the highly visible CEO of a much loved brand, the large-scale public reaction to his death was surprising. The world sat up and took notice. Apple Stores had their shop front windows filled with Post-it notes upon which admirers had scrawled tributes.

Writing in the *Guardian* newspaper in Britain, Tanya Gold commented shortly after Jobs' passing that the deification of Apple's co-founder was the company's greatest marketing achievement to date. '[Steve Jobs] is now,' she wrote, 'just a little too late to enjoy it, the world's most famous man, one pixel short of saviour.' Not simply content to love consumer products, Gold suggested, it is now an imperative that when the people who created them die, users are required to enter a mode of spiritual decline themselves. Almost immediately after Jobs had stepped down as Apple CEO on 24 August, tributes had started pouring out from the various people he had worked with and known over the years. Aware that it was only a matter of time before his death, 'Bad Steve' stories were quickly reconfigured as 'Passionate Steve' stories. Google's vice-president of engineering, Vic Gundotra, spoke about how

Jobs had once phoned him early on a Sunday morning in 2008 to berate him over choosing the wrong shade of yellow for an interface element of Google's iPhone app. Following Jobs' death exactly six weeks later, there was a second wave of Jobsmania. Virgin America, the only airline based in Silicon Valley, named one of its fleet after the CEO, emblazoning Stewart Brand's 'Stay Hungry, Stay Foolish' sentiment on the side of an Airbus A320 in tribute. A Hong Kong company announced its plans to release a poseable Steve Jobs figurine for $99.99 plus shipping, complete with 'a number of accessories to help users replicate any number of famous Jobs poses'. (The concept was later ditched following 'immense pressure' from lawyers representing both Apple and the Jobs estate.) Meanwhile, Marc Jacobs, the designer who had created the Apple co-founder's famous three-piece rimless spectacles in 1998, was inundated with requests for identical frames, to the point where his company had to place customers on a three-month waiting list.

In a sense it was symbolic that this should be the post-Steve era. At his last ever keynote event in March 2011, where Jobs appeared painfully thin but mentally as sharp as ever, he talked about the ushering in of a 'post-PC world'. The idea behind it was simple. Instead of Apple's digital hub revolving around the personal computer, from this point on it would instead operate around the iCloud: a network online storage platform, which replaced Apple's disastrous MobileMe subscription service. For iOS 5, the latest iteration of Apple's mobile operating system, the company had combed through iOS and identified every feature which assumed or required a PC, and retooled it to rely instead on the cloud. As journalist Leander Kahney of the Apple news website *Cult of Mac* wrote:

> *With iOS 5, Apple stores all of your data — your*
> *mail, your calendar, your address book, your photos,*

your music, your ebooks, even your Doodle Jump
save games — in the iCloud. iTunes Match hurls
your complete music collection onto Apple's servers,
available to download anywhere and anytime without
pulling out your Apple Connector cable. Meanwhile,
Wi-Fi Syncing makes sure that if your iPhone or iPad
does need to talk to your PC, it can do so just by being
plugged into a wall socket and within a stone's throw
of your PC.

The iCloud would be officially announced later in 2011, at the Apple Worldwide Developers Conference. 'The question is has Apple severed iOS's innate tether to the PC, or will iOS 5 be remembered as a smaller interim step towards the post-PC world Steve so presciently envisioned?' Kahney asked. 'We've been playing with iOS 5 for months. Here's what we think: by gum, Apple's done it.' If Apple genuinely was indeed heading into a brave new world in which the PC was no more central than the iPod, the iPhone or the iPad which is certainly less important than the first two) then it would be one that wasn't inhabited by Steve Jobs, the man whose life had been dedicated to turning the personal computer into a consumer product.

The question, of course, is where Apple goes from here. By the time of Jobs' passing he had already been succeeded as Apple CEO by Timothy Donald Cook, a man five years Jobs' junior. Tim Cook is unlike Steve Jobs in many ways. Although he was born in 1960 and thus also qualifies as a part of the baby-boomer generation, he grew up a world away from the countercultural high-tech haven of Silicon Valley, in the small town of Robertsdale, Alabama. The son of a shipyard worker,

Cook majored in industrial engineering at Auburn University, received a business degree from Duke in 1988 and then joined IBM, where he stayed for the next twelve years, working in PC logistics. From IBM he jumped to Compaq Computer, filling the post of vice-president of corporate materials. Cook had only been at Compaq for six months when he heard, in 1998, that Apple was looking for a new head of operations. The previous person in the position had stayed just three months under Jobs before throwing in the towel. If Jobs was eager to hire someone quickly simply in order to relieve himself of a bit of extra work, however, he wasn't letting on. Prior to bringing Cook aboard, Jobs had rejected a string of other highly qualified operations managers for the role. One executive from Compaq Computer lasted a mere five minutes in his interview before Jobs got up and walked out on him. The executive in question had only just finished explaining that he liked to collect barbers' chairs when he wasn't working.

Cook was markedly different, both to the other candidates and to Jobs himself. Where Jobs had dropped out of college, Cook earned his MBA by going to night school. Where Jobs proudly boasted that Apple was a company in which spreadsheets were a thing of the past, Cook tirelessly obsessed (and continues to obsess) over figures and graphs in a way that Jobs never had the patience for. Within months of joining Apple, Cook had managed to drastically reduce Apple's on-hand inventory, comprising both components and finished products; bringing it down from $400 million the previous December to just $78 million. The move earned him the nickname the 'Attila the Hun of inventory'. Where Jobs would yell and foam at the mouth, Cook was, and remains, soft-spoken but firm. Where Jobs told biographer Walter Isaacson, 'I'm willing to go thermonuclear war' in order to destroy Google's Android platform, Cook has been outspoken in his desire to litigate less than his predecessor and to 'settle rather than battle' with competitors. Among the first decisions Cook made upon assuming the position of CEO was to implement a charitable matching scheme, so that Apple would mirror employees' personal charitable contributions up to $10,000 per year. The program resulted in $2.6 million in donations within its first two months in existence. (By contrast, one of Jobs' first initiatives upon returning to Apple in 1997 was to cancel all of the company's philanthropic programs as a cost-saving measure.) Could it be that under Cook Apple will become that most countercultural of paraxodes: the friendly corporation?

Several people, in fact, have speculated that it was Cook's differences to Jobs, rather than his similarities, that enabled him to ascend through the ranks as Jobs' apparent protégé. There are people at Apple far more like Jobs than Cook was. Scott Forstall, Apple's senior vice-president of iOS software, for example, possesses Jobs' attention to product detail – and even went so far as to purchase the same model of silver Mercedes coupé that Jobs drove to show where his inspirations lay. Cook, on the other hand, succeeded because he was willing to take on the tasks that Jobs himself had little interest in.

But the two men did share things in common. Fellow workaholics, Cook has shown the kind of dedication to Apple that Jobs demanded, routinely answering emails at 4.30 in the morning. He also *got* Apple's culture in a way that not everyone does. This was evidenced early in 2009, a week after Jobs announced that he would be taking a six-month leave of absence from Apple for medical reasons, when Cook presided over a scheduled conference call with investors and Wall Street analysts to discuss Apple's last-quarter earnings for the previous year. Not unexpectedly, the first question asked was about the possibility of Cook replacing Jobs as CEO in the event that the company co-founder was unable to return to fulfil his duties. It was a point that Cook could very well have brushed off, as much for political reasons as for anything else. Instead he took the opportunity to deliver a stirring – and seemingly impromptu – speech about Apple's community and culture, which harked back not just to the Apple Values of the company's early years, but to the kind of rhetoric that wouldn't have looked out of place in a Stewart Brand-penned editorial in *The Whole Earth Catalog*:

> *We believe that we are on the face of the earth to make great products, and that's not changing. We are constantly focusing on innovating. We believe in the simple, not the complex. We believe that we need to own and control the primary technologies behind the products that we make, and participate only in markets where we can make a significant contribution. We believe in saying no to thousands of projects, so that we can really focus on the few that are truly important and meaningful to us. We believe in deep collaboration and cross-pollination of our groups, which allow us to innovate in a way that others cannot. And frankly, we don't settle for anything less than excellence in every group in the company, and we have the self-honesty to admit when we're wrong and the courage to change. And I think, regardless of who is in what job, those values are so embedded in this company that Apple will do extremely well. And ... I strongly believe that Apple is doing the best work in its history.*

□ □ □

Steve Jobs did not want to leave Apple as a company in which people ask 'What would Steve do?' in any given situation.

Apple's executives must learn not to do this; a challenging ask in an industry in which even those outside Apple consciously try and model themselves after Jobs, with varying degrees of success. Jobs himself had no great preciousness for the past. When he returned to Apple in 1997 he gave away hundreds of boxes containing vintage Apple materials – including blueprints, user manuals and company shirts – to the Silicon Valley Archives at Stanford University, informing everyone that Apple was no longer going to live on former glories. From that point on, he told people, Apple would look only forward.

Of course, when someone like Jobs, who was among the most visible CEOs in the world and a person many see as directly corollary to (and sometimes, misreportedly, wholly responsible for) Apple's success, things become more challenging. Many business analysts have pointed to cases involving similar once-dominant companies, headed up by immensely charismatic individuals, and noted how these went into a steep period of steady decline following that person's departure. Three such examples would include Polaroid, Sony and Disney; all companies, incidentally, that Jobs greatly admired. As noted, all were, in the words of sociologist Max Weber in his 1947 book *The Theory of Social and Economic Organization*, governed by individuals 'set apart from ordinary men and treated as endowed with supernatural, superhuman, or at least specifically exceptional powers or qualities'. In each case the passing of the company's charismatic founder (respectively Edwin Land at Polaroid, Akio Morita at Sony and Walt Disney at Disney) led to an interlude during which the companies coasted on past success, before the inevitable fall from grace occurred. Will the same happen at Apple, much as it did when Jobs left the company for the first time in 1985? This remains to be seen, but it would seem too early to write it down as an inevitability.

In the same way that Apple-watching has become a recognisable media occupation in its own right, so, too, has predicting the company's forthcoming demise for a certain percentage of those reporters. There is even a tongue-in-cheek webpage, the Apple Death Knell Counter, dedicated to chronicling these incorrect aspersions as they arise from various journalists, analysts, pundits and business executives. (At time of writing, Apple had been declared moribund no less than fifty-nine times since April 1995.)* While history would dictate that, at some point, Apple's influence will certainly wane, there is no reason why this need be in the foreseeable future.

So far all evidence suggests the contrary, in fact. During the first half of the 2012 fiscal year, Apple generated more revenue, at $85.83 billion, than it did for the entire fiscal 2010. Of course it is going to take more than a few successful quarters to show that Tim Cook is the kind of long-term visionary that Apple needs. What is going to be fascinating to watch play out is Apple's move the next time the company needs to make the kind of industry-redefining manoeuvre that it did when Jobs decided to launch the iPod or iPhone. Today around three-quarters of Apple's revenue come from iOS devices (that's the iPad and the iPhone). Will Cook possess the Jobsian chutzpah to shake up the Apple system to that extent?

What will more assuredly happen is that whatever steps Apple takes from this point on, they will be compared – both in and outside the company – to Jobs' run as CEO. Given that Jobs' fourteen-year stint running Apple is about as good a turnaround story as one could hope for in business, this is a daunting challenge. If Apple's products succeed from here, they will be compared to the astronomical successes Jobs had with the iPod, the iPhone and the iPad. If they fail – or if they contain elements

---

* The webpage in question can be found at http://www.macobserver.com/tmo/death_knell/

that appear clunky or unfinished – they will be met with snipes, noting how they would never have been allowed out of the door were Jobs still at the company helm. Apple has built up a degree of goodwill from years of providing industry-redefining tools, but it has also never been a company to sit on its laurels.

One key to Apple's future success links back to another hippie dream as appropriated by the cyber-libertarians. This is the ecological model of the machine; an idea that owes it roots to both Buckminster Fuller's 1968 *Operating Manual for Spaceship Earth* and Richard Brautigan's dream for a world in which machines and animals can live together in perfect harmony. In Kevin Kelly's book *Out of Control: The New Biology of Machines, Social Systems and the Economic World* the *Wired* writer demonstrates how the hippie notion of a natural balance to the earth can fit in with today's technological world:

> *The tangled flow of manufactured materials from machine to machine can be seen as a networked community – an industrial ecology. Like all living systems, this interlocking human-made ecosystem tends to expand, to work around impediments, and to adapt to adversity. Seen in the right light, a robust industrial ecosystem is an extension of the natural ecosystem of the biosphere … Stuff circles from the biosphere into the technosphere and back again in a grand bionic ecology of nature and artefact.*

Apple, for its part, has been able to create an effective product ecosystem in a way that few companies are ever able to do. Sales of the iPhone drive use of the App Store, which in turn drives music sales, which drive sales of the iPhone, which then drive sales of the iPad, which drive sales of music, which drive sales of apps, which drives sales of the Macbook Air, which drives sales of apps from the App Store, which drives sales of the iPhone, and so on *ad infinitum*. This ecosystem is enhanced by positive feedback, with each element supporting the enhanced use of every other part. As each increases, it bootstraps the increased use of all other parts. The genius of Apple's ecology is that its products ask users to buy into a system which, once learned, is difficult to step outside of. After all, once the controls for an iPod have been ingrained in the 'implicit' memory (which governs everything a person can do without thinking) why trade the product for a Microsoft Zune or another brand name when it comes time to upgrade? And if an iPhone and iMac can sync together in perfect harmony, why buy a smartphone or personal computer which cannot?

Jobs' dream of a self-sustaining ecosystem went beyond just product lines, however. He also wanted to create a company that could sustain itself without him at its head. In 2008, as his health worsened, Jobs hired the dean of the Yale School of Management, Joel M. Podolny, to create what he referred to as Apple University: an alternative training school that would ensure that Apple's values were not lost after Jobs stepped down from his post at Apple. Podolny was not a run-of-the-mill professor. Becoming head of Yale's graduate business school in 2005 at the age of thirty-nine, Podolny's revamped – controversial – curriculum had been full of esoteric courses on subjects such as 'creativity and innovation'.

The question is whether Jobs has been successful in this, and whether all of the uniquely talented individuals he groomed as his various lieutenants will be able to work together in the harmonious way that he hoped. 'If you're looking at Apple today you can't worry about a lack of talent, because they've got talent up the wazoo,' says former Apple engineer Michael Lopp. 'The thing that I worry about is that what Steve did was to align some very powerful personalities, with some very big egos. You look at the senior staff of Apple today and you've got all these amazing people that were kept in check by a guy who had an even bigger ego than any of them. What I worry about is, in the absence of that greater ego, do you [end up returning] to that world of fiefdoms that existed before Steve came back?'

This, of course, was a problem when Jobs left Apple for the first time. Although (his standing in the Macintosh team aside) he was not considered particularly key to the company's success, Jobs was able to imbue Apple with his personality in such a way that the organisation still reflected aspects of the brash, young Steve Jobs when the older, less brash version returned over a decade later. 'In many corporations, the culture can come to resemble the leader and take on aspects of his culture – and this is especially true at Apple,' says cyber-libertarian John Perry Barlow. 'You had an institution that became secretive and competitive and ideological in a way that was directly correlative with Steve.' It was this challenge that Jobs was struggling to overcome towards the end of his life. As Apple board member Al Gore told Walter Isaacson: 'The context for Apple is changing dramatically ... It's not hammer-thrower against Big Brother. Now Apple's big, and people see it as arrogant ... [Steve was] still adjusting to it. [He was] better at being the underdog than being a humble giant.'

Going all the way back to the original '1984' Macintosh commercial, Apple has always been a brand with a certain degree of knowing superiority. As Lev Grossman wrote in a 2005 *Time* magazine article entitled 'How Apple Does It':

> *It's almost eerie: Apple employees all like one another,*
> *and they have a strong sense that they are the chosen*
> *of the earth, and they're not going to be a jerk about it,*

but all others who dwell on this mortal coil are missing
out by not working here.

The company possesses what former Apple marketing director Michael Mace refers to it as the 'snob factor'. 'By using Apple products, you were telling yourself and the world that you were a superior form of human being,' he says. 'You knew better than anyone else, and you were part of a special elite above everybody else. You could kind of smirk at anybody else, so there was definitely a level of self-satisfaction involved. We were selling life validation products to the smartest, most creative segment of the population. It's a gorgeous example of how you can match market segment with company culture, product design and advertising – and how freakishly powerful that is when you get all of those factors right.' Alan Randolph, a former Apple accessory designer who lives in San Francisco, backs him up: 'Apple flatters our own mythologies, our secret vision of ourselves as sophisticated and popular with dozens of friends who cannot wait to hear from us. Jobs [knew] you cannot touch us more deeply than that. We are, at the end of the day, very simple vulnerable squishy creatures, and Apple does pretty up even the dullest lives.'

Despite its increasing ubiquity, Apple has thus far managed to avoid the kind of backlash that saw Microsoft labelled the 'Evil Empire' in the 1990s while at the height of its powers. For all its positioning as a company straight out of the Peace and Love generation, however, Apple's reputation as a company to deal with in business is far from utopian. As it has grown larger and more successful the 'we can do anything' hippie dream took on a decidedly more dystopian 'I can do anything' mentality.

For example, when Apple first decided that it wanted to use the name iPhone, it discovered that Cisco Systems, the networking giant which employs 10,000 more people than Apple, already owned the trademark. The name had been registered back in 1996; one year before Jobs rejoined Apple and two before Ken Segall came up with the term iMac. In short, there was no doubting that Cisco had come up with the name themselves. But because the company had yet to release a product under that name, Apple assumed that the term, trademarked or otherwise, was still up for grabs. When Cisco heard about this they threatened litigation. Apple went ahead and announced the iPhone. Cisco sued. The suit was settled so that Apple kept the name and the two companies agreed to cooperate with one another. An easy truce was reached. Three years later Apple announced that it wanted to use the term iOS for its iPhone operating system. Cisco balked. The company had been referring to its own Internet operating system as IOS for almost twenty years. 'No way,' they said, shocked at Apple's audacity. A second agreement was reached. Jobs announced the renamed iOS when he debuted the iPhone 4 in June 2010.

Of course, it's one thing to act like this in business, another to parade this kind of behaviour in front of customers. In its 'Get a Mac' television campaign, which ran between 2006 and 2009, Apple managed to achieve the impossible: make people feel sorry for Microsoft. Featuring comedian John Hodgman as the embodiment of the repressed, buttoned-down PC and Hollywood actor Justin Long as the younger, hipper Mac, the spots revolved around the Macintosh scoring endless victories against its overly fuddy-duddy rival. What at first was amusing quickly became not unlike watching Muhammad Ali having a sparring session with the unpopular fat kid at school. In an article for *Slate* magazine, writer Seth Stevenson referred to the campaign as 'mean spirited' and asked whether 'smug superiority (no matter how affable and casually dressed) [was] a bit off-putting as a brand strategy'. The dichotomy was heightened yet further when the campaign made its way to the United Kingdom, with Long and Hodgman replaced by comedy duo Mitchell and Webb. In the fifteen UK-specific advertisements, David Mitchell stars as the PC while Robert Webb channels the Mac. As well-executed as the individual ads were, the problem – escaping the notice of precisely no one familiar with the two actors – was that in their established roles in the sitcom *Peep Show*, Mitchell plays the likeable, if neurotic underdog, while Webb portrays a self-adoring poseur. As such, as the *Guardian*'s Charlie Brooker pointed out, the audience comes away feeling that 'PCs are a bit rubbish yet ultimately lovable, whereas Macs are just smug, preening tossers'.

It's a fine line to walk.

□ □ □

Today it is increasingly difficult to think of Apple as a countercultural entity. Although my interest in the company primarily relates to its links with the sixties and seventies hacker-hobbyists and techno-hippies – when the very idea of building a computer for the people was a radical notion – I continue to return to Apple today, a company as full of paradoxes as at any point in its history. Naturally, the foremost paradox is: how can the world's most valuable high-tech brand become the establishment without becoming the Establishment? Despite its self-loathing during the first half of the 1990s, in many ways it was easier to accept Apple as truly revolutionary at a time when, ironically, it wasn't doing very much that was revolutionary at all. To follow the hippie countercultural hypothesis through to its logical conclusion, one can either stay young, free and nonconformist, or grow old and sell out. At the start of Chapter 13 I laid out these two options as presented in Norman Mailer's 1957 essay 'The White Negro', which did as much to shape the burgeoning countercultural aesthetic as anything. 'One is Hip or one is Square ...' Mailer writes, in sentiments not dissimilar to those echoed forty years later by Rob Siltanen in Apple's 'Think Different' campaign. 'One is a rebel or one conforms, one is a frontiersman in the Wild West of American night life, or else a Square cell, trapped in the totalitarian tissues of American society, doomed willy-nilly to conform if one is to succeed.' It is for this reason that musicians like Jim Morrison and Jimi Hendrix, who passed away at the age of twenty-seven, will forever be seen as young, idealistic rebels, while certain parties will cluck about Mick Jagger still performing at the age of sixty-nine, or the fact that you can today hire Bob Dylan to perform at 'your corporate event, private party, fundraiser, or club'. In the nineties, Apple might have been declining in almost every aspect of business – and sometimes consciously striving to avoid doing the very thing which made it desirable as a brand – but it was still an underdog, and there is nothing the counterculture loves more than an underdog.

Nowadays that couldn't be further from the truth. 'Apple is getting to the point where it can't keep growing,' says former employee George Crow. 'They dominate their markets so much that if they can't find new places to go then they cannot continue to expand at the speed that they have been doing. The other challenge is that as the management turns over and new people come in – if we're lucky they'll be indoctrinated to the Apple culture. But if we're not so lucky Apple will just turn into a basic US corporation that's [only] interested in the bottom line for the next quarter. That's probably the single most important thing that Steve brought back to Apple: the fact that he was not interested in the next quarter; he was interested in conquering the world. I think Tim Cook shares that vision and I know that Jony Ive does. What I don't know is how they're going to be able to maintain that vision in the long term ... I hope that they surprise me.'

To others, the sheer scale of the current Apple makes it difficult to consider it any kind of rebel. 'They're just too big,' says Michael Grothaus, a former Apple employee who worked for the company in the 2000s, when I put the question to him about Apple's countercultural status in today's world. 'They own the tablet market; they own the MP3 player market – even though that's practically insignificant today; they're pretty damn good when it comes to the smartphone market ... I mean, what exactly are you countering? They *are* the culture now.'

This might be exactly the point. The Apple ideology today – a geeky, countercultural mix of hackers united against a monolithic technocratic elite, has gone mainstream. After all, as Thomas Frank writes in his book *What's the Matter with Kansas*:

> *Counterculture is so commercial and so business-friendly today that a school of urban theorists thrives by instructing municipal authorities on the fine points of luring artists, hipsters, gays, and rock bands to their cities on the ground that where these groups go, corporate offices will follow.*

But Apple has at least one thing going for its revolutionary credentials. Its rise to prominence has come not from following other companies and attaching a bit of revolutionary zeal to the marketing, but, rather, from a repeated propensity for going against the popular wisdom. High-tech and other media companies can largely be divided into the two distinct categories of 'wave makers' and 'wave riders'. A wave maker blazes its own trail, and in the process fundamentally affects the behaviour of both consumers and other companies. A wave rider is content to go with the flow. In Silicon Valley there are many examples of the latter, but a proportionately small number of the former. Apple is among the former. Time and again the company has shown an ability – and a burning desire – to reinvent itself under a new guise, even when this has seemed to run counter to the direction in which pundits see the industry heading.

A candy-coloured computer that looked, in Jobs' words, more like something you'd want to 'lick' than something you would want to conduct serious business on ran against everything analysts thought was true about the PC market. iTunes democratised music distribution, and in doing so eschewed more popular subscription-based means of listening to music online in favour of an unbundling of albums into its individual tracks. The iPhone completely flipped what people expected from a smartphone – making it not so much a phone with computing functions, but a computer that could make phone calls – and as such helped revolutionise the mobile-communications industry. Meanwhile, Apple's advertising campaigns have repeatedly (and very cleverly) presented the company's products not as the latest *le gadget* in order for mass consumers to 'keep up with the Joneses' but, rather, as individualistic devices designed for a select few freethinking dissidents. Squares need not apply. By doing this Apple has built up a wider customer base than ever. After all, deep down doesn't everyone want to rock the boat?

Apple is banking on the fact that you do.

AUTHOR INTERVIEWS

(conducted 2011–12)

Michael Albaugh, John Alfano, Dennis Allison, William Arms, Ron Avitzur, Richard Barbrook, John Perry Barlow, Jules Bloomenthal, Stewart Brand, Jim Burger, Dick Cavett, Dennis Cohen, George Crow, David Curbow, Paul Dali, Martin Darbyshire, Steve Demeter, Alfred DiBlase, Tim Dierks, John Draper, Bill Dresselhaus, Paul N. Edwards, Craig Elliott, Ron English, Hartmut Esslinger, Lee Felsenstein, Tobey Fitch, David Fradin, Bob Frankston, Philip Gray, Clive Grinyer, Michael Grothaus, Andy Hertzfeld, Patrick Holleran, Michael Holley, Bruce Horn, Dean Hovey, Rob Janoff, Michael Jay, Ted Kaehler, Mitch Kapor, Bill Kelley, David Kelley, Daniel Kottke, Bob Lash, Marc LeBrun, Michael Lopp, Paul Lutus, Michael Mace, Jerry Manock, John Markoff, Matthew Medeiros, Richard Melmon, Mike Murray, Kee Nethery, Frank O'Mahoney, Alan Oppenheimer, Rich Page, Joseph Patane, Ken Perlin, Peter Phillips, Jim Reekes, Greg Robbins, Philip Roybal, John Sculley, Ken Segall, Hank Shiffman, Rob Siltanen, Brad Silverberg, Alvy Ray Smith, David Smith, Walter Smith, Marty Spergel (email), Richard Stallman, Mitch Stein, Craig Tanimoto, Larry Tesler, Bruce Tognazzini, Jim Warren, Randy Wiggington, Lance Williams, Dave Wilson, Larry Yaeger.

BIBLIOGRAPHY

printed and electronic sources

Ahl, David. 'The First West Coast Computer Faire'. *The Best of Creative Computing*, Vol. 3, 1980.

— and Betsy Staples. 'Woz and Us'. *Creative Computing*, September 1982.

Amerika, Mark. 'Countdown to Ecstasy: the disappearance of the interface'. Amerika On-Line #6, 22 December 1997. Available at: http://www.heise.de/tp/artikel/3/3145/1.html

Arlidge, John. 'Father of Invention'. *Observer*, 21 December 2003.

Arreguín-Toft, Ivan. *How the Weak Win Wars*. Cambridge University Press, 2005.

Atkinson, Bill, and Andy Hertzfeld. Computer History Museum interview with Grady Booch, 8 June 2004.

Avitzur, Ron. 'The Graphing Calculator Story', 2004. Available at: http://www.pacifict.com/Story/

Barbrook, Richard. *Imaginary Futures*. Pluto Press, 2007.

—. 'The Hi-Tech Gift Economy'. December 1998. Available at: http://www.imaginaryfutures.net/2007/04/19/the-hi-tech-gift-economy-by-richard-barbrook/

—and Andy Cameron. 'The Californian Ideology (Extended Mix)'. September 1998. Available at: http://www.hrc.wmin.ac.uk/theory-californianideology-main.html

Belk, Russell and Gülnur Tumbat. 'The cult of Macintosh'. *Consumption, Markets and Culture*, Vol. 8, No. 3, September 2005.

Bird, Brad. Interview with *Readymade*. March 2005.

Biskind, Peter. *Easy Riders, Raging Bulls*. Simon & Schuster, 1998.

Brand, Stewart. 'Spacewar'. *Rolling Stone*, 7 December 1972.

—. 'Keep Designing'. *Whole Earth Review*, 46, May 1985.

—. *The Media Lab*. Viking, 1987.

—. 'We Owe It All To The Hippies'. *Time*, 145, special issue, Spring 1995.

Braunstein, Peter and Michael William Doyle (ed.). *Imagine Nation*. Routledge, 2002.

Brooks, David. *Bobos in Paradise*. Simon & Schuster, 2000.

Buxton, Bill. *Sketching User Experiences*. Morgan Kaufmann, 2007.

Catmull, Edwin Earl. 'A Subdivision Algorithm for Computer Display of Curved Surfaces'. University of Utah dissertation, December 1974.

—. Interview with Gardiner Morse. *Harvard Business Review*, August 2002.

Chen, Brian X. *Always On*. Da Capo Press, 2011.

Cortada, James. *The Digital Hand*. Oxford University Press, 2004.

Cringely, Robert X. 'Next Question'. *InfoWorld*, 17 October 1988.

—. *Accidental Empires*. Viking Penguin, 1992.

Daly, James. 'Of Sailing Ships and Sales Dips'. *Computerworld*, 1 October 1990.

Dediu, Horace. 'Apple sold more iOS devices in 2011 than all the Macs it sold in 28 years'. *Asymco*, 16 February 2012. Available at: http://www.asymco.com/2012/02/16/ios-devices-in-2011-vs-macs-sold-it-in-28-years/

Deleuze, Gilles. *Cinema 1*. University of Minnesota Press, 1986.

—. *Cinema 2*. University of Minnesota Press, 1989.

Deutschman, Alan. *The Second Coming of Steve Jobs*. Broadway, 2000.

Donovan, Tristan. *Replay*. Yellow Ant, 2010.

Drucker, Peter F. *The Effective Executive*. Harper & Row, 1967.

Eco, Umberto. 'The Holy War'. *Espresso*, 30 September 1994.

Ellul, Jacques. *The Technological Society*. Knopf, 1964.

English, Ron, Carlo McCormick and Colin Moynihan. *Popaganda*. Soft Skull Press, 2001.

Esslinger, Hartmut. *A Fine Line*. Jossey-Bass, 2009.

—. Computer History Museum interview with Barry Katz, 20 April 2011.

Feenberg, Andrew. *Between Reason and Experience*. MIT Press, 2010.

Fisher, Lawrence. 'Computer venture confirmed by Perot'. *New York Times*, 31 January 1987.

Floch, Jean-Marie. *Visual Identities*. Continuum, 2000.

Flynn, Laurie. 'Apple tells buyers not to expect low-cost Mac'. *InfoWorld*, 30 June 1989.

Frank, Thomas. *The Conquest of Cool*. University of Chicago Press, 1997.

—. *What's the Matter with Kansas?* Henry Holt and Co. 2004.

Freiberger, Paul, and Michael Swaine. *Fire in the Valley*. McGraw-Hill, 1999.

Friedman, Ted. *Electric Dreams*. New York University Press, 2005.

Fuller, Buckminster. *Operating Manual for Spaceship Earth*. Southern Illinois University Press, 1969.

Gere, Charlie. *Digital Culture*. Reaktion, 2002.

Gibson, William. *Neuromancer*. Ace, 1984.

Gladwell, Malcolm. *Outliers*. Little, Brown and Company, 2008.

—. 'Creation Myth'. *New Yorker*, 16 May 2011.

—. 'The Tweaker'. *New Yorker*, 14 November 2011.

Glaser, Eliane. *Get Real*. Fourth Estate, 2011.

Gross, Doug. 'Still relevant after decades, the Beatles set to rock'. *CNN Entertainment*, 4 September 2009.

Haraway, Donna. *Simians, Cyborgs and Women*. Routledge, 1991.

Hayden, Steve. 'Working with Steve Jobs was "challenge of a lifetime"'. *Ad Age*, 26 August 2011.

Heath, Joseph, and Andrew Potter. *The Rebel Sell*. Capstone, 2006.

Hertzfeld, Andy. *Revolution in the Valley*. O'Reilly Media, 2004. (See also his website, folklore.org)

Hiatt, Brian. 'Bono on Steve Jobs' Rock and Roll Spirit'. *Rolling Stone*, October 7 2011. Available at: http://www.rollingstone.com/music/news/exclusive-bono-on-steve-jobs-rock-and-roll-spirit-20111007

Hiltzik, Michael. *Dealers of Lightning*. HarperBusiness, 2000.

Illingworth, Monteith. 'George Coats: Toast of the Coast'. Available at: http://www.cyberstage.org/archive/cstage12/coats12.htm

Isaacson, Walter. *Steve Jobs*. Simon & Schuster, 2011.

Ive, Jonathan. Interview with British Council Design Museum, 2007.

Jastrow, Robert. *The Enchanted Loom*. Simon & Schuster, 1981.

Jenkins, Henry. *Convergence Culture*. New York University Press, 2006.

Jobs, Steve.

— Akst, Daniel. 'Unbundles of Joy'. *New York Times*, 11 December 2005.

— and Dan'l Lewin. Interview with Peter Denning and Karen Frenkel. Communications of the ACM, April 1989.

—. Grossman, Lev. 'How Apple does it'. *Time*, 16 October2005.

—. Interview with Jeff Goodell. *Rolling Stone*, 16 June, 1994.

—. Interview with Jeff Goodell. *Rolling Stone*, 3 December 2003.

—. Interview with Bill Gates and Brenten Schlender. *Fortune*, 26 August 1991.

—. Junod, Tom. 'Steve Jobs and the Portal to the Invisible'. *Esquire*, October 2008.

—. Langer, Andy. 'The God of Music?' *Esquire*, July 2003.

—. Moore, Geoffrey. *Crossing the Chasm*. HarperBusiness, 1999.

—. Negroponte, Nicholas. *Being Digital*. Knopf, 1995.

—. Schefter, Jim. 'Apple's Lisa'. *Popular Science*, June 1983.

—. Smithsonian oral history interview with Daniel Morrow, 20 April 1995.

—. Stanford commencement address, 12 June 2005.

Johansen, Jon Lech. 'iPhone Independence Day'. Available at: http://nanocr.eu/2007/07/03/iphone-without-att/

Kahney, Leander. *Inside Steve's Brain*. Portfolio, 2008. (See also his website, cultofmac.com)

— 'Evil/Genius'. *Wired*, April 2008.

Kay, Alan. 'A personal computer for children of all ages'. Proceedings of the ACM National Conference, Boston, August 1972.

Kelly, Kevin. *Out of Control*. Addison-Wesley, 1994.

—. *New Rules for the New Economy*. Viking, 1998.

Kirkpatrick, David. 'The Second Coming of Apple ...'. *Fortune*, 9 November 1998.

—. *The Facebook Effect*. Virgin Books, 2010.

Klein, Naomi. *No Logo*. Flamingo, 2000.

Koning, Ben, and Anneke Metz. *Images of America: Sunnyvale*. Arcadia, 2011.

Kushner, David. *Masters of Doom*. Random House, 2003.

Lanier, Jaron. *You Are Not a Gadget*. Penguin, 2010.

Lappin, Joan. 'Motorola's Zander has real trouble now'. Forbes. com, 20 February 2007. Available at: http://www.forbes. com/2007/02/20/motorola-dell-zander-pf-ii-in_jl_0220soapbox_ inl.html

Lashinsky, Adam. *Inside Apple*. John Murray, 2012.

Lasseter, John. Interview with Aubrey Day. *Total Film*, 3 June 2009.

—. Interview with John Young. *Entertainment Weekly*, 16 June 2011.

Leach, Edmund. *A Runaway World?* Oxford University Press, 1968.

Leary, Timothy. *Chaos and Cyberculture*. Ronin, 1994.

Leonard, Andrew. 'Let My Software Go!' *Salon*, 30 March 1998.

Levine, Robert. *The Power of Persuasion*. Wiley, 2003.

Levy, Steven. 'Hackers in Paradise'. *Rolling Stone*, 15 April 1982.

—. *Hackers*. Doubleday, 1984.

—.'The Whiz Kids Meet Darth Vader'. *Rolling Stone*, 1 March 1984.

—. *Insanely Great*. Viking Publishing, 1994.

—. *In the Plex*. Simon & Schuster, 2011.

Lew, Julie. 'House's Cave is Apple of Steve Wozniak's Eye'. *Chicago Tribune*, 7 April 1990.

Linzmayer, Owen W. *Apple Confidential 2.0*. No Starch Press, 2004.

Ludlow, Peter (ed.). *Crypto Anarchy, Cyberstates and Pirate Utopias*. MIT Press, 2001.

Lyotard, Jean-François. *The Postmodern Condition*. University of Minnesota Press, 1984.

Mace, Scott, and Tom Shea. 'Tech-powered concert puts tech fair in the shade'. *InfoWorld*, September 1982.

Mailer, Norman. 'The White Negro'. *Dissent*, Summer 1957.

Malone, Michael S. *The Valley of Heart's Delight*. John Wiley & Sons, 2002.

Markoff, John. *What the Dormouse Said*. Viking Penguin, 2005.

— and Paul Freiberger. 'Visit with Cap'n Software, Forthright Forth Enthusiast'. *InfoWorld*, October 1982.

Mason, Paul. *Why It's Kicking off Everywhere*. Verso, 2012.

McGregor, Douglas. *The Human Side of Enterprise*. McGraw-Hill, 1960.

McLuhan, Marshall. *Understanding Media*. McGraw-Hill, 1964.

Meikle, Graham. *Future Active*. Routledge, 2002.

Meyers, Cynthia B. 'Psychedelics and the Advertising Man: the 1960s "Countercultural Creative" on Madison Avenue'. *Columbia Journal of American Studies,* Vol. 4. No. 1, 2000.

Miller, Brian, and Mike Lapham. *The Self-Made Myth*. Berrett-Koehler, 2012.

Mitnick, Kevin D., with William Simon. *Ghost in the Wires*. Little, Brown and Company, 2011.

Montessori, Maria, and Anne E. George. *The Montessori Method*. Frederick A. Stokes, 1912.

Moore, Gordon. 'The Future of Integrated Electronics'. Fairchild Semiconductor internal publication, 1964.

—. 'Cramming more components onto integrated circuits'. *Electronics*, Vol. 38, No. 8, 19 April 1965.

More, Thomas, and Robert M. Adams. *Utopia: A New Translation*, Backgrounds, Criticism. Norton, 1975.

Moritz, Michael. *Return to the Little Kingdom*. Duckworth Overlook, 2009. (Originally published, excluding prologue and epilogue, as *The Little Kingdom*. William Morrow, 1984.)

Morozov, Evgeny. *The Net Delusion*. Allen Lane, 2011.

Nachbar, Jack, and Kevin Lause. *Popular Culture*. Bowling Green State University Popular Press, 1992.

Naughton, John. *A Brief History of the Future*. Phoenix, 2000.

Nelson, Theodore H. *Computer Lib; Dream Machines*. Rev. edn, Tempus Books, 1987.

Nocera, Joseph. 'The Second Coming of Steven Jobs'. *Esquire*, December 1986.

Noer, Michael. 'The Stable Boy and the iPad' *Forbes*. 8 September, 2010.

Nusselder, André. *Interface Fantasy*. MIT Press, 2009.

Orwell, George. *Nineteen Eighty-Four*. Secker & Warburg, 1949.

Paik, Karen. *To Infinity and Beyond!* Chronicle Books, 2007.

Parker, Rachel. 'Canon invests $100 million in NeXT'. *InfoWorld*, 19 June 1989.

Perot, Ross. Transcript of address at National Press Club, 17 November 1988.

Perry, Charles. *The Haight-Ashbury*. Random House, 1984.

Pink, Daniel. *Free Agent Nation*. Warner Business Books, 2001.

Pollack, Andrew. 'Businessland endorses takeover for $54 million'. *New York Times*, 5 June 1991.

Ponting, Bob. 'RenderMan imaging gets vendor support'. *InfoWorld*, 27 February 1989.

Prochnik, George. *In Pursuit of Silence*. Doubleday, 2010.

Raskin, Jef. Interview with Alex Pang, 13 April 2000. Available at: http://library.stanford.edu/mac/primary/interviews/raskin/trans.html

—. 'Book of Macintosh' papers.

Reich, Charles. *The Greening of America*. Random House, 1970.

Rheingold, Howard. *Smart Mobs*. Perseus Publishing., 2003.

Rose, Frank. *West of Eden*. Viking Penguin, 1989.

Rosenbaum, Ron. 'Secrets of the Little Blue Box'. *Esquire*, October 1971.

Roszak, Theodore. *The Making of a Counter Culture*. Faber, 1970.

—. *From Satori to Silicon Valley*. Don't Call it Frisco Press, 1986.

Rubin, Michael. *Droidmaker*. Triad, 2006.

Saxenian, AnnaLee. *Regional Advantage*. Harvard University Press, 1996.

Schaefer, Peter, and Meenakshi Gigi Durham. 'On the Social Implications of Invisibility ...' *Critical Studies in Media Communication*, Vol. 24, No. 1, March 2007.

Schlender, Brenton. 'How Steve Jobs Linked Up With IBM,' *Fortune*, 9 October 1989.

Schwartz, Peter, and Peter Leyden. 'The Long Boom'. *Wired*, July 1997.

Scott, Michael. Interview with Jay Yarow. *Business Insider*, 24 May 2011.

Segall, Ken. *Insanely Simple*. Portfolio Penguin, 2012.

Shone, Tom. *Blockbuster*. Simon & Schuster, 2004.

Shorris, Earl. 'Love is Dead'. *New York Times Magazine*, 29 October 1967.

Sloan, Alfred P. *My Years with General Motors*. Doubleday, 1964.

Smith, Douglas, and Robert C. Alexander. *Fumbling the Future*. iUniverse, 1999.

Stewart, James B. *Disney War*. Simon & Schuster, 2005.

Stross, Randall E. *Steve Jobs & the NeXT Big Thing*. Atheneum/ Macmillan, 1993.

Thompson, Hunter S. *Fear and Loathing in Las Vegas*. Random House, 1971.

Turkle, Sheery. *Life on the Screen*. Simon & Schuster, 1995.

—. *The Second Self*. MIT Press, 2005.

—. *Alone Together*. Basic Books, 2011.

Turner, Fred. *From Counterculture to Cyberculture*. University of Chicago Press, 2006.

Vogelstein, Fred. 'Weapon of Mass Disruption'. *Wired*, February 2008.

Waters, John K. *Blobitecture*. Rockport Publishers, 2003.

Weber, Max. *The Theory of Social and Economic Organization*. Free Press, 1947.

Whyte, William Hollingsworth. *The Organization Man*. Simon & Schuster, 1956.

Wiener, Norbert. *The Human Use of Human Beings*. Houghton Mifflin, 1950.

Wittkower, D. E. (ed.). *iPod and Philosophy*. Open Court, 2008.

Wolf, Gary. 'The Next Insanely Great Thing'. *Wired*, February 1996.

Wolfe, Tom. *The Electric Kool-Aid Acid Test*. Farrar Straus Giroux, 1968.

—. 'The Tinkerings of Robert Noyce'. *Esquire*, December 1983.

Wozniak, Steve, with Gina Smith. *iWoz*. Norton, 2006.

—. 'Homebrew and How the Apple Came To Be'. Available at: http://www.atariarchives.org/deli/homebrew_and_how_the_apple.php

—. Interview with Rupert Neate. *Daily Telegraph*, 6 October 2008.

Young, Jeffrey. *Steve Jobs*. Scott, Foresman, 1988.

— and William Simon. *iCon*. John Wiley & Sons, 2005.

Zittrain, Jonathan. *The Future of the Internet and How to Stop It*. Yale University Press, 2008.

Žižek, Slavoj. *The Sublime Object of Ideology*. Verso, 1989.

INDEX

n represents entries found in footnotes